Equation of State of Uranium Dioxide

C. Ronchi · I. L. Iosilevski · E. S. Yakub

Equation of State of Uranium Dioxide

Data Collection

with 19 figures and 55 tables

Springer

Dr. Claudio Ronchi
Head of the Materials Research Department
European Commission
Joint Research Centre
Institute for Transuranium Elements
P.O. Box 2340
76125 Karlsruhe
Germany
e-mail: ronchi@itu.fzk.de

Professor
Igor Lvovitch Iosilevski
Moscow Institute of Physics and Technology
Dolgoprudny, Moscow Region, 141700
Russia
e-mail: ilios@orc.ru

Professor
Eugene Solomonovich Yakub
Medical State University
Odessa
Ukraine
e-mail: eugene.yakub@unive.odessa.ua

Library of Congress Control Number: 2004109050

ISBN 3-540-22122-0 Springer-Verlag Berlin Heidelberg New York

Springer-Verlag is a part of Springer Science+Business Media
springeronline.com

Typesetting: Digital data supplied by authors
Cover-design: design & production, Heidelberg
Printed on acid-free paper 2/3020 xv - 5 4 3 2 1 0 -

To the raising generations:
to our children and grandchildren

Foreword

In the beginning of the 1990's, in the course of the events which were rapidly changing the political configuration of the East European countries, the crisis which invested the vast research apparatus of the former Soviet Union was entailing consequences whose dimension and depth were immediately realized by the international scientific community.

In the same years, however, the most important branch of nuclear energy research and development, in particular that concerning fission reactor, was worldwide undergoing a substantial reduction due to a variety of decisional situations. Yet, paradoxically, it was a very good fortune that a number of concerns on the future of nuclear research were shared by East- and West-European scientists, especially those who were working in advanced fields. In fact, the only hope for coping with an uncertain future was to erect bridges between similar institutions and employ safeguarding tactics linked to a long term collaboration strategy.

A decade later, this proved to be a winning decision, since the revival of nuclear energy is presently starting from a basis of common intentions and a network of established cooperation, whose seeds are to be searched in those initial, individual efforts.

I must also acknowledge the clairvoyance of international organizations like *INTAS*, which made it possible to launch a number of joint projects, whose objectives were evaluated on the basis of their scientific relevance, independently of the political priorities of the individual countries. I enjoyed these exchanges that were both lively and productive. This monograph is to a great extent the result of one of these projects.

At that time, as director of the High Energy Density Laboratory of the Russian Academy of Sciences, I realized that the know-how of some of our specialists could be conveyed to investigate the extreme states of nuclear materials during the most severe hypothetical reactor accidents - a domain where in Europe and USA most studies had been stopped after the decision of not further pursuing the Fast Breeder development.

Specialists from laboratories in Russia, Ukraine and Uzbekistan participated in this project, coordinated by the Joint Research Centre of the European Commission, with the objective of constructing a comprehensive equation of state of the nuclear fuel up to the critical point.

I should like to express here my congratulations to the authors of this book and to the entire staff of the *INTAS* 93-66 project for their success, a success that demonstrates that good cooperative research can be effectively performed even – or perhaps only - without the heavy administrative and managerial armament which characterizes most of the current international projects.

Prof. Vladimir E. Fortov
Russian Academy of Sciences, Moscow, March 2004

Preface

In 1993, a collaboration was started between the Joint Research Centre of the European Commission, Institute for Transuranium Elements (*ITU*) and the Russian Academy of Sciences, High Energy-Density Laboratory (now Institute for High Energy Density, *IHED*) aimed at constructing an Equation of State (*EOS*) for liquid and gaseous UO_2, which should fully reproduce a consolidated thermodynamic database for this compound and, further, provide trustworthy extrapolations up to the thermodynamic critical point. The framework of this project was provided by the International Association for the Promotion of Co-operation with Scientists from the Independent States of the Former Soviet Union (*INTAS*), which made it possible to establish administrative channels between the European Union and different East European Countries.

The quest for an Equation of State, which describes in a unified and self-consistent way the thermodynamic properties of the liquid and vapour phases of uranium dioxide, has had a long history of over 30 years. Yet, given the high melting point of UO_2, the portfolio of experimental data, which could be drawn upon to construct the *EOS* empirically, was initially rather limited and, accordingly, theoretical work was undertaken in the attempt to extrapolate what little data there was to experimentally inaccessible regions of the phase diagram, and to predict the values of the parameters characterising the critical point. The results of this study, as well as some matter concerning its development, constitute the subject of this monograph.

Before starting the work, a number of specialists were asked to participate in debates and discussions in view of choosing an appropriate method to construct an equation of state, which should be able to satisfy important requirements that could not be considered in the previous attempts. However, the principal objective of the project was not to merely add a new, more sophisticated model to the collection of the existing ones, but rather to define a general approach, with a rigorous selection of basic hypotheses, as well as a critical analysis of the significance of the sub-models and approximations involved. Actually, this discussion started from the ground level of thermodynamics and accompanied the project development through a series of working meetings, seminars and conference presentations. Interesting documentation and information was finally accumulated, which was worth being preserved from dispersion.

In fact, though the new elaborated *EOS* is marking a significant progress with respect to the previous models, the last word has certainly not yet been said on this subject. Therefore, as important as the results obtained, is the method and approach used, and how the same line of thought can be corroborated and improved by considering experimental data that might be made available in the future. This was the reason why this monograph has been finally edited, presenting not only the final results, but also a description of the various steps, including some inconclusive ones, whose failure was however able to evidence, in the examined context, not only the merits but also the weakness and the limits of some classical approaches.

This book contains a description of the work carried out between 1995 and 2000, starting from the year when results were first presented at international conferences and published in scientific magazines. In this perspective, the matter has been exposed in such a way as to guide the reader through a sequence of questions, whose answers led to the eventual formulation of the searched equation of state. The reported mathematical treatment, though necessarily abridged, is in essence exhaustive. As for the numerical calculation methods adopted and implemented in the *SAHA* computer code system, no explicit mention is made here, as they are presently of common use in thermodynamic equilibrium calculations.

The function of this monograph is twofold. First, it is intended to provide numerical data for nuclear engineering applications; secondly, to furnish the tools for a possible extension of the model to more complex systems. For this reason, some matter is presented in a didactic form, starting from basic concepts, which might be self-understanding for specialists. On the other hand, for most readers these introductory instructions may not provide a sufficient platform for discerning the meaning of each detail of the various subjects, so that additional reading of fundamental literature, indicated in the references, might be necessary. Within these limits, however, it is hoped that the monograph is reasonably complete and self-containing.

A team of specialists actively participated in this project with specific work, as well as with constructive criticisms and suggestions. An invaluable contribution was provided by the late and unforgettable Lev V. Gurvich and, after him, by Vladimir S. Yungman (*IVTAN*, Moscow) on thermodynamic data of uranium oxides, by Alexander M. Semenov (Power Engineering Institute, Moscow) on atomic and molecular parameters and by Gerard J. Hyland (University of Warwick, United Kingdom) on thermophysical properties. Their contributions are here summarised in an Appendix. The extensive computer calculations have been carried out by Victor K. Gryaznov at the Institute of Chemical Physics of Chernogolovka, Russia. This latter was more than a routine work, since without Gryaznov's devoted and efficient labours this project would have never been successful.

Finally, special thanks are due to Vladimir E. Fortov, who initiated this international collaboration and, though holding the high office of Minister of Research in the Russian government, supervised this work with attention and stimulating discussions.

Karlsruhe, Moscow
April, 2004

Claudio Ronchi
Igor Iosilevski
Eugene Yakub

Contents

1 Introduction 1
1.1 Reactor Accident Analysis and Fuel Equation of State 1
1.2 The Role of Equation of State 2
1.3 Equation of State for Liquid UO_2: Historical 4
1.4 Summary of the New Equation of State Features 13
1.5 General Notations 15

2 Governing Equations and Fundamental Formulae 17
2.1 Introduction 17
2.2 The Concept of Equation-of-State 20
2.3 Model Equations of State 22
2.3.1 Ideal Gas 22
2.3.2 Ideal Reacting Gas 24
2.3.3 Hard Bodies 24
2.3.4 Soft Spheres 26
2.3.5 Van der Waals' Model 27
2.3.6 Lennard-Jones Fluid 29
2.3.7 Charged Spheres 30
2.3.8 Composite Models 31
2.4 Physical and Chemical Models in Thermodynamics of Reacting Fluids 33
2.4.1 The Concept of Composition 33
2.4.2 Neutral Models of $UO_{2\pm x}$ 34
2.4.3 Ionic Models 36
2.4.4 MIX Models 37
2.5 Conclusions 38

3 Ionic Models for Liquid Urania 41
3.1 Restricted Primitive Ionic Model 41
3.2 Extended Ionic Model 45

3.3 Local Equations of State for the Liquid Phase: General Requirements ... 49
3.4 Improved Restricted Primitive Ionic Model ... 49
3.5 Compatibility Conditions ... 51
3.6 Determination of the Coulomb Contribution ... 52
3.7 Conclusions ... 54

4 Gas-Liquid Coexistence in Uranium Dioxide ... 59
4.1 General Conditions of the Phase Equilibrium ... 59
4.1.1 Single-Component Fluids ... 61
4.1.2 Chemically Reactive Systems without Ionisation ... 61
4.1.3 Congruently Coexisting (Azeotropic) Compositions ... 62
4.1.4 Extremal Properties of the Thermodynamic Functions in the Azeotropic Points ... 63
4.1.5 Systems of Charged Species ... 63
4.1.6 Chemically Reacting Fluids with Ionisation ... 64
4.2 Calculation of the Equilibrium Composition and Thermodynamic Functions ... 65
4.3 General Structure of the Liquid-Vapour Phase Boundaries in $UO_{2\pm x}$ ... 67
4.4 Equilibrium Properties and Composition of $UO_{2\pm x}$ ... 70
4.4.1 Liquid Phase at Low Temperatures ... 70
4.4.2 Vapour Phase without Ionisation ... 71
4.4.3 Partially Ionised Vapour Phase ... 72

5 Application of the Chemical Model within the van der Waals Approximation ... 75
5.1 Non-Congruent Evaporation over $UO_{2\pm x}$ ... 75
5.1.1 Assumptions ... 75
5.1.2 Calibration ... 76
5.1.3 Results ... 78
5.2 Oxygen Potential ... 86
5.2.1 Simplified Calibration Procedure ... 86
5.2.2 Calibration in the Case of Non-Congruent Evaporation ... 87
5.3 Composition of the Liquid Phase ... 89
5.4 Discussion ... 92
5.5 Justification of the MIX Models ... 95

6 New Equation of State for Fluid Uranium Dioxide Based on Thermodynamic Perturbation Theory ... 99
6.1 Thermodynamic Perturbation Theory ... 100
6.1.1 Simple Fluids ... 101
6.1.2 One-Fluid Approximation ... 104
6.2 Fluids Composed of Molecules with Anisotropic Interaction ... 106
6.2.1 Averaged Diameter of Non-Spherical Molecules ... 107

6.2.2 Effective Molecular Diameter 109
6.2.3 Estimation of the Anisotropy Parameter 110
6.3 Conclusions 111

7 Thermodynamic Properties of UO_2, as Predicted by the New Equation of State 113
7.1 Summary of the Model Features 113
7.2 Calibration 114
7.3 Validation of the INTAS-99-EOS 116
7.3.1 Density 118
7.3.2 Total vapour pressure 119
7.3.3 Heat capacity and thermal coefficients 121
7.4 Non-Congruent Equilibrium and Critical Point 123
7.4.1 Oxygen potential of liquid UO_2 124
7.4.2 Enthalpy on the vapour-liquid boundary 125
7.4.3 Equilibrium composition and non-ideality effects 127
7.4.4 Azeotropic compositions 128
7.4.5 Critical point in FCE approximation 130
7.4.6 Critical point under non-congruent evaporation 131
7.5 Concluding Remarks 133

A Appendix 135
A.1 Thermophysical Properties of UO_2 135
A.1.1 Density of UO_2 135
A.1.2 Volume Changes on Melting 137
A.1.3 Thermal Expansion 137
A.1.4 Elastic Properties and Adiabatic Compressibility 138
A.1.5 Heat Capacity of Molten UO_2 140
A.1.6 Enthalpy and Entropy of Fusion 140
A.1.7 Vapour Pressure over Nominally Stoichiometric UO_2 Near Melting 142
A.1.8 The Congruently Vaporising Compositions of Urania 144
A.2 Individual Components. Tables of Thermodynamic Functions 145
A.3 Estimated Molecular and Ionic Interaction Constants 176
A.3.1 Polarisability of Atomic O and U 178
A.3.2 Polarisability of U_mO_n Molecules 179
A.3.3 Dispersion Constants 180
A.3.4 Dipole Moments of U_mO_n Molecules 180
A.3.5 Sizes of U_mO_n Molecules 183
A.3.6 Conclusions 186
A.4 Thermodynamic Tables 187

References 349

Index 355

1

Introduction

1.1 Reactor Accident Analysis and Fuel Equation of State

For more than fifty years, uranium dioxide was world-wide used as the main fuel for two generations of nuclear power plants. Chosen first as the most convenient uranium compound for easy fabrication of sintered pellets, UO_2 displayed in the long run an excellent combination of thermophysical properties as well as an uncommon resistance to burn-up and radiation damage. Furthermore, some features of the dioxide crystallographic structure (fluorite face centred cubic), in conjunction with the electronic properties of uranium, allowing for different oxidation states, enable solid solutions, MO_2, to be formed with all existing fissile elements, in particular, with plutonium and thorium. Furthermore, mixed oxides (*MOX*) with variable oxygen stoichiometry can be obtained, so that the oxygen chemical activity may be adjusted in the fabricated fuel to avoid chemical interaction with metallic cladding. For this reason, uranium-plutonium *MOX* could be successfully adopted also for fast breeder reactors. In this application, the fuel is submitted to a severe thermal regime, with centreline pellet temperatures of up to near the melting point, and gradients of the order of 500 K/mm . Though under these conditions the fuel undergoes almost complete in-pile restructuring, the MO_2 rods still exhibit an excellent chemical and mechanical performance up to very high burn-ups. It was, indeed, in the context of the safety studies of liquid-metal cooled fast breeder reactors (*LMFBR*) that the properties of the oxide fuel were invoked to describe hypothetical core disruptive accidents where extremely high temperatures far above the melting point are reached. In fact, the worst conceivable accident sequence in a nuclear reactor can be conjectured in a *LMFBR* scenario, starting from a supercritical reactivity excursion, initiated, for example, by a loss of the liquid sodium coolant. Such an excursion entails a rapid release of energy, which - if large enough - can result in the explosive disassembly of the completely molten reactor core. Upon melting, the uranium-plutonium oxide fuel and its cladding collapse to the bottom of the core, and, should the compaction of fissile atoms lead to a supercritical configuration, even more energy will be released in the fuel. This results in even higher temperatures, and - in consequence of the inertia of the molten fuel (and core debris) which prevents instantaneous adjustment to the rapid input of energy - also in high pressures. Both act to terminate the transient - the high temperature expansion of the fuel reducing the macroscopic nuclear cross-sections (and hence the fission rate), whilst the pressure pulse ultimately results in the hydrodynamic dispersal of the fuel. This hypothetical sequence of events may occur within a very short time, and the fission energy released during the reactivity ramp - which is a measure of the gravity level of the accident - results from the competition between the lifetime of prompt and delayed neutrons, and the quasi-adiabatic rate of expansion of the core, governed, in the most conservative case, by the vapour pressure of the molten fuel.

In the context of the current *LMFBR* technology, this type of accident is highly improbable, however, the frightening scenario of an explosive core disassembly at temperatures at which also refractory materials melt and vaporise within seconds or even fraction of seconds, has produced a strong impact in the public debate on peaceful uses of nuclear energy.

At present, the objectives of the nuclear technology developments are aiming at understanding and preventing less severe reactor accidents, in which however, core melting is contemplated as well as radionuclide dispersion by fuel vaporisation. Even in this scenario, the high temperature thermodynamic properties of molten fuel represent a key issue in the evolution and consequences of the accidents. On the one side, licensing authorities are demanding trustworthy information on the physical processes taking place during these hypothetical accidents, on the other one, engineers and physicists must admit that certain predictions are to a large extent based on conjectures and estimates. Small scale simulation tests of reactor core behaviour under high and fast power excursions have been conducted worldwide, but a realistic assessment of the reactor materials properties, and in particular of fuel at temperatures far above the melting point still remains a challenge. In the last three decades, in spite of the uncertain and fluctuating context of the nuclear energy development, groups of specialists have not ceased to work on this problem, which is centred on the definition of a thermodynamic equation of state of the fuel applicable from the melting up to the critical point.

1.2 The Role of Equation of State

The pressure-volume-temperature relation is provided by the *thermal* equation of state (*EOS*), $P = P(T, V)$, in which the pressure P, temperature T, and volume, V, of the given system enter. However, like the internal energy U, the volume is not fixed by the *thermal EOS*, but requires, instead, a *caloric* equation of state, such as is afforded, for example, by the heat capacity at constant volume, $C_V(T)$. It is thus the equations of state of the fuel which provide the essential link between the nuclear aspects of a supercritical excursion and the dynamic response of the molten core; $T(U,V)$ controls the nuclear fission cross-sections *via* the Doppler coefficient, whilst $P(U,V)$ controls the time-dependence of the fuel-mass geometry *via* hydrodynamics. The reliability of reactor safety analyses is thus contingent on the accuracy to which the equations of state of the fuel are known - in particular, in the liquid-vapour coexistence region, which extends to the critical point - a most important invariant point on the phase diagram of any material.

As mentioned before, fuels of different compositions may be concerned. In particular, in fast reactors the fuel is normally a mixed oxide of uranium and plutonium. The description of the high temperature thermodynamic behaviour of *MOX* is obviously more complicated than for uranium oxide. Yet, the difference in the extrapolated thermodynamic properties of UO_2 and MO_2 at temperatures far above the melting point are very likely much smaller than the uncertainties of the current theoretical models. Therefore, the first objective should be to obtain a sufficiently accurate description of UO_2, the simplest and most known compound[1].

In the context of the above mentioned *LMFBR* reactivity excursions, any increase of hydrostatic pressure in the molten core contributes to an earlier termination of the power ramp. From this point of view, an accident occurring in a fresh core is expected to be more severe than in a core at high burn-up. As for the effect of the fuel composition, under anoxic conditions, *MOX* is slightly more volatile than uranium dioxide. Therefore, a fast reactor core composed of enriched UO_2 does probably provide the case of the highest possible energy insertion during core disruption.

Despite considerable effort over the past thirty years, the high-temperature phase diagram of the U-O system in the vicinity of stoichiometric UO_2 was still very incomplete in the middle of the 90's, due primarily to three features:

- *the very high melting temperature* ($\sim 3150\ K$), which, until the advent of rapid laser heating techniques, not only restricted the acquisition of experimental data, but also posed problems for the containment of the sample, with the attendant possibility its significant contamination,
- *the rather wide range of stoichiometries* over which urania, $UO_{2\pm x}$, can exist as a single phase system, and the related problem of controlling the stoichiometry of any *given* sample throughout the course of an experiment,
- *the non-congruency of the melting and vaporisation* of the material under conditions of thermodynamic equilibrium.

[1] Furthermore, the fuel irradiated at high burn-ups contains significant amounts of fission products, some of which are gaseous or highly volatile. In this case, under temperature ramps above the melting point, pressures are created in addition to the equilibrium vapour pressure of the matrix. In a first approximation, the pressures of components not reacting with the matrix can be evaluated independently.

Given, however, the recent advances both in high temperature experimental technology and in the theoretical understanding of binary systems of variable composition, the possibility of further progress presented itself, involving not only further refinements, but also, more importantly, the possibility of exploring hitherto experimentally inaccessible regions of the *phase diagram.*

1.3 Equation of State for Liquid UO_2: Historical

Thirty years ago, when the first attempts were being made to construct an *EOS*, which describes in a unified and self-consistent way the thermodynamics of the coexisting liquid and vapour phases of urania, the portfolio of experimental data, which was then available to act as empirical constraints on such an *EOS*, was rather limited. In an attempt to fill this lacuna, much attention and effort was accordingly directed towards theoretical modelling in order to permit what little data there was to be extrapolated to experimentally inaccessible areas of the phase diagram - in particular, the critical region; the results obtained prior to *INTAS* 93-66 project are summarised in Table 1.1.

For the first 10 years, attempts toward obtaining values of the critical parameters were based on purely empirical procedures - notably the law of rectilinear diameters (*LRD*) and the principle of corresponding states (*PCS*). The values of the critical temperature, T_C, obtained by such methods are described in Table 1.2, and varied between 6400 K and 10000 K. It was eventually realised, however, that little significance could be attached to such values, since the procedures used to predict them either lacked any physical basis (as is the case with the *LRD*), or had no demonstrable validity (as is the case with the *PCS*) for non-congruently vaporising materials like urania. Accordingly, a less empirical approach based on Eyring's significant structure theory (*SST*) was initiated in 1974 (see Table 1.3), and progressively developed and refined, until by 1987 it was not only able to treat two-phase equilibria involving non-stoichiometric molten $UO_{2\pm x}$, but also incorporate the experimentally well-established [131] existence in the vapour phase - even over stoichiometric $UO_{2.00}$ - of molecular species other than UO_2(g), notably UO(g), UO_3(g) - as well as O(g), O_2(g) and U(g). For the first time, it was thus possible to consider, theoretically, the non-congruency of the vaporisation of molten urania in terms of a model which treated the liquid and vapour phases in a unified way - an attractive feature not shared by some earlier essays [37] which, through their description of the liquid and vapour phases using quite separate models, did not fully respect the coexistent equilibrium of the two phases. For the case of stoichiometric UO_2, a critical temperature around 10000 K was finally predicted [39] - a value close to that given [76] by the alternate perturbed hard sphere (*PHS*) approach, which is essentially a refinement of the classical van der Waals' treatment of two-phase equilibria. The proximity of these two values of T_C must, however, be considered fortuitous, since within the *PHS* approach the vaporisation is necessarily congruent, the

Table 1.1. Synopsis of the Main Published Works on the Equation of State of UO_2

Year	Author	Method of Evaluation	Predicted Critical Point Parameters[(a)] T_C	P_C	V_C	ρ_C	z_C	Input Data
1964	Meyer and Wolfe [103]	*PCS*	7300	194.0	85.5	3.01	0.27	Melting point(*MP*), density, vapour pressure
1965	Miller [104]	*LRD*	9115	124.6	170	1.59	0.27	*MP*, density, vapour pressure
1966	Menzies [102]	*PCS*	8000	200.0	90	3.16	0.27	Saturated vapour pressure(*VP*)
1966	Robbins [118]	*PCS*	5500-10000	100-280	96-120		0.2-0.4	density, heat of fusion, *VP*
1974	Booth [17]	*PCS*	6723	142.3	99	2.74	0.251	*MP*, density, *VP*, enthalpy of solid
1974	Gillan [50]	*SST*	6960	108.4		1.64	0.308	*MP*, density, *VP* of UO_2, molecular data
1974	*ibid.*	*SST*	9930	146.9		1.63	0.308	*MP*, density, different *VP* data
1976	Fischer *et al*[42]	*SST*	7560	122.6		1.66	0.316	*VP*, enthalpy of solid UO_2, molecular data (spectroscopic)
1976	Kapil [83]	*PCS*	6744	142.2		2.74	0.25	*MP*, density, linear expansion, *VP*
1977	Browning *et al*[22]	*PCS*(1)	6723	140.4	98.7		0.25	
		PCS(2)	6820	138.0	98.4		0.24	
1977	Browning *et al*[22]	*SST*(1)	7320	125.6	152		0.32	*VP*
		SST(2)	8840	142.4	158		0.31	*MP* data, liquid density, and expansion
1978	Finn *et al*[37]	*LRD*	7365	705		2.99	1.00	Liquid density, oxygen potential , liquid C_V, calculated partial pressures
			6400			3.27	0.27	

(continued)

Table 1.1. Synopsis of the Main Published Works on the Equation of State of UO_2 (end)

Year	Author	Method of Evaluation	Predicted Critical Point Parameters[(a)] T_C	P_C	V_C	ρ_C	z_C	Input Data
1979	Fischer [38]	*SST*	7970	178[(b)]	130	2.08	0.349[(b)]	O/M=1.992, solid-state defect energies, solid-state data near T_m; spectroscopic data on UO(g) and UO_2(g) molecules
		SST	7990	182[(b)]	129	2.10	0.353[(b)]	O/M=1.911
1982	Dharmadurai [30]	*RK-EOS*	7000	113	178	1.52	0.346	*VP* in equilibrium with UO_2(l)
1985	Mistura *et al*[105]	*PHC-EOS*	7567	141	156	1.73	0.350	*VP* and liquid molar volume at T_m
1987 (1989)	Fischer [39]	*SST*	10600	158[(b)]	173	1.56	0.310[(b)]	*VP* at $T = 2150\ K$ and over liquid; geometrical mean of experimental partial pressures of UO and UO_3; molecular spectroscopic data.

<u>**Notes:**</u>
[(a)] T_C *in K*, P_C *in MPa*, V_C in $10^{-6}\ m^3kg^{-1}$, ρ_C in $10^3 kg\ m^{-3}$, $z_C = P_C V_C / MRT_C$, $M = 0.27 kg\ mol^{-1}$
[(b)] only contributions from U-bearing species are included.
PCS = Principle of Corresponding States; *LRD* = Law of Rectilinear Diameters; *SST* = Significant Structure Theory; *RK-EOS* = Redlich-Kwong Equation of State; *PHC-EOS* = Perturbed Hard Core Equation of State.
Only the 1987 work of Fischer includes UO(g), UO_2(g) and UO_3(g) molecules - UO_3(g) having been *omitted* from his 1979 work. Apart from the 1974 and 1976 *SST* contributions (which treated UO_2(g) as a bent molecule), all other *SST*-based calculations have assumed a *linear* UO_2(g) molecule; all *SST* attempts (*except* 1974) have included electronic contributions in the UO_2(g) partition function.

only molecular species admitted being UO_2, in either phase; furthermore, the degree to which it can be considered realistic to treat a system such as molten UO_2 - which in its solid state is appreciably ionic - as composed of neutral molecules of UO_2 might seem slight, especially in view of the appreciable ionic conductivity exhibited by other binary molten systems having the same solid-state fluorite structure [29]. However, there exists an equivalence theorem, based on the works of Stillinger [128] and Reiss [117] which asserts that the thermodynamic properties of a binary molten salt are actually the same as those of a hypothetical non-electrolytic liquid (of the same overall density and temperature) composed of electrically neutral, identical 'molecules' of size equal to the distance of closest approach of a charged cation and an anion in the real (binary) liquid; these hypothetical molecules are considered to interact non-Coulombically by a short-range potential, identical to that to which the cations and anions in the actual liquid are subject. The asserted dependence of thermodynamic properties only on the sum of the cation and anion radii (rather than on their individual diameters) represents a *PCS* [117] - but one more subtle than that already referred to and criticised. The importance of this equivalence is, of course, that it reconciles a finite electrical conductivity (which molten UO_2 most probably has) with a description of the liquid in terms of seemingly neutral molecules. The much better overall agreement with available thermodynamic data which is obtained [76] by assuming such hypothetical spherical molecules, rather the linear form which real UO_2 molecules actually have [105], strongly suggests that the hypothetical, non-electrolytic model might form the basis of a novel approach to the *EOS* appropriate to the liquid-vapour coexistence region of the phase diagram of uranium dioxide.

The desirability of such an approach which is rooted more strongly in the liquid than is the *SST*, for example (in which the liquid is described in terms of the partition functions of the solid and gaseous phases), is indicated by certain fundamental difficulties which continued to plague even the most refined version [39] of the *SST* approach - notwithstanding heroic attempts at a realistic description of the multi-species vapour phase over a non-stoichiometric UO_2 – notably:

- certain pathologies exhibited by the derived *EOS* with respect to the composition of the vapour phase, particularly in the vicinity of the critical point where, over $UO_{2.00}^{(liq)}$, the predicted value of $(O/U)^{(vap)} \neq 2$,
- the fact that the very concept of a critical point - in the sense that it is used in the case of a mono-atomic substance - is actually ill-defined in the case of a binary system with variable stoichiometry, where the only well-defined extrema of such a system are the so-called *cricondentherm* and *cricondenbar* [151].

The former defines the highest temperature attainable on the saturation curve for which $\partial T/\partial P = 0$, whilst the latter defines the highest pressure attainable on the boiling curve for which $\partial P/\partial T = 0$; beyond these extrema, where the phenomena of retrograde condensation and vaporisation occur, lies the critical point, the precise location of which is defined by the intersection of this $P-T$ section of the phase diagram with the $T-x$ section (x =O/U-2), at the point where $T(x)$ is maximum. These boiling and saturation curves define the extent of the two-phase region, and are the $P-T$ equivalent of the more familiar (liquid-vapour) two-phase region usually depicted in the $T-x$ plane.

In connection with experiments to determine the vapour pressures as functions of temperature, it is clearly essential to establish which curve corresponds to the prevailing experimental conditions. Evidently, experiments based on the classical boiling point method [16] (where a specimen, heated in a quasi-stationary way on a relatively slow timescale, evaporates against an external inert gas whose pressure can be varied) refer to the boiling curve (which separates stoichiometric liquid from a hyperstoichiometric vapour phase), whilst the conditions obtained on the saturation curve (which separates a stoichiometric vapour from a liquid phase whose surface is hypostoichiometric) are identical to those of the so-called forced-congruently vaporising mode (*FCM*), which is realised in experiments utilising very rapid laser heating (on a sub-millisecond time scale), the sample here being allowed to evaporate freely into vacuum [108]. Under these conditions, an initially stoichiometric sample is virtually instantaneously vaporised with preservation of its bulk stoichiometry, whilst the surface itself is rendered hypostoichiometric - the rate of evaporation being much greater than the rate at which oxygen can be replenished from the bulk by diffusion [20].

The T-dependence of the heat capacity, $C_P(T)$, of molten UO_2 predicted by the *SST* does not even qualitatively conform to the experimental $C_P(T)$ data which extends up to 8000 K, subsequently obtained at *ITU* [119]. This extensive set of $C_P(T)$ data, which was published in 1993, constitutes not only a most valuable contribution to the caloric *EOS* of UO_2, but also acts a powerful constraint on any acceptable thermal *EOS*, to which the *SST*-based *EOS*, for example, apparently does not conform.

The development of *EOS* for UO_2, presented in this monograph progressed through several stages. A relatively simple *EOS* based on a pure ionic model was first devised. Background, method of calibration and resulting expressions of the thermodynamic quantities are presented in Chapter 3, following a general discussion of the available models and approaches presented in Chapter 2. The pure ionic model had obviously a restricted success, since it is unable to treat the non-congruent evaporation of uranium dioxide. It also fails, at higher temperatures, to reach the vicinity of the vapour-liquid critical point. However, the experience gained on this way was important in developing an improved *EOS* model, with a first attempt to define simple physical atomic interactions having meaningful parameters expressed in terms of known ionic radii. To resolve further problems encountered in applying this *physical* model more sophisticated *chemical* models of fluid UO_2 have been invoked. The general conditions of the vapour-liquid phase equilibrium and their implementation within this more complicated context are discussed in Chapter 4. Given the problems associated with pure ionic and *SST*-based approaches, in particular in the liquid-vapour coexistence region, and the absence of any obvious way of resolving them, it was decided to base the theoretical efforts on a model which essentially generalises the single substance, neutral molecule approach already mentioned (and validated by the Stillinger-Reiss theorem), by including:

- molecular species other than just UO_2 - notably, UO and UO_3 in reactive equilibrium with UO_2,
- their associated singly charged ions (UO^+, $UO_2^{\pm}$, UO_3^-),
- electrons and other elemental species (U, O, O_2), together with their associated monovalent ions (U^+, O^-, O_2^-), and
- possible neutral dimers, such as U_2O_n with n = 2, 3, 4, 5.

It should be pointed out that a description of the liquid state in terms of such units is not necessarily inconsistent with the conventional *ionic* description of the bonding in the solid - if for no other reason than the stability of the O^{2-} ion is contingent on the Madelung lattice potential. Between these species the following interactions are admitted:

- short-range repulsion calculated in the *hard-spheres mixture* approximation, neglecting any difference between the radii of a given species and its associated ion,
- short-range attractions, calculated in the van der Waals' approximation, neglecting any difference between neutral and charged species, and assuming an equal attraction between all U-bearing species, and a much weaker, but still equal attraction between all combinations of O and O_2, and between either O or O_2 and any U-bearing molecule, and
- the Coulomb interaction between charged species, calculated within a variant of the modified Debye-Hückel approximation.

This further step in the development of the *EOS* was much more cumbersome but the context of the starting assumptions proved to be well grounded on principles of evidence and much more successful than those adopted in the past (Chapter 5). The results presented demonstrate the ability of a *chemical* model to reproduce the oxygen potential as a function of temperature as well as the most specific features of non-congruent evaporation of UO_2. At the same time, however, essential drawbacks were found; in particular the van der Waals - like model was still unable to reproduce the experimental temperature dependence of the heat capacity.

To overcome this difficulty a modification based on the thermodynamic perturbation theory (*TPT*) was developed. The final version of the *EOS* described in Chapter 6 is a generalisation of the approach first proposed by Barker and Henderson [10] for a Lennard-Jones fluid. This generalisation was also implemented in two stages. In the first one, the simple version of *TPT* (actually the model of a Lennard-Jones fluid) was utilised, in the second stage a more sophisticated version was constructed by taking into account the non-central character of the interactions of complex molecules in a dense fluid.

The calibration of this model and validation of its predictions with experimental data are presented in Chapter 7. Calibration of the model was performed by using as input selected thermodynamic quantities. Different choices were, obviously, possible (recommended values of pertinent thermophysical properties of urania are presented in Appendix A.1), but only few quantities, with the respective experimental uncertainties, were providing sufficiently strong constraints for unambiguous calibration. The optimal input was found to be: *i*) the density of molten UO_2, *ii*) the partial pressure of $UO_2(g)$ at the melting point, and *iii*) the value of $(O/U)^{(vap)}$ at $T = 3400\ K$.

Furthermore, elaboration of reactive equilibria considered in the model, requires a specific database containing the thermodynamic functions of the relevant gaseous species, together with the ionisation energies and electron affinities of certain atoms and molecules. These data are presented in Appendix A.2.

Calibration is implemented by treating molecular radii and van der Waals' interaction constants as *free* parameters, whose fitted values must be ultimately in reasonable agreement with measured molecular quantities. Accordingly, a detailed, independent assessment of these molecular parameters is presented in Appendix A.3.

To test the predicative accuracy of the model, experimental data on thermophysical quantities related to the *EOS* are required at temperatures well above the melting temperature. The $C_P(T)$ data (constituting the caloric *EOS*) obtained in *ITU* are extremely valuable in this respect. The situation with the thermal *EOS* is, however, less satisfactory, despite the extensive experimental attempts over the past 20 years to acquire reliable $P_S(T)$ data up to the highest possible temperatures. These temperatures vary from experiment to experiment. Those attained by using laser pulse heating - probably the most reliable technique – are based on the classical boiling point method [16]. Yet, the boiling temperature of UO_2 is relatively low ($\sim 5000\ K$), and well below the estimated critical temperature (of which there is no indication in the new $C_P(T)$ data of molten UO_2 which extends up to 8000 K). Pressure measurements at higher temperatures were attained either in-pile [21] with low measurement accuracy[2], or from very fast laser heating (microsecond pulses). These latter are unfortunately plagued with contributions to the total vapour pressure from ions thermo-ionically emitted from the heated surface, an artefact that establishes the intrinsic application limit of this technique. Additionally, in all these experiments vaporisation was produced in the *FCM*, where the vapour pressure is significantly lower than under thermodynamic equilib-

[2] In-pile vaporisation experiments are to date more informative than they were prior to the acquisition of $C_P(T)$ data of molten UO_2 [119]. In fact, in these experiments only the deposited energy, H, and the generated total pressure, P, could be measured, but not the sample temperature. A conversion of $P(H)$ into $P(T)$ was therefore necessary. This conversion, which was initially performed through a complex numerical procedure, can now be obtained with much greater accuracy.

Table 1.2. Inter-comparison of Purely Empirical Theoretical Approaches for Estimating the Critical Parameters of Urania

	Law of Rectilinear Diameters (*LRD*)	**Principle of Corresponding States (*PCS*)**
Range of T_c	$6400 - 9115\ K$	$5000 - 10000\ K$
Basic Principles	$1/2(\rho^{(liq)} + \rho^{(vap)}) = a - bT$	Universal relationship in *reduced* variables
Predicted Properties	Critical point data	Critical point data, *EOS*, (transport properties)
Calculation Procedure	Intersection of $\rho^{(liq)}$ and $\rho^{(vap)}$ curves	Application of universal functions in their reduced state within the *same* class of materials
Input Data	Liquid expansion data, vapour pressure data, and density at melting	*T, P, ΔH...,* mainly near melting
Validity of Results	Dependent on the range of extrapolation of the input data, and gives only a first estimate of the range of the critical parameters	Dependent on the availability of data of *similar* substances; gives a first approach to *EOS*
Advantages	Easy to apply	Simple empirical relationship. Can be deduced from statistical mechanics under certain restrictive conditions.
Problems	Availability of input data close enough to T_c. *No theoretical basis*	Availability of data on similar substances; this is *particularly* so with UO_2!

Table 1.3. Inter-comparison of Less Empirical Theoretical Approaches for Estimating the Critical Parameters of Urania

	Significant Structure Theory (*SST*)	**Perturbed Hard Spheres (*PHS*)**
Range of T_c	6960 – 10000 K	7000 – 9934 K
Basic Principles	Liquid partition function taken as a geometrical average of solid and gas partition functions	Perturbation theory expansion about the hard sphere potential
Predicted Properties	Critical point data, *EOS*, transport properties	Critical point data, *EOS* and certain transport properties
Calculation Procedure	Construction of the partition functions. Double tangent technique to determine phase equilibrium and pressure	Calculation of free energy from first few terms of a pertinent series. Calculation of thermodynamic quantities
Input Data	Melting temperature, volume expansion upon melting, vapour pressure data, heat of fusion	Pair potential for pure substances and interaction for mixtures. Boiling point, thermal expansion coefficient
Validity of Results	Dependent on input data and validity of model, especially for ionic systems.	Depends on how close the real potential is approached. Additive assumption of pair potential
Problems	Limited validity of the model	Accuracy of the pair potential and hard sphere correlation function
Advantages	Analytic form of free energy based on a liquid model, involving reasonable amounts of computing time	Based on same physical principles as van der Waals theory. Exact treatment of hard core. Treats attractive part and soft core of potential

rium (characterised by non-congruent vaporisation). This difficulty identifies an additional key role for the thermal *EOS*, namely that of permitting conversion from *FCM* vapour pressure values to those corresponding to *global* equilibrium. Without the possibility of such a conversion, information yielded by laboratory experiments employing rapid, laser-induced evaporation into vacuum would be irrelevant for predicting real reactor conditions arising from reactivity excursions.

1.4 Summary of the New Equation of State Features

In the following Chapters, the construction of the equation of state is described step by step, by following a line of though, which started from a simple scenario, and gradually encompassed new aspects or details to attain an adequately comprehensive model.

The final model is based on the thermodynamic perturbation theory (*TPT*) modified in order to account for the specific properties of UO_2. It describes, in a unified formalism, a multi-component mixture of chemically reactive, strongly interacting molecules and atoms, both neutral and charged. The description of dense liquid UO_2 is based on the demonstrated formal correspondence of the partition function of ionic liquids to that of liquids consisting of virtual molecular clusters (Stillinger-Reiss [128, 117] theorem). In such a model, detailed information is needed to describe the contributions of the internal degrees of freedom of complex (bounded) particles, *e.g.*, energies of ionisation and dissociation, excitation energy levels, *etc.*. On the other hand, however, the offered advantage is that the renormalized effective interaction of free (unbounded) particles is relatively weak, and, furthermore, arbitrary stages of ionisation and different structures of the particles can be composed in the frame of a unified calculation procedure. In principle, once a proper definition of the effective interaction of free particles is achieved, and the bound states are properly restricted, the model can continuously describe the wide variety of states ranging from the presumably highly ionised liquid at very high temperatures, to the *cold* vapour represented by a neutral molecular mixture. Along with free UO_2, a number of 'clusters', like U_2O_4, U_2O_2, U_2O_3, U_2O_5, *etc.* , as well as free atoms and molecules of uranium and oxygen (U, U_2, O, O_2), and ions (UO_2^+, UO^+, U^+, UO_3^-, UO_2^-, O^-) are included in the picture as distinct species with own partition functions, repulsion parameters (intrinsic volumes) and short-range attraction. This model can be considered as sufficiently realistic for dense vapours as well as dense, expanded liquids in the high-temperature region of phase coexistence. The liquid phase is essentially described as a set of conventionally distinguishable clusters (mainly with one uranium in the centre surrounded by several oxygen atoms), which are allowed to freely exchange oxygen and/or electrons. The general conditions for the free-energy minimum correspond to a system of equations for the chemical potentials, or may alternatively be expressed in the form of mass action law. Once explicit expressions for the free energy are known, this minimisation becomes a mere numerical problem. The equilibrium composition is evaluated, and, consequently, all thermodynamic functions can be deduced from the *thermal* and *caloric* equations of state, which are self-consistently defined as first derivatives of the free energy. It is worth remarking that, in spite of the additional, implicit dependence of the free energy on temperature and density (through the variable equilibrium composition) its first full derivatives, *e.g.*, pressure, entropy and chemical potentials, are equal to the corresponding partial derivatives of the free energy for fixed composition, at the free energy minimum. However, wherever the second derivatives are concerned, (*e.g.*, for the calculation of heat capacity, compressibility, *etc.*), this additional dependence of the free energy must be taken into account *explicitly*. Therefore, these quantities must be calculated via a direct numerical differentiation. The method applied to obtain the results reported here, was developed by V.K. Gryaznov and is described in Ref. [58].

The results of the final calculations of thermodynamic properties are discussed in detail in Chapter 7 and presented in an extended set of tables in Appendix A.

1.5 General Notations

List of Symbols

A	–	Electronic affinity	$J,\ eV$
a	–	van der Waals attraction parameter	$J\ m^3\ kmol^{-1}$
C_V	–	Isochoric heat capacity	$J\ kmol^{-1}K^{-1},\ J\ kg^{-1}K^{-1}$
C_P	–	Isobaric heat capacity	$J\ kmol^{-1}K^{-1}$
D_i	–	Dissociation energy if i-th species	$J\ kmol^{-1}$
d	–	Effective particle diameter	$m, \text{Å} \equiv 10^{-10}m$
U	–	Internal energy	$J\ kmol^{-1},\ J\ kg^{-1}$
F	–	Helmholtz free energy	$J\ kmol^{-1}$
G	–	Gibbs free energy	$J\ kmol^{-1},\ J\ kg^{-1}$
H	–	Enthalpy	$J\ kmol^{-1},\ J\ kg^{-1}$
I	–	Ionisation potential	$J,\ eV$
k	–	Boltzmann constant	$1.38\ 10^{-23}J\ K^{-1}$
L	–	Latent heat of vaporisation	$J\ kmol^{-1},\ J\ kg^{-1}$
m_i	–	Mass of i-th species	$kg\ kmol^{-1}$
N_i	–	Number of i-th species	$kmol$
N	–	Total number of particles	$N = \sum_i N_i,\ kmol$
n	–	Number density	$n = N/\mathcal{V}$
n_i	–	Number density of i-th species	$n_i = N_i/\mathcal{V}$
N_a	–	Avogadro number	$6.023\ 10^{26}\ kmol^{-1}$
P	–	Pressure	$Pa,\ bar \equiv 10^5 Pa$
R	–	Ideal-gas constant	$8.314\ J\ kmol^{-1}K^{-1}$
r_D	–	Debye radius	$m, \text{Å} \equiv 10^{-10}m$
S	–	Entropy	$J\ kmol^{-1}K^{-1},\ J\ kg^{-1}K^{-1}$
T	–	Temperature	K
$\mathcal{V}$	–	Volume	m^3
V	–	Molar volume	$V = \mathcal{V}/N,\ m^3\ kmol^{-1}$
V_s	–	Speed of sound	$m\ s^{-1}$
x	–	Non-stoichiometry parameter	$x = \mathrm{O/U} - 2$
x_i	–	Mole fraction of i-th species	$x_i = N_i/N$
z	–	Compressibility factor	$z = PV/RT$
Z_i	–	Charge of i-th species, in units of $\lvert e \rvert$	$\lvert e \rvert = 1.602\ 10^{-19}C$
α_T	–	Isobaric volume expansion	K^{-1}
β	–	Inverse temperature ($\beta = 1/kT$)	J^{-1}
$\beta_T(\beta_S)$		Isothermal (adiabatic) compressibility	Pa^{-1}, bar^{-1}
Γ_D	–	Coulomb non-ideality parameter	$\Gamma_D = d/r_D$
ε, σ	–	Lennard-Jones potential parameters	J,Å
η	–	Packing fraction	$\eta = {}^1\!/_6 \pi n d^3$
μ_i	–	Chemical potential of i–th species	$J\ kmol^{-1}$
ρ	–	Density	$\rho = \sum_i m_i n_i, kg\ m^{-3}$
$\Phi(r)$	–	Interaction potential	J

Abbreviations

BC – Boiling Curve
CCP – The point of Maximal Pressure on the *BC* (cricondenbar)
CCT – The point of Maximal Temperature on the *SC* (cricondentherm)
CCX – The point of Maximal O/U-ratio on the *BC*
CCE – Conditions of Chemical Equilibrium
CMC – Congruently Melting Compositions
CMINX – The point of Minimal O/U-ratio on the *SC*
CP – Critical Point
CPE – Conditions of Phase Equilibrium
CRF – Chemically Reacting Fluid
CVC – Congruently Vaporising Compositions
DHLL – Debye-Hückel Limiting Law
DHSE – Debye-Hückel Approximation for Charged Hard Spheres
EOS – Equation of State
FCE – Forced Congruent Phase Equilibrium
FCM – Forced Congruent Evaporation Mode
HS – Hard Sphere
IM – Ionic Model
M – Metals (U, Pu, *etc.*)
MD – Molecular Dynamics
MIX – Mixed Model
MOX – Mixed Oxide Fuels
MSA – Mean Spherical Approximation
MSAE – Mean Spherical Approximation + Energy Equation
MSAV – Mean Spherical Approximation + Virial theorem
PCP – Pseudo Critical Point
PCS – Principle of Corresponding States
RPIM – Restricted Primitive Ionic Model
SC – Saturation Curve
SSA – Single Substance Approximation
TPT – Thermodynamic Perturbation Theory
VdW – Van der Waals

ANL – Argonne National Laboratory, USA
IHED – Institute for High Energy Density, Russian Academy of Sciences
INSC – International Nuclear Safety Center, ANL, USA
INTAS – International Association for East-West Scientific Co-operation
ITU – Institute for Transuranium, Joint Research Centre, European Commission
IVTAN – Institute for High Temperatures, Russian Academy of Sciences
MIPT – Moscow Institute of Physics and Technology

2

Governing Equations and Fundamental Formulae

2.1 Introduction

One can consider as a chemically reacting fluid (*CRF*) a fluid in which the chemical bonds between atoms or ions may break up and reform, or ionisation and recombination occur. Formation of ion pairs in ionic liquids also may be regarded as a chemical reaction. There are many kinds of chemical bonding, ranging from very strong covalent bonds to weak hydrogen bonds, and many types of ions and charged complexes may exist in solution. Therefore, this definition is general enough to include a wide spectrum of fluids of very different nature, from dense dissociating hydrogen or nitrogen, to associated liquids, like hydrogen fluoride or water, dense non-ideal plasma, ionic melts, *etc.*. All *CRF*'s display a few specific features in their macroscopic behaviour. The most well known is the abnormal rise with temperature of the specific heat, $C_P(T)$, or of the thermal and electrical conductivity. Another specific feature is a non-monotonic dependence on T of the heat of vaporisation, $\Delta H^{(vap)}(T)$, which can often exhibit maxima as, *e.g.*, in the case of the reactions $N_2O_4 \leftrightarrow 2\ NO_2$ or $(HF)_6 \leftrightarrow 6\ HF$. This phenomenon is closely related to non-congruent vaporisation, *i.e.*, to differing compositions of the coexisting vapour and liquid. Note that thermodynamic effects related to the chemical reaction rates might be very significant, because the chemical bonding (and ionisation) energies are large compared with kT.

The influence of strong intermolecular interactions on the equilibrium of chemical reactions is another important point. The equilibrium constants measured in the gas and in the liquid phases may differ by several orders in magnitude. The heat of reaction in the liquid phase may also be very different from that in the gas phase. Both effects are strongly coupled, and this entails complications in the theoretical description. An inadequate treatment may well lead to wrong conclusions: *e.g.*, the macroscopic description of liquid-phase reactions leads in many cases to large $\Delta H^{(liq)} - \Delta H^{(vap)}$ differences, as if the temperature dependence of the equilibrium constant $K(T)$ be different in the gas and in the liquid phase, whilst $\Delta H^{(liq)} - \Delta H^{(vap)}$ is larger than expected from considering the typical intermolecular interaction energies. Hence, it has sometimes been concluded that there exist some as yet unknown specific interactions between molecules in the liquid phase, which lead to such unusual macroscopic effects. Often, however, these effects can be more easily explained by taking into account the thermal expansion of the liquid; the essential value of the volume change in the reaction may indeed become very important here, as it can be an order of magnitude larger than the estimated values of the change of the intrinsic volumes of the reacting molecules.

A general explanation of these phenomena was already obtained in statistical mechanical theory of reacting fluids [143] (see also the review in Ref. [146] and references cited therein) and will be summarised below. In several, more complicated cases the real gas-phase molecules, embedded into a dense liquid, are dissociated into ions. Even if they remain in a molecular form, the change of their molecular parameters, such as vibration frequencies, may be significant. Many infrared or Raman spectroscopic measurements have been made in the liquid phase [8], which show that only for simple non-polar diatomic molecules such as H_2, N_2, only a slight shift in optically active vibration frequencies (from 1 to 10 cm^{-1} in non-polar solvents) is observed. Rotational and electronic degrees of freedom in the liquid phase are also affected by intermolecular interactions. All this requires a detailed comparison of the predictions made on the basis of simplified models (*e.g.*, partition functions calculated for individual molecules) with existing experimental data.

The thermodynamic behaviour of the simplest *CRF*, for instance the homonuclear dimerising fluids having the only reaction $A_2 \rightleftharpoons 2\,A$ (A = H, N, F, ...), is presently well understood. The atoms with open electronic shells, frequently occurring in liquids, reveal some specific interaction features, such as the dependence of the interaction energy on the interatomic distance and the mutual orientation of electronic spin and orbital momenta. Thus, in the simplest case of two hydrogen atoms, two possible orientations of spins (parallel or anti-parallel) and two interaction curves $^1\Sigma$ and $^3\Sigma$ exist. The interaction of two nitrogen atoms, on the other hand, is characterised by four ($^1\Sigma$, $^3\Sigma$, $^5\Sigma$ and $^7\Sigma$) potential curves, whilst fluorine-fluorine requires eighteen curves *etc.*. Some of them correspond to very deep wells (strong chemical bonding), while others are purely repulsive. Another essential feature of *CRF* is the strong non-additivity of the interactions, leading to the formation of chemical bonds, *e.g.*, the well-known effect of saturation of the chemical bonding.

As a material, which can also exist as a non-stoichiometric compound, uranium dioxide generally vaporises non-congruently, with different compositions in the vapour and in the liquid phase. The vapour phase contains neutral and ionised molecular species (*e.g.*, UO_2, UO, UO_3, UO^+ *etc.*,), whilst the liquid phase is supposed to consist of ions such as U^{4+} and O^{2-} [125]. It should be emphasised, however, that the structure of the liquid state has still not been definitively established, and the problem appears to be particularly complex. For instance, based on this hypothesis the results of molecular dynamics (*MD*) simulations of liquid UO_2 give much less acceptable results than for solid state (see, *e.g.*, [11, 87, 125]).

In view of constructing an equation of state valid up to the critical point, we are forced to choose an universal model, applicable both to the gas and to the liquid phase. Yet, for the proposed problem, it is necessary to extend the traditional approach by taking into account:

- the wide temperature interval from melting up to the critical point, with densities from very dilute ideal gas up to the dense liquid;
- the non-stoichiometry of $UO_{2\pm x}$;
- the presence of neutral and charged species in the vapour phase;
- the chemical reactivity of gaseous species.

From the logical (but not historical) viewpoint we may subdivide all existing *EOS* into three groups.

1. Theoretical *EOS*. These are based on microscopic potential models and are derived by statistical mechanical theories or *ab initio* quantum mechanical calculations. Normally, such *EOS*' have no fitting parameters or the parameters are those of the underlying potential model. The most widely known is the Carnahan-Starling *EOS* for hard spheres [23]. Though the direct practical applicability of such *EOS* is very limited by the poor reality of the underlying models, they play an important role in the construction of semi-empirical *EOS*.
2. Empirical *EOS*. Many empirical formulae having little or no connection to the real laws of intermolecular interaction have been proposed since the pioneering work of van der Waals [135]. Empirical *EOS*' have been widely used for reconciliation, interpolation and compact representation of large collections of experimental data for a number of important substances like water, hydrocarbons, *etc.*. Usually such *EOS*' have many fitting constants and almost no scaling or extrapolation properties. The only applicability of empirical *EOS*' derived for one substance for prediction of properties of other ones is only justified by the *Principle of Corresponding States* (*PCS*).
3. Semi-empirical (model) *EOS*.
   Such *EOS*' employ the approximate analytical forms of theoretical *EOS*' for simplest molecular models (or a combination of them) to describe and predict thermodynamic functions of real substances. Their applicability essentially depends on the availability of (even very limited) experimental data on some basic properties (density, vapour pressure, *etc.*). These data are normally used for the determination of few essential *EOS* parameters. This procedure (calibration) is usually based on a previous selection of essential parameters and on the determination of their effect on different properties (sensitivity study). The accuracy of the predictions essentially depends on the appropriateness of the model chosen for the description of

the structure and on the specific features of the real inter-particle forces in the material. The *EOS* must be verified by additional experimental data (not used in the calibration procedure) or by comparison with other independent predictions.
Semi-empirical *EOS* models usually are more appropriate for extrapolation purposes and can be applied for prediction of thermodynamic properties beyond the explored range of temperatures and pressures. The model *EOS* have been exploited for prediction of high-pressure and high-temperature behaviour of complex strongly non-ideal multicomponent systems with chemical reactions and ionisation, like detonation products, shock compressed gases and fluids, non-ideal plasma, dissociating reactor coolants, *etc.* [58, 143].

Owing to the requirements described at the end of the preceding section the last type of *EOS* is the most suitable for fluid urania in the widest interval of temperatures and pressures.

2.2 The Concept of Equation-of-State

The thermodynamic state of an arbitrary single-phase M-component system in thermal, mechanical and chemical equilibrium is defined by the full set of $M+2$ thermodynamic parameters of state, *e.g.*, by temperature T, volume $\mathcal{V}$, and numbers of molecules for all components $N_1, \ldots N_M$. If this set of parameters is given, and the appropriate thermodynamic potential, the Helmholtz free energy $\mathcal{F}$, is known as a function of all the parameters of state:

$$\mathcal{F} = \mathcal{F}\left(T, \mathcal{V}, N_1, \ldots, N_M\right) \tag{2.1}$$

then all other thermodynamic properties (thermodynamic functions) of the system can be evaluated by differentiation of the free energy, $\mathcal{F}$, with respect to the parameters of state[1] [70].

The first derivatives with respect to the volume, $\mathcal{V}$, and temperature, T, give the *thermal*:

$$P = -\left(\frac{\partial \mathcal{F}}{\partial \mathcal{V}}\right)_{T, N_1, \ldots, N_M} \tag{2.2}$$

[1] All thermodynamic functions denoted by *calligraphic* symbols ($\mathcal{F}, \mathcal{U}, \mathcal{S}, \mathcal{H}, \mathcal{V}$, *etc.*) are extensive quantities, *e.g.*, proportional to the size of the system. In practice, it is more convenient to use *intensive* quantities calculated per unit mass (specific values), or molar ones (per *kmol*, or molecule, ion, *etc.*). We shall use mostly the latter kind of variables, *e.g.*, free energy $F = \mathcal{F}/N$, entropy $S = \mathcal{S}/N$, internal energy $U = \mathcal{U}/N$, *etc.*, as well as molar (specific) volume, V, number densities $n_i = N_i/\mathcal{V}$), *etc.*. These quantities along with pressure P, temperature T, and chemical potentials μ_i are intensive variables, whose values do not depend on the size of the system.

and the *caloric EOS*:

$$U = -T^2 \left(\frac{\partial F/T}{\partial T}\right)_{V,N_1,\ldots,N_M} . \tag{2.3}$$

Therefore, any subsequent mention of *EOS* should be understood as a reference to Eq.(2.1).

Differentiation of Eq.(2.1) with respect to N_i gives *chemical potential* of the i-th specie:

$$\mu_i = -\left(\frac{\partial \mathcal{F}}{\partial N_i}\right)_{T,\mathcal{V},N_{j\neq i}} . \tag{2.4}$$

The subscript here indicates that all N_j numbers, except N_i, are kept constant.

The entropy is defined either by the combination of free and internal energy $S = (U - F)/T$ or directly by the partial derivative of the *EOS*:

$$S = -\left(\frac{\partial F}{\partial T}\right)_{V,N_1,\ldots,N_M} . \tag{2.5}$$

If r reversible chemical reactions:

$$\sum_i \nu_i^{(\alpha)} \mathrm{A_i} = 0, \qquad \alpha = 1, \ldots, r \tag{2.6}$$

occur in the system, then only $M - r$ out of the N_i, $i = 1, \ldots, M$ parameters of state are independent since r additional constraints:

$$\sum_i \nu_i^{(\alpha)} \mu_i = 0, \qquad \alpha = 1, \ldots, r \tag{2.7}$$

are imposed. Here $\nu_i^{(\alpha)}$ are stoichiometric coefficients of the i-th component in the α-th reaction. Eqs.(2.6) are essentially equivalent to the requirement of the minimum of the Helmholtz free energy, and express the equilibrium of the α-th reaction proceeding in the system [70]. Due to these conditions, all thermodynamic functions expressed via *first* derivatives of the Helmholtz free energy (pressure, P, Eq.(2.2), internal energy, U, Eq.(2.3), entropy, S, Eq.(2.5)) are unaffected by the chemical reactions. The same is true for all thermodynamic functions which are linear combinations of F, U, P, and S, for instance, enthalpy $H = U + PV$ and Gibbs free energy $G = H - TS$.

This is not the case for all other thermodynamic functions, which depend on the *second* derivatives of the Helmholtz free energy, namely for:

– the isochoric heat capacity

$$C_V = \left(\frac{\partial U}{\partial T}\right)_V \equiv -T \left(\frac{\partial S}{\partial T}\right)_V , \tag{2.8}$$

– the isobaric heat capacity

$$C_P = \left(\frac{\partial H}{\partial T}\right)_P \equiv -T\left(\frac{\partial S}{\partial T}\right)_P, \tag{2.9}$$

– the isothermal (β_T) and adiabatic (β_S) compressibilities

$$\beta_T = -\frac{1}{V}\left(\frac{\partial V}{\partial P}\right)_T, \tag{2.10}$$

$$\beta_S = -\frac{1}{V}\left(\frac{\partial V}{\partial P}\right)_S, \tag{2.11}$$

– the volume thermal expansion

$$\alpha_T = \frac{1}{V}\left(\frac{\partial V}{\partial T}\right)_P, \tag{2.12}$$

– the speed of the sound, *etc.*.

All these properties are affected by the chemical reactions, *e.g.*, the isobaric heat capacity consists of the so-called 'frozen', $C_P^{(f)}$, and 'reactive', $\Delta C_P^{(r)}$, components:

$$C_P = C_P^{(f)} + \Delta C_P^{(r)}, \tag{2.13}$$

where the frozen component of the isobaric heat capacity is evaluated as the slope of the enthalpy at a constant fluid composition:

$$C_P^{(f)} = \left(\frac{\partial H}{\partial T}\right)_{P,N_1,\dots,N_M}. \tag{2.14}$$

The reactive contribution, $\Delta C_P^{(r)}$, depends on the composition and on the heats of the chemical reactions. In some cases this contribution is much greater than the frozen counterpart [71].

2.3 Model Equations of State

2.3.1 Ideal Gas

The simplest model *EOS* corresponds to the monatomic ideal gas, *i.e.*, to a system of non-interacting particles[2] (atoms or ions). The *EOS* for such systems is well known from statistical mechanics [89]:

$$\mathcal{F}^{(id)} = -kT\sum_{i=1}^{M} N_i \ln\left[\frac{e\,Q_i}{n_i}\left(\frac{m_i kT}{2\pi\hbar^2}\right)^{3/2}\right], \tag{2.15}$$

[2] Note that this model is applied frequently not only to dilute gases, where its validity is evident, but also to dense systems as an addition to *excess* contributions, accounting for particle interactions in composite *EOS*'.

where m_i is the mass if the i-th specie, n_i is its number density, $\hbar = h/2\pi$ is the reduced Planck constant, k is the Boltzmann constant, and

$$Q_i = \sum_k g_k \, e^{-E_k/kT}; \tag{2.16}$$

is the internal partition function of the i-th species; g_k is the degeneracy of the k-th electronic energy level E_k, the summation is here performed over all E_k.

By applying Eqs.(2.2) and (2.3) one gets the *thermal*

$$P^{(id)} = nkT \ , \tag{2.17}$$

and the *caloric EOS* for the ideal gas:

$$E^{(id)} = {}^3\!/_2 \, kT + \frac{\sum\limits_i N_i \langle E_i \rangle}{\sum\limits_i N_i} \ , \tag{2.18}$$

where

$$\langle E_i \rangle = \frac{\sum\limits_k g_k \, E_k \, e^{-E_k/kT}}{\sum\limits_k g_k \, e^{-E_k/kT}} \tag{2.19}$$

are internal contributions from electronic excitations.

The chemical potential of the i-th species in the ideal-gas model is:

$$\mu_i^{(id)} = -kT \, \ln \left[\frac{Q_i}{n_i} \left(\frac{m_i kT}{2\pi\hbar^2} \right)^{3/2} \right] . \tag{2.20}$$

All the thermodynamic functions of the ideal gas can be easily evaluated if the internal partition function Eq.(2.16) is known [63]. The internal energy, enthalpy and heat capacities of the ideal gas at a given composition depend on temperature only, whilst the free energy, entropy and chemical potential depend also on volume (pressure). This additional volume dependence is rather trivial and can be easily separated from the temperature dependence. For instance, the chemical potential may be written as:

$$\mu_i^{(id)} = \mu_i^{(id)} (T) - kT \, \ln \, n_i \ , \tag{2.21}$$

where

$$\mu_i^{(id)} (T) = -{}^3\!/_2 \, kT \, \ln \left(\frac{m_i kT}{2\pi\hbar^2} \right) - kT \, \ln Q_i . \tag{2.22}$$

2.3.2 Ideal Reacting Gas

The model of an ideal, chemically reacting gas is usually conceived[3] as being similar or even the same as that described in the previous section, except that the concentrations n_i of some (or all) species depend on temperature and, possibly, pressure (density), owing to the reversibility of the chemical reactions. We can substitute the expressions Eq.(2.21) in the conditions Eq.(2.7) of the equilibrium reactions Eq.(2.6), and transform these conditions into the equation

$$\prod_{1\le i\le M} n_i^{\nu_i^{(\alpha)}} = K_\alpha^{(id)}(T), \quad \alpha = 1, \ldots, r \tag{2.23}$$

known as *mass action law* [89] with temperature-dependent *equilibrium constants*:

$$K_\alpha^{(id)}(T) = \exp\left\{\sum_{i=1}^{M} \nu_i^{(\alpha)} \frac{\mu_i^{(id)}(T)}{kT}\right\}. \tag{2.24}$$

The set of equilibrium concentrations n_i, $i = 1, \ldots, M$ can be found as a solution of the non-linear set of Eqs.(2.23) at each temperature and density (pressure) by taking into account the normalization condition $\sum_i n_i = n$.

The first derivatives of the equilibrium constants give the ideal-gas heats of reactions:

$$\Delta H_\alpha^{(id)}(T) = kT^2 \frac{\partial \ln K_\alpha^{(id)}}{\partial T}. \tag{2.25}$$

The values of the ideal-gas constants Eq.(2.24) and reaction heats Eq.(2.25) can be calculated if the internal partition functions Eq.(2.16) are known or are tabulated in handbooks [9, 63].

2.3.3 Hard Bodies

The model of hard spheres is the simplest and the most commonly used hard-body molecular model representing the short-range repulsive forces acting between all particles (atoms, ions and molecules). According to this model, two particles do not interact until the distance, r, between their centres is greater than a certain limit d, characterising the size of a particle. This hard-sphere diameter, d, is the only parameter of the model. At distances $r < d$ the repulsion forces are so strong that such states do never realise. The Helmholtz free energy of a hard-sphere system may be written as a sum[4]:

---

[3] Rigorously, this model is self-contradictory because an event of chemical reaction cannot be thought without invoking some strong inter-particle forces responsible for the formation and breaking of chemical bonds, electron excitation, or ionisation. Even if such events are extremely rare and do not contribute essentially to the mean potential energy, this model should be applied with caution, and always in combination with a particular non-ideality model.

[4] The superscript $^{(R)}$ is assumed to denote any contribution of repulsive forces.

$$\mathcal{F} = \mathcal{F}^{(id)} + N\Delta F^{(R)}_{HS}, \tag{2.26}$$

where $\mathcal{F}^{(id)}$ is expressed by Eq.(2.15), and $\Delta F^{(R)}_{HS}$ is the excess contribution to the free energy due to the hard-sphere repulsion. This excess free energy has the following form:

$$\frac{\Delta F^{(R)}_{HS}}{RT} = f(\eta), \tag{2.27}$$

where $\eta = {}^1\!/_6 \pi n d^3$ is the *packing fraction, i.e.*, the portion of the volume occupied by hard bodies. The corresponding thermal equation of state in terms of the compressibility factor $z = P/nRT$ can be written as:

$$z(\eta) = 1 + \eta \frac{d\,f(\eta)}{d\eta}. \tag{2.28}$$

Many attempts to develop the explicit expression for $f(\eta)$ have been undertaken since the first work of van der Waals [135] who proposed the simplest concept of excluded volume: $z(\eta) = (1-4\eta)^{-1}$. Nowadays, computer simulation methods provide very precise data for the excess compressibility factor of hard spheres which is well represented by the simple formula[5] due to Carnahan and Starling [23]:

$$z_{HS} = \frac{2}{3}\left(\frac{PV}{RT}\right)^{PY/C}_{HS} + \frac{1}{3}\left(\frac{PV}{RT}\right)^{PY/V}_{HS} = \frac{1+\eta+\eta^2-\eta^3}{(1-\eta)^3},$$

that leads to the following equations for the excess free energy and compressibility factor:

$$f(\eta) = \frac{4\eta - 3\eta^2}{(1-\eta)^2}, \tag{2.29}$$

$$\Delta z_{HS} = \frac{4\eta - 2\eta^2}{(1-\eta)^3}. \tag{2.30}$$

A more general model of hard convex bodies was studied by Boublik [19], who proposed the following formula for the one-component fluid:

$$z_{HB}(\eta,\gamma) = \frac{1+\eta(\gamma-2)+\eta^2(1-\gamma+\frac{\gamma^2}{3})-\eta^3\frac{\gamma^2}{9}}{(1-\eta)^3}. \tag{2.31}$$

The value of the parameter γ depends on the molecular shape. Larger values of γ correspond to elongated molecules. For $\gamma = 3$, Eq.(2.31) reverts to the hard-sphere Eq.(2.30).

The generalisation of the Carnahan-Starling formula for a mixture of spheres having different diameters is [99]:

[5] This expression was derived as an average between two analytical expressions provided by the virial theorem $^{(V)}$ and the compressibility $^{(C)}$ equation based on the exact solution of the integral Percus-Yevick $^{(PY)}$ equation [142, 132].

$$\frac{\Delta F_{HS}^{(R)}}{RT} = X\frac{\eta}{(1-\eta)^2} + 3Y\frac{\eta}{1-\eta} + (X-1)\ln(1-\eta)\ , \tag{2.32}$$

$$X \equiv \left(\overline{d_i^{\,2}}\right)^3 \left(\overline{d_i^3}\right)^{-2};\quad Y \equiv \overline{d_i^2}\,\overline{d_i}\left(\overline{d_i^3}\right)^{-1};\quad \overline{d_i^k} \equiv \sum_i N_i d_i^k \Big/ \sum N_i;\ k = 1,2,3.$$

2.3.4 Soft Spheres

The common feature of the hard-body models is a zero non-ideality contribution to the caloric *EOS*. The in reality smoother short-range repulsion of atoms and ions produces an increase in the internal energy with temperature and density. This effect, excluded in hard bodies, is taken into account in the soft-sphere models.

The simplest way to include the effect of softness in the short-range interatomic repulsion is to replace the potential of hard spheres by the inverse-power repulsion. If the repulsive energy decreases proportionally to the m-th power of the interatomic distance, the excess free energy depends on both temperature and density in the form:

$$\Delta F_{SS}^{(R)} = RT\ f_{SS}(nT^{-3/m}). \tag{2.33}$$

This specific dependence entails the following relationship between excess caloric and thermal non-ideality contributions:

$$\Delta U_{SS}^{(R)} = \frac{3RT}{m}\Delta z_{SS}^{(R)}. \tag{2.34}$$

Within the soft-sphere model, the increase of the internal energy with temperature is proportional to the excess compressibility factor $\Delta z = z - 1$ and inversely proportional to the hardness of repulsion. If $m \to \infty$, the soft-sphere model reverts to that of hard spheres.

An additional simplification is possible at large $m \gg 1$. Function f_{SS} at the right side of Eq.(2.33) may be replaced by a correspondingly reduced hard sphere free energy Eq.(2.27) with an effective soft-sphere diameter:

$$d\,(T) = d_o\,(T_o/T)^{1/m}. \tag{2.35}$$

Consequently, the packing fraction becomes temperature-dependent:

$$\eta = {}^1\!/_6\pi\, n\, d_o^3\,(T_o/T)^{3/m}\ . \tag{2.36}$$

Barker and Henderson [10] introduced another definition of temperature-dependent effective molecular size. For molecules interacting via a central potential $\Phi\,(r)$ they found:

$$d_{BH}(T) = \sigma - \int_0^\sigma \exp\left(-\frac{\Phi(r)}{kT}\right)dr\ , \tag{2.37}$$

where σ is the smallest (short-distance) root of the equation $\Phi(\sigma) = 0$. Levesque and Verlet [95] represented the computed values of $d^*_{BH}(T^*) = d_{BH}/\sigma$ by the simple analytical fit:

$$d^*_{BH} = \frac{0.9955 + 0.168\,T^*}{1 + 0.1958\,T^*}. \tag{2.38}$$

Here $T^* = kT/\varepsilon$ is the reduced temperature. A more sophisticated version of the soft-sphere model was proposed by Andersen, Chandler and Weeks [6]. Instead of the inverse-power repulsion they used the repulsive part of the Lennard-Jones potential and calibrated the effective diameter to fit the first peak of the static structure factor. Here the effective diameter becomes both temperature and density dependent. We refer for details to the original paper [6] and to a comprehensive review [150].

All definitions of the soft-sphere diameter predict its slow decrease with increasing temperature, what affects the caloric *EOS* by providing a positive contribution to the internal energy. Therefore, all the proposed temperature functions have positive second derivatives with respect to T, what corresponds to a positive non-ideality contribution to the isochoric heat capacity $\Delta C_V = C_V - C_V^{(id)} > 0$.

2.3.5 Van der Waals' Model

At high temperatures and densities the repulsive forces between atoms and molecules provide the main contribution to the non-ideality corrections to the thermodynamic functions. At intermediate temperatures and densities, *e.g.*, near the gas-liquid critical point, the short- and intermediate-range intermolecular attraction becomes an equally important part of the particle interaction.

The first successful attempt to take into account both repulsive and attractive forces was made by van der Waals in 1873. His classic thermal *EOS* (*VdW-EOS*):

$$\left(P + \frac{a}{V^2}\right)(V - b) = RT \tag{2.39}$$

includes along with the repulsion parameter $b = 4\eta V$ (in terms of the hard-sphere model), the attraction parameter a.

The equation for the Helmholtz free energy can be written as a sum of three terms:

$$F = F^{(id)} + \Delta F^{(R)}_{VdW} + \Delta F^{(A)}_{VdW}, \tag{2.40}$$

where the superscripts $^{(R)}$ and $^{(A)}$ refer to repulsive and attractive forces, respectively.

The repulsion contribution in Eq.(2.40) is essentially the same as in the hard-sphere model Eq.(2.29), but the approximation corresponding to the original *VdW-EOS* Eq.(2.39) gives:

$$\Delta z^{(R)}_{VdW} = \frac{4\eta}{1-4\eta}, \tag{2.41}$$

and the excess repulsion free energy is:

$$\Delta F^{(R)}_{VdW} = -RT \ln(1-\eta) \, . \tag{2.42}$$

Eqs.(2.42) and (2.29) have the same first-order term in expansion on powers of η, whilst the hard-core repulsion contribution becomes much larger at higher densities. Among the numerous attempts to improve the *VdW-EOS*, the most popular are those based on replacement of Eq.(2.41) by Δz from Eq.(2.30) (see, *e.g.*, Haar and Shenker [65]).

The approach proposed by Mistura, Magill and Ohse for UO_2 [105] is also based on a modified *VdW-EOS*:

$$z = z_{HB}(\eta, \gamma) - \frac{a}{VRT}, \tag{2.43}$$

where $z_{HB}(\eta, \gamma)$ is the compressibility factor of Boublik defined by Eq.(2.31).

The key point of the *VdW* model is the contribution of the attractive forces, which is negative and proportional to the number of interacting molecular pairs:

$$\Delta P^{(A)}_{VdW} = -a\,n^2 . \tag{2.44}$$

The positive constant, a, in Eqs.(2.39) and (2.44) is the measure of the intensity of the attractive forces. The quadratic dependence of the excess pressure on density in Eq.(2.44) corresponds to the first-order correction (at the level of the second virial coefficient) in the group expansion of pressure of non-ideal gas [70]. Attractive forces give a negative and temperature-independent contributions to the caloric *VdW-EOS*:

$$\Delta U^{(A)}_{VdW} = -a\,n \, , \tag{2.45}$$

though there is no contribution to the isochoric heat capacity:

$$(\Delta C_V)_{VdW} \equiv 0 \, . \tag{2.46}$$

It is worth remarking that the *VdW-EOS* is open to a straightforward generalisation to multicomponent mixtures.

More sophisticated approaches like the *thermodynamic perturbation theory* (*TPT*) [10, 6] are essentially the more rigorous extension of the van der Waals' ideas, and their main success is related to a more accurate evaluation of the attractive contribution in realistic potential models[6].

[6] More detailed discussion of *TPT-EOS* is presented in Chapter 6.

2.3.6 Lennard-Jones Fluid

The *EOS* for one-component fluid composed of particles interacting via the additive potential:

$$\Phi_{LJ}(r) = 4\varepsilon\left[\left(\frac{\sigma}{r}\right)^{12} - \left(\frac{\sigma}{r}\right)^{6}\right] \tag{2.47}$$

(*LJ*-fluid) has been the object of study in many computer simulations and theoretical approaches since it is believed to suitably represent simple fluids (Ar, Kr, Xe, *etc.*). Monte Carlo and molecular dynamics simulations provide enough material for the construction and calibration of empirical *EOS'* (see Ref.[150] and references therein), $P^* = P^*(\eta, T^*)$, expressed in terms of reduced pressure, P^*, packing fraction, $\eta = {}^1\!/_6 \pi\, n\, \sigma^3$, and reduced temperature, $T^* = kT/\varepsilon$. These *EOS* describe the thermodynamic $P - V - T$ surface of the *LJ*-fluid with high accuracy in a broad interval of T^*. Although few real substances (condensed noble gases) exhaust the immediate applicability of such a model, its significance is wide-ranging. An attractive idea is to use the *LJ*-fluid as a reference system for more complex fluids on the basis of *PCS*, or as a *reference system* in *TPT*. In the last case, the analytical *EOS* for pure *LJ*-fluid is not sufficient. One needs first its binary distribution function to evaluate the first-order perturbation corrections or at least some moments of this distribution function.

Another important application of the model *LJ*-fluid is the generalisation of its *EOS* for multicomponent fluid mixtures. The simplest approximation to evaluate the properties of a fluid mixture is the *one-fluid approximation.* The *EOS* is written as a sum of ideal-gas contributions and the excess free energy of a fictitious single-component fluid whose effective molecular parameters are combinations of individual molecular constants. Certain additional rules for combining individual parameters of particle interactions in single effective values σ_{MIX} and ε_{MIX} [71, 85, 101] are invoked, like:

$$\sigma^3_{MIX} = \sum_j x_j\, \sigma^3_{jj}\,, \tag{2.48}$$

$$\varepsilon_{MIX} = \frac{1}{\sigma^6_{MIX}} \sum_{i,j} x_i x_j\, \sigma^6_{ij}\, \varepsilon_{ij}\,, \tag{2.49}$$

where $\sigma_{ij} = \frac{1}{2}(\sigma_{ii} + \sigma_{jj})$ and $\varepsilon_{ij} = (\varepsilon_{ii}\varepsilon_{jj})^{1/2}$ are empirical combination rules used to estimate the potential parameters of interactions between molecules of different sorts, via known potential interactions parameters between individual molecules of the same sort. The choice of combination rules like Eqs.(2.48) and (2.49) has normally no cogent reasons, and is much dependent on the success in the description of the given fluid mixture. However, the successful applications of one-fluid approximations are normally limited to fluids having nearly spherical molecules of approximately equal sizes.

The most promising application of this model is its use as a reference fluid for construction of *EOS*' based on *TPT* approaches. In fact, it offers a better alternative to empirical approximations like the one-fluid model, and will be described in more details in Chapter 6.

2.3.7 Charged Spheres

The first successful attempt to evaluate the Coulomb contribution to *EOS* for fluid composed of charged particles is the well-known Debye-Hückel theory of dilute electrolytic solutions [89]. According to the *Debye-Hückel limiting law* (*DHLL*) at low concentrations of ions the Coulomb contribution to the Helmholtz free energy is written as:

$$\Delta F^{(C)}_{DHLL} = -\frac{kTV}{12\pi r_D^3}, \tag{2.50}$$

where

$$r_D = \left(\frac{e^2}{\epsilon_0 kTV} \sum_{i=1}^{M} N_i Z_i^2 \right)^{-1/2} \tag{2.51}$$

is the *Debye radius* [34]. Eqs.(2.50) and (2.51) do not involve any characteristic size of the interacting particles and correspond to the pure Coulomb interaction potential:

$$\Phi^{(C)}(r_{ij}) = \frac{e^2 Z_i Z_j}{4\pi\epsilon_0 r_{ij}}, \tag{2.52}$$

where $\Phi^{(C)}(r_{ij})$ is the potential energy of electrostatic interaction between two point charges eZ_i and eZ_j placed at a distance r_{ij}; e is the elementary charge, and ϵ_0 is the dielectric permittivity of free space (dielectric constant $\epsilon_0 = 8.85\ 10^{-12} F m^{-1}$). Eq.(2.50) is valid in the limit of very small ratios of the characteristic molecular size d to the Debye radius (Coulomb parameter):

$$\Gamma_D \equiv \frac{d}{r_D} \ll 1. \tag{2.53}$$

More sophisticated versions of Debye-Hückel theory were developed for higher concentrations of charged particles:

- *Improved Debye-Hückel approximation for charged hard spheres* (*DHSE*)[33]. This approximation, when combined with an equation of energy, provides[7] the following formula for the Coulomb contribution to the excess Helmholtz free energy:

$$\Delta F^{(C)}_{DHSE} = \frac{kT}{24\eta} \left[\ln(1+\Gamma_D) - \Gamma_D + \frac{\Gamma_D^{\,2}}{2} \right], \tag{2.54}$$

  where η is the packing fraction.

[7] The expression for the Helmholtz free energy within any approximate microscopic theory of structure and properties, depends on the basic equation which it is derived from [70].

- *Mean-Spherical Approximation* (*MSA*) [140]. This is a generalisation of the Percus-Yevick (*PY*) approximation in the theory of integral equations for pair correlation functions of charged hard spheres. From the energy (*MSAE*) and virial (*MSAV*) equations the following formulae are attained:
  - *MSAE*: *MSA* approximation + Energy equation

$$\Delta F^{(C)}_{MSAE} = \frac{kT}{72\,\eta}\left[2\,(1+\Gamma_D)^{3/2} - 2 - 6\,\Gamma_D - 3\Gamma_D{}^2\right]; \tag{2.55}$$

  - *MSAV* : MSA approximation + Virial theorem

$$\Delta F^{(C)}_{MSAV} = \frac{kT}{72\,\eta}\left[\frac{4}{\Gamma_D{}^2}\ln\left(\frac{\sqrt{1+2\Gamma_D}+1}{2}\right) - \frac{2\sqrt{1+2x}-1-\Gamma_D}{\Gamma_D{}^2}\right]. \tag{2.56}$$

It should be noted that Eqs.(2.54) to (2.56) fit the *DHLL* for $\Gamma_D \to 0$, and give almost identical results for $\Gamma_D \ll 1$. The expansion of the function:

$$\varphi\,(\Gamma_D) = \frac{12\pi r_D^3}{V}\,\frac{\Delta F^{(C)}}{RT} \tag{2.57}$$

corresponding to Eqs.(2.50), (2.54), (2.55), and (2.56) in power series has the form:

$$\varphi\,(\Gamma_D)_{DHLL} = -1;$$

$$\varphi(\Gamma_D)_{DHSE} = -1 + \frac{3}{4}\Gamma_D - \frac{3}{5}\Gamma_D{}^2 + \ldots;$$

$$\varphi(\Gamma_D)_{MSAV} = -1 + \frac{1}{2}\Gamma_D - \frac{5}{12}\Gamma_D{}^2 + \ldots;$$

$$\varphi(\Gamma_D)_{MSAE} = -1 + \frac{3}{4}\Gamma_D - \frac{3}{4}\Gamma_D{}^2 + \cdots\ .$$

At high Coulomb non-ideality $\Gamma_D \gg 1$, the above formulae, except for the *DHLL*, predict a qualitatively similar behaviour, but with very different values of $\varphi\,(\Gamma_D)$.

2.3.8 Composite Models

In the general case the interaction between particles includes a number of different, physically important contributions both repulsive and attractive. Aside from the common short-ranged repulsion (van der Waals attraction and Coulomb electrostatic interactions), some systems demonstrate the importance of short ranged saturative covalent forces responsible for formation of chemical bonds, as well as hydrogen bonding, the interaction of molecular dipoles, quadrupoles, *etc.*. Some of them are strongly angular-dependent, and cannot be described by a central-force model. A general extension of the above expressions for the Helmholtz free energy for simple systems Eqs.(2.26) and (2.40), which we will call the *composite model*, is written as:

$$F = F^{(id)} + \Delta F^{(R)} + \Delta F^{(A)} + \Delta F^{(C)} + \dots \,, \tag{2.58}$$

where the superscripts $^{(\mathrm{id})}$, $^{(\mathrm{R})}$, $^{(\mathrm{A})}$, $^{(\mathrm{C})}$ stand for ideal gas, short range repulsion, attraction and Coulomb contributions, respectively.

The additivity of different contributions can be in no way regarded as self-evident. All kinds of particle interactions affect the static structure of the fluid and each contribution of a particular kind of forces in Eq.(2.58) depends on the presence and the intensity of others. Even the separation of the ideal-gas contribution, so usual for statistical mechanical and chemical approaches [6], is an approximation valid only for central forces. The shift and broadening of spectral lines and bands in optical spectra of molecules illustrate very well the effect of interactions between intermolecular forces and internal degrees of freedom in the condensed phase [8, 143].

Not all kinds of forces can be regarded as structure-forming factors. The most important in this sense are the short-ranged repulsive forces. Therefore, in each term of approximate additive form should be present at least a common parameter, *e.g.*, the effective size(s) of particles. The intensive Coulomb forces also affect the spatial distribution of charged species, and hence the presence of Coulomb parameter(s) is desirable in other contributions. Unfortunately, the state of the art in statistical mechanical theory of complex dense Coulomb systems is far from being satisfactory, and this aspect is not adequately considered in practical applications.

Another point, which must be mentioned in connection with the approximate additive form of the composite model, is the implicit interdependence of all terms in a case of the dense, chemically reacting fluid (*CRF*). This relationship is due to common equilibrium composition which is present in all the terms of Eq.(2.58). This requirement provides an additional important link between all contributions to the composite model.

The problem of the equilibrium composition will be discussed below in more detail. We shall here present only the general expressions for the chemical potentials and equilibrium constants.

According to Eq.(2.4), the chemical potential for the i-th species is:

$$\mu_i = \mu_i^{(id)} + \left(\frac{\partial \Delta \mathcal{F}}{\partial N_i}\right)_{T,\mathcal{V},N_j} . \tag{2.59}$$

Correspondingly, we obtain the equilibrium constant for an arbitrary reaction by substituting the general Eq.(2.58) into Eq.(2.7):

$$K_\alpha = K_\alpha^{(id)}(T)\, \exp\left\{\sum_{i=1}^{M} \nu_i^{(\alpha)} \left(\frac{\partial \Delta \mathcal{F}}{\partial N_i}\right)_{T,\mathcal{V},N_j}\right\} . \tag{2.60}$$

Here $K_\alpha^{(id)}(T)$ is the ideal-gas equilibrium constant Eq.(2.24). Note that the real equilibrium constant Eq.(2.60) depends not only on temperature and density but also on the equilibrium composition of the *CRF*. The corresponding heat of reaction, which determines the shift of the chemical equilibrium with temperature, can also be split into an ideal and an excess part:

$$\Delta H_\alpha = kT^2 \left(\frac{\partial \ln K_\alpha}{\partial T} \right)_P = \Delta H_\alpha^{(id)} + \Delta H_\alpha^{(ex)} . \tag{2.61}$$

The effect of non-ideality in the chemical equilibrium is characterised by the *reaction volume*:

$$\Delta V_\alpha = -RT \left(\frac{\partial \ln K_\alpha}{\partial P} \right)_T , \tag{2.62}$$

and can be also evaluated from the *EOS* Eq.(2.58).

2.4 Physical and Chemical Models in Thermodynamics of Reacting Fluids

2.4.1 The Concept of Composition

Essentially, there are two approaches in the microscopic theory of *CRF*. A first one, the *physical model*, where only structural elements are atoms or ions. In this approach, all kinds of chemical bonding appear as a consequence of the specific particle interaction, and are not treated explicitly; consequently, no concept of composition (except of elemental one) is introduced. This treatment is formally rigorous, at least within the framework of the classic mechanical description, and is free of arbitrary assumptions related to the chemical bonding. Physical models, however, encounter some difficulties when applied to a real, dense *CRF*. No useful calculation method had been developed so far for fluids in the framework of a physical model (except direct computer simulations) and, accordingly, very crude analytical approximations are usually used here. The second, unsatisfactory feature of the physical models is the inability to predict concentrations of any compound species formed in the liquid phase.

The second approach, (also known as the *mixture approach*), is intended to construct a *chemical model* of the system to be represented. Under certain aspects it appears at a first sight preferable. Here the basic concept is the subdivision of all microscopic states of atoms into 'bonded' and 'unbonded' parts. One of the first successful attempts to develop a coherent chemical model was made by Hill in his well-known *physical cluster theory* [70]. In fact he was the first to set the problem of subdividing the phase space of interacting groups of particles into 'bonded' and 'unbonded' parts. *CRF*'s are treated as mixtures of free atoms (or ions) as well as of all possible molecules or complexes, which can be formed. The law of mass action for a given *EOS* model combined with a rule for selection of species (model of composition) defines the equilibrium composition of the mixture. The main advantage of the chemical model is the possibility to define the set of species in a way that includes both very strong attractive valences and Coulomb forces in the *internal* interactions of the species formed in the mixture. Sometimes this is also possible in a physical model, which actually may be considered as a limiting case of chemical model with one sort of elemental or composite particles. For example, the equilibrium structure of dense dissociating hydrogen [147] or a concentrated electrolyte [128] can be effectively described in terms of atomic [147] or ionic [128] pairs in a coherent manner. This point, which makes the essence of the Stillinger-Reiss theorem [117, 128], gives the possibility to describe the macroscopically neutral electrolyte as a fluid composed of neutral microscopic particles - for instance, of bonded ionic pairs.

On the other hand, the chemical model has the disadvantage that it needs some arbitrary definition of chemical bonding. This problem was extensively discussed in the past [124, 122, 144]. The conclusion was that, whatsoever self-consistent definition of bonding is applied, the resulting thermodynamic properties are always the same. However, the calculated composition depends on the definition of bonding, which must be suitable for comparison with experimental data. An alternative approach to this problem is based on the variational principle [144]. Subdivision of the phase space into 'bonded' and 'unbonded' parts was considered in Ref. [144] as a variational problem for each specific reference system. Another possible way, at least for the simplest cases of dimerising atomic systems, is to follow a rigorous, based on the integral equations for the conventional interatomic correlation functions [145].

Liquid uranium dioxide may be regarded as a complex *CRF*. The specific features of such a liquid are related to the type of interactions between atoms or ions. There is only one viable approach, which is suitable for developing an equation of state over a wide interval of temperatures, namely, the mixture approach. However, for comparison, an alternative approach, based on a physical model valid in a restricted interval of temperatures and densities is employed and commented in Chapter 3.

2.4.2 Neutral Models of $UO_{2\pm x}$

In accordance with the general idea inherent in the chemical approach that ionic liquids – in particular, liquid uranium dioxide - may be successfully described by different physically equivalent set of components, a variety of appropriate models may be considered. For many of them exploratory calculations were performed [77].

- **Physical neutral model (*NM-0*).** This is the simplest mono-molecular description of the real multi-component U-O system as being composed only of UO_2 molecules [105].
  Merit: No concept of composition required.
  Limitations: The concepts of independent chemical potentials of oxygen and uranium cannot be introduced.

The structure of phase boundaries of the model must necessarily be different from the real ones.

- **Restricted Neutral Model(*RNM*)**. The simplest approximation of the chemically reacting U-O system, is assured consisting of three molecular species only: UO, UO_2, and UO_3.
  Merit: The oxygen chemical potential can be introduced as a distinct parameter, strongly dependent (in the liquid phase) on the interaction parameters of the three species involved

$$\mu_{\mathrm{O}}=\mu_{\mathrm{UO_2}}-\mu_{\mathrm{UO}}=\mu_{\mathrm{UO_3}}-\mu_{\mathrm{UO_2}} , \tag{2.63}$$

  so that non-congruent phase equilibrium in UO_2 may be considered in a rough approximation.
  Limitation: Absence of atomic and molecular oxygen is not essential for the chemical picture of the liquid phase at low temperatures, but it is of primary importance for the vapour phase. Absence of oxygen makes the parameters of the saturated vapours of uranium dioxide at $T \gg T_m$ and $(O/U)^{vap} \gg 2.0$ non-realistic. For example, at $T = T_m \approx 3120\ K$ and $O/U > 2.25$, the oxygen mole fraction[8] in the vapour is $x_o > 0.02$.
- **Restricted Primitive Neutral Model**. Similarly to the well-known *restricted primitive model* of the conventional electrolyte theory, one may introduce a model including only three neutral species: UO, UO_2 and UO_3 with equal parameters of repulsion and attraction. The only possible chemical reaction is:

$$2\mathrm{UO_2} \rightleftarrows \mathrm{UO}+\mathrm{UO_3}. \tag{2.64}$$

  The contribution of this reaction to the caloric *EOS* may be taken into account, while the shift of the equilibrium constant of this reaction in the liquid, in comparison with those in the vapour, is absent, so that the gas-liquid phase transition in this model is always congruent.
- **Neutral Model (*NM-I*)**. A restricted set of neutral species (O + O_2 + U + UO + UO_2 + UO_3) is considered in this model, not only for the vapour, as usually done [54, 108], but also for the liquid phase.
  Merit: This represents a realistic model for the non-congruent equilibrium vapour pressure far from the critical region, where the Coulomb non-ideality of charged species must be taken into account.
  Limitations: Absence of charged species, which are important even at low temperature (also in the vapour [98]).
  Possibility of a strong chemical coupling (binding) of molecules UO_n into complexes such as U_2O_2, U_2O_3, U_2O_4, U_2O_5, *etc.*, is not taken into account. It should be noted that hypothetical formation of chains and clusters U_iO_j, with i and $j \gg 1$ is quite plausible at high temperatures, and in the dense vapour, and, obviously, also in the expanded liquid close to the critical region. This effect may influence the location of the critical

[8] See the discussion in Section 5.2

point and affects properties (such as the oxygen potential) , which are very sensitive, in this region, to the details of the particle interactions[9].

- **Neutral Model (*NM-II*).** Same as *NM-I*, but with addition of complex molecules of the type U_2O_2, U_2O_3, U_2O_4, and U_2O_5.

2.4.3 Ionic Models

- **Restricted Primitive Ionic Model (*RPIM*).** Stoichiometric composition of two, Coulombically interacting ionic species: U^{4+} and O^{2-} is adopted. Both species have the same size.
- **Primitive Ionic Model (*IM-I*).** Same as *RPIM* except for different diameters of hard-sphere for short-range repulsion of different ionic species. Both models are referred to as *IM-I* model.
  Merit: The *RPIM* model is a standard ionic model in the theory of the non-ideal plasma [140]. Numerous theoretical applications (integral equations, computer simulations) provide exhaustive information on the properties of this model.
  Good analytical equations of thermal and caloric properties at moderate non-ideality are available.
  The *IM-I* models can have limited success when used for calculation of properties of dense stoichiometric liquid uranium dioxide at low temperatures.
  The chemical potential of UO_2 can be defined as[10]:

$$\mu_{UO_2} = \mu_{U^{4+}} + 2\,\mu_{O^{2-}}\,. \tag{2.65}$$

  The model can be used for calculation of phase coexistence in $UO_{2.00}$ when combined with a separate or *local EOS* for the vapour phase.
  Limitations: Only stoichiometric liquid phase ($O/U = 2.00$) can be treated. No independent chemical potentials for components. Only one parameter (ionic size), entailing low flexibility.
  Oxygen chemical potential cannot be introduced. *IM-I* cannot be used for the calculation of non-congruent evaporation.
  Moreover, absence of possible reactions of electronic exchange between ions confines the validity of *IM-I* to low temperatures. In fact, it is known that the ionisation degree is meaningful not only for the vapour phase, but also for the liquid. It is also known that in the vicinity of the critical

[9] The hypothesis has been made [123] concerning the possible coexistence of two different kinds of ionic liquids: a normal, compact one at low temperature and high density, and a *chain-structured* friable one in a hot expanded state close to the critical point.

[10] One may consider a similar model composed of ions U^{2+} and O^-. It is known that the magnitude of the ion charge in a given substance is an effective parameter. In particular, in many real oxides the ion labelled as O^{2-} have an effective charge of $Z = 0.9 \ldots 1.1$. [141].

point for many ionic liquids the ionisation degree decreases with increasing temperature [61].

- **Extended Pure Ionic Model (*IM-II*)**. Ions other than those of *IM-I* are taken into consideration[11]. For example, U^{5+}, U^{4+}, U^{3+}, *etc.*, as well as O^{2-} and O^{-}.
Merit: Electron and oxygen chemical potentials can be defined, as:

$$\mu_e = \mu_{U^{4+}} - \mu_{U^{5+}} , \tag{2.66}$$

$$\mu_O = \mu_{O^{2-}} + \{\mu_{U^{5+}} - \mu_{U^{3+}}\} . \tag{2.67}$$

IM-II can be used for calculation of the non-congruent evaporation when it is combined with a separate model for the gas phase. A realistic molecular composition of the vapour can be reproduced, at least in principle, for example, with the use of a direct numerical simulation (like *MD*-simulation [125]) of actual complex formation. This approach is, however, very complicated and hardly applicable.
- **Improved Restricted Ionic Model (*IM-III*)**. The same as IM-I, but additional short-range van der Waals attraction between ions is included. The importance of short-range dipole-dipole and dipole-quadruple dispersion attractive forces was demonstrated in numerous Monte Carlo and molecular dynamics computer simulations of ionic fluids.
Merit: The model is well founded and more flexible than *IM-I.* It can be successfully used in combination with additional constraints to reproduce important properties of ionic fluids.
Limitations: Only stoichiometric liquid phase ($O/U = 2.00$) can be considered. Calibration is more complicated, and the empirical separation of van der Waals contributions from the total attractive energy of liquid is a rather ambiguous and not indisputable procedure.

2.4.4 MIX Models

The most general type of the models for a mixture are called here *MIX* models. The composition in this type of models includes both neutral and charged particles. The principal advantage of *MIX* models is their flexibility, the seamy side is usually the very complicated calibration procedure as well as risks of ambiguity in the set of parameters found.

- ***MIX-I*** - It includes the following species: O, O_2, U, UO, UO_2 UO_3, U_2O_2, U_2O_3, U_2O_4 , U_2O_5, U^+, UO^+, UO_2^+, UO_2^-, UO_3^-, O^-, O_2^-, and e^-. This model may be considered sufficiently realistic for dense vapours and expanded liquids in the high-temperature region of phase coexistence. The essential assumption in its application to the liquid phase is the existence of

[11] It should be noted that electrons are implicitly present as a phantom component in this model.

Table 2.1. Compositions Adopted in the Model *EOS*'

Model	*EOS*	O/U	UO_2	UO UO_3	U O	O_2	$U^{(4+)}$ $O^{(2-)}$	$U^{(5+)}$ $U^{(3+)}$	$U^{(2+)}$ $O^{(-)}$	$UO_2^{(\pm)}$ $UO_3^{(\pm)}$
Neutral	*PHC* [105]	2.0	√							
	Simplest *VdW*	2.0	√							
	Simplest *TPT*	2.0	√							
Ionic	*IM-I*	2.0					√		√	
	IM-II	2±x					√	√	√	
	IM-III	2.0					√			
MIX	*VdW-DH-HSM*	2±x	√	√	√	√			√	√
	INTAS-99-*EOS*	2±x	√	√	√	√			√	√

a set of conventionally distinguishable clusters, mainly with one uranium centre and several oxygen constituents, where free exchange of oxygen and electrons can take place. The main problem here is the self-consistent definition of the effective potentials of the various interactions.

- ***MIX-II***- Besides the composition included in model *MIX-I*, the possibility of multi-stage ionisation is taken into account, *i.e.* :

$$MIX - II = MIX - I + \{U^{6+}, U^{5+}, U^{4+}, U^{+3}, U^{2+}, O^{2-}\}. \qquad (2.68)$$

The main purpose of this model is a unique, wide-ranging description of the uranium dioxide, which exhibits a continuous transition from the rare molecular gas phase to the dense cold liquid state with presumably highly ionised ionic structure. This model is widely used in the theory of the *strongly coupled plasma (SCP)*, a state which has been intensively studied in the past time in the context of the *Gas-Core Nuclear Reactor* design [58].

2.5 Conclusions

Several equations of state are tested and discussed in the following Chapters, starting from (inadequate) simplest ideal systems of neutral particles or purely ionic systems, and proceeding towards more articulated equations, encompassing several aspects of the atomic interactions.

In Table 2.1 we present the correspondence between these *EOS*' and the composition models discussed in the Section 2.4.

The scheme in Fig. 2.1 shows how basic models of interatomic forces have been combined to construct different types of equation of state for liquid and gaseous uranium dioxide. Three types of forces are considered: the short-range repulsive and attractive ones, whose origin resides in the interaction of electronic and nuclear charges, and the long-range Coulomb forces, which are mainly created by the cumulative atomic charges. All these types of forces with

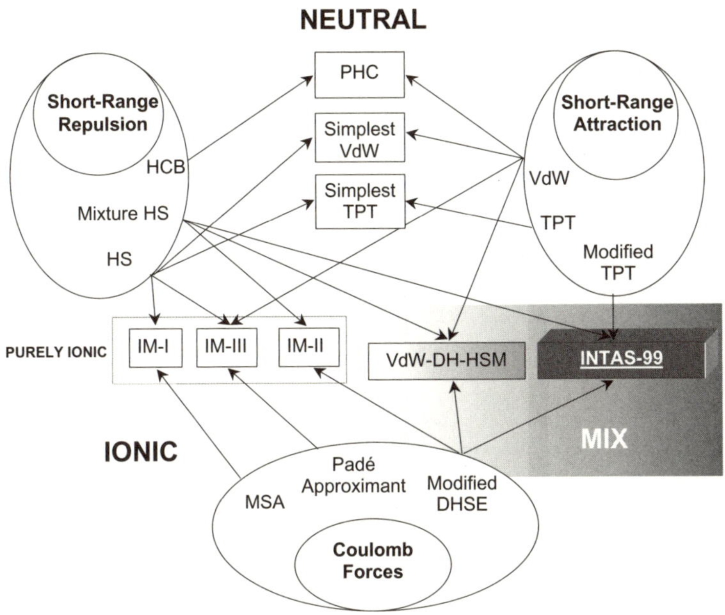

Fig. 2.1. The basic models of interatomic forces and different types of *EOS*' for uranium dioxide

their related potentials assume an increasingly complex spatial dependence for higher atomic densities, so that their functional expressions require further corrections; these are indicated in the figure in the respective elliptical boxes with acronyms explained in the Chapter 1. The models used to describe the contributions of these interactions to *EOS* are discussed in Section 2.3.

The equation of state labelled *INTAS*-99-*EOS* represents the most comprehensive and realistic model which could be presently devised for fluid uranium dioxide. The results of extensive calculations, reported in the Tables at the end of this monograph, have been obtained by applying this equation.

3

Ionic Models for Liquid Urania

Several substances in the condensed state are essentially governed by electrostatic forces and hence can be described, at least in principle, by the simplest Coulomb potential. Yet, though the short to medium range interactions can be evaluated with a consistent mathematical treatment, the long range effects, increasingly important at higher ion densities, are much more problematic. Actually, a general solution of this problem is still wanted, and only restricted applications are available. Despite these limits, ideal ionic fluids are exerting a particular fascination for the important reason that they are fully determined by the mere definition of their charges, and no other parameter is required to specify their thermodynamic quantities at any temperature. Furthermore, explicit relationships between some of them can be established, having an absolutely general validity for all these substances.

This Chapter is dedicated to the discussion of pure ionic models, with the purpose of explaining their properties in contexts where they are taken to represent partial aspects of more complex thermodynamic systems.

3.1 Restricted Primitive Ionic Model

Instead of a multi-component mixture of different types of ions we consider here only a two-component mixture of ions of species U^{4+} and O^{2-}. Electroneutrality implyes $n_{U^{4+}} = 2\,n_{O^{2-}}$, where n_α are the corresponding number densities of ions. The interactions are assumed to be pair-wise additive:

$$U_N = \frac{1}{2}\sum_{\alpha,\beta}\sum_{i,j}\Phi_{\alpha\beta}(r_{ij}), \tag{3.1}$$

where U_N denotes the total interaction energy. The interaction potential takes into account only two kinds of forces: short-range repulsion and Coulomb forces:

$$\Phi_{\alpha\beta}(r_{ij}) = \Phi^{(R)}_{\alpha\beta}(r_{ij}) + \Phi^{(C)}_{\alpha\beta}(r_{ij}), \tag{3.2}$$

The first term is here represented by the hard–sphere potential:

$$\Phi^{(R)}_{\alpha\beta}(r) = \begin{cases} +\infty, & r < \sigma_{\alpha\beta} \\ 0\ , & r > \sigma_{\alpha\beta}, \end{cases} \tag{3.3}$$

and the Coulomb interaction $\Phi^{(C)}_{\alpha\beta}(r_{ij})$ is represented by Eq.(2.52).

All the ionic diameters are the same ($\sigma_{\alpha\beta} = d$) so that only one adjustable parameter is available.

This model has been extensively studied in computer simulations [150]. A property of prime importance, which must be reproduced by *any* model is the density $\rho^{(liq)}(T_m)$ at the melting point. For $UO_{2.00}$ the value $\rho^{(liq)}(T_m) = 8870\ kg\ m^{-3}$ is well established [36]. At the melting temperature the ambient pressure in the liquid can be replaced by zero without any loss of precision, therefore, the only constraint to be fulfilled is:

$$P_{UO_{2.00}}(T_m, \rho^{(liq)}) \cong 0. \tag{3.4}$$

The effective diameters fitted from this equation and some thermodynamic properties calculated at melting temperature on the basis of the Carnahan-Starling formula Eq.(2.29), are respectively presented in columns 3 (*MSAE*) and 5 (*MSAV*) of Table 3.1 for different approximations of the Coulomb contribution, Eqs.(2.55) and (2.56). The comparison with experimental data and even with computer simulations show that this one-parametric *IM-I* model has little success.

Only the speed of sound, V_s, is predicted in reasonable agreement with the experiment, this might by due to coincidental error cancellation. Although the value of the effective diameter of ions is in reasonable accordance with the known ionic sizes, the prediction of entropy and thermal coefficients is poor. From these comparisons one can see that there are two additional sources of inaccuracy:

- The uncertainty in the analytical representation of the Coulomb contribution to free energy of ionic liquid, and
- the uncertainty in the values of the energies of formation of the ions U^{4+} and O^{2-}.

In fact, the formation energies of U^{4+} and O^{2-} ions incorporate the ionisation potentials of uranium and the electron affinity of the oxygen ions, $A(O^{2-})$.

According to Ref.[133] the sum of the first four ionisation potentials of uranium is:

$$6.12 + 11.45 + 17.92 + 31.12 = 66.6\ (eV). \tag{3.5}$$

We shall refer to this data set as to *Set 1*. According to another data set [35](*Set 2*), this sum is:

$$6.19 + 11.6 + 19.8 + 36.7 = 74.3\ (eV). \tag{3.6}$$

The value of the electron affinity is reasonably well defined only for atomic oxygen:

Table 3.1. Restricted Primitive Ionic Model *IM-I* for UO_2 at T_m. Different Approximations for the Coulomb Contribution

	Model:	*DHSE-MSAE*$^{(a)}$	*MSAE*	*DHSE-MSAE*$^{(b)}$	*MSAV*	*MD* [125]	*Exp.*
Ionic diameter, d,	$10^{-10}\ m$	1.98	2.43	2.35	2.43		
Entropy, S,	$10^3 J\ kg^{-1}K^{-1}$	1.18	0.38	0.63	0.22		1.18
Enthalpy, H,	$10^3 J\ kg^{-1}$	-6340	-14800	-8430	-1000		-6360
Gibbs Potential, G,	$10^3 J\ kg^{-1}$	-1003	-1603	-1040	-1069		-1003
Heat Capacity, C_P,	$10^3 J\ kg^{-1}K^{-1}$	0.388	1.07	0.885	0.916	0.360	0.450
Speed of Sound, V_S,	$10^3\ m\ s^{-1}$	1.26	3.09	2.55	2.14		1.85
Compressibility, β_T,	$10^{-10}\ Pa^{-1}$	8.31	1.45	2.15	3.22	8.13	4.80
Expansivity, α_T,	$10^{-4}K^{-1}$	1.15	0.90	1.03	1.40	0.02	1.03

$^{(a)}$ Modified *DHSE* approximation Eq.(3.9) $\alpha = 0.925$, fitted from experimental S, ρ and G values, by using *Set 1* (Eq.(3.5), p.42) for the uranium ionisation potentials;
$^{(b)}$ *Ibid.* $\alpha = 1.25$, by using the second set of ionization potentials data (Eq.(3.6));
All energies are calculated relative to the energy of an atomic ideal gas at $T = 0\ K$

$$A(O^-) = E(O) - E(O^-) \cong 1.47\ (eV). \tag{3.7}$$

The electron affinity of O^-, however, is much more uncertain. The energy necessary to attach two electrons to an oxygen atom $A(O^{2-}) = E(O)$ - $E(O^{2-})$ was estimated to be between *zero* and 6.5 eV, the latter value being based on the outdated Ref.[18]. Accordingly, for one ion U^{4+} and two ions O^{2-} the total uncertainty in the energy of formation is approximately $\sim$ 20 eV. In view of this uncertainty, we used the tabulated *entropy* of liquid $UO_{2.0}$ as appropriate additional constraint in the parameters fitting procedure, because the entropy is independent of this uncertainty. According to Ref.[63] :

$$S_{UO_{2.00}} \cong 1.18\ kJ\ kg^{-1}K^{-1}\ (\text{at } T_m, \text{liquid}). \tag{3.8}$$

This additional constraint is important if one wants to describe the caloric properties of UO_2 correctly. To satisfy two conditions of Eq.(3.4) and Eq.(3.8) simultaneously at least two adjustable parameters in the model are required. A second parameter can be found by considering the inaccuracy source mentioned above. Numerous comparisons between the predictions of computer simulations and of the integral-equation theory [72] show that the simulated values of energy within *RPIM* are usually between the predictions of *DHSE* and *MSAE* approximations. The Helmholtz free energy of liquid UO_2 (modified *DHSE* approximation) can be written in the following form:

$$F = F^{(id)} + \Delta F^{(R)}_{HS} + \alpha\, \Delta F^{(C)}_{DHSE} + (1-\alpha)\, \Delta F^{(C)}_{MSAE}, \tag{3.9}$$

where α may represent the additional fitting parameter (which is *not* necessarily $0 < \alpha < 1$). The conditions Eq.(3.4) and Eq.(3.8) should now fix the values of the two adjustable parameters: the hard-sphere diameter, d, in $\eta = {}^1\!/_6 \pi n d^3$, Eq.(2.27), and the fitting parameter α.

Note that one can also impose the values of the Gibbs free energy and of the enthalpy[1] of liquid UO_2 by alternatively using the equilibrium vapour pressure of the liquid as a fitting parameter. This requires the setting of the UO_2 *partial* pressure to $P_{UO_2} \cong 2.3\ kPa$ instead of the experimental value of the total *non-congruent* vapour pressure $\sim 5\ kPa$ [63].

The fitting procedure described above is only successful when combined with ionisation potentials *Set 1* along with the value $A(O^{2-}) \sim +5\ eV$ for the electron affinity of O^-. The results are presented in Table 3.1 (columns 2 and 4 - *DHSAE-MSAE*) and in Table 3.1. Note that a similar fitting in the context of the modified *MSAE* approximation is doomed to fail, and is proved to be impossible for any other variant of the *virial* versions of the approximations discussed above in Section 2.4.

One can first conclude that the two-parametric modified *DHSE* approximation of the *pure ionic IM-II* model is sufficiently flexible to attempt a general description of the system. Many of the differential thermodynamic

[1] Provided that the energies of U^{4+} and O^{2-} formation are accurate enough.

Table 3.2. Restricted Primitive Ionic Model *IM-I*. Modified *DHSE* Approximation. Calculated Thermodynamic Parameters at Selected Temperatures

Temperature, K		3120	4000	5000	6000
Volume, $V^{(liq)}$,	$10^6 m^3 mol^{-1}$	30.4 [(a)]	33.4	36.7	40.0
Entropy, $S^{(liq)}$,	$10^3 J\ kg^{-1} K^{-1}$	1.183[(a)]	1.278	1.361	1.427
Enthalpy, $H^{(liq)}$,	$10^3 J\ kg^{-1}$	-6343 [(a)]	-6006	-5633	-5270
Compressibility, β_T,	$10^{-1} MPa^{-1}$	8.3	8.65	9.11	9.63
Expansivity, α_T,	$10^{-4} K^{-1}$	1.15	1.00	0.89	0.81
Speed of sound, V_S,	$m\ s^{-1}$	1260	1300	1333	1360
Heat Capacity, C_P,	$J\ kg^{-1} K^{-1}$	386	378	368	360

[(a)] the parameters were fitted to reproduce this value at T_m

quantities of liquid uranium dioxide can be reproduced by fitting the two parameters of the model to non-thermodynamic quantities whose definition does not involve second derivatives. Fitting is very sensitive to the choice of the *form* of the Coulomb corrections Eq.(2.54) and Eq.(2.56). Thermodynamic quantities obtained by extrapolating the modified *DHSE* approximation to high temperatures are presented in Table 3.1.

One must keep in mind that, though the approximations Eqs.(2.54) to (2.56) lead to some improvement of the simplest *DHLL* model, giving *positive* deviations from this limiting law in the Helmholtz free energy, they increasingly differ for higher values of Γ_D. This is illustrated in Fig. 3.1, where functions $\varphi(\Gamma_D)$ Eq.(2.57) are plotted for different approximations.

To provide a better approximation for large Γ_D values, it is necessary to invoke all available experimental data[2] on liquid UO_2 for the model calibration.

3.2 Extended Ionic Model

A more sophisticated version of the ionic model (*IM-II*) was proposed in Ref.[80] to describe the liquid-vapour equilibrium of non-stoichiometric liquid $UO_{2\pm x}$, by taking into account possible formation of additional ionic species: U^{5+}, U^{3+}, and O^-. The *EOS* within *IM-II* model retains the same form of Eq.(3.9) where instead of Eq.(2.29), used in the primitive model, the more general Eq.(2.32) for a mixture of hard spheres is utilized. This extension is equivalent to assuming possible *electronic exchanges*

[2] Note that the parameterisation of the semi-empirical *EOS'* depends also on the adopted value of the equilibrium oxygen pressure in the vapour phase, which enters in the calibration procedure.

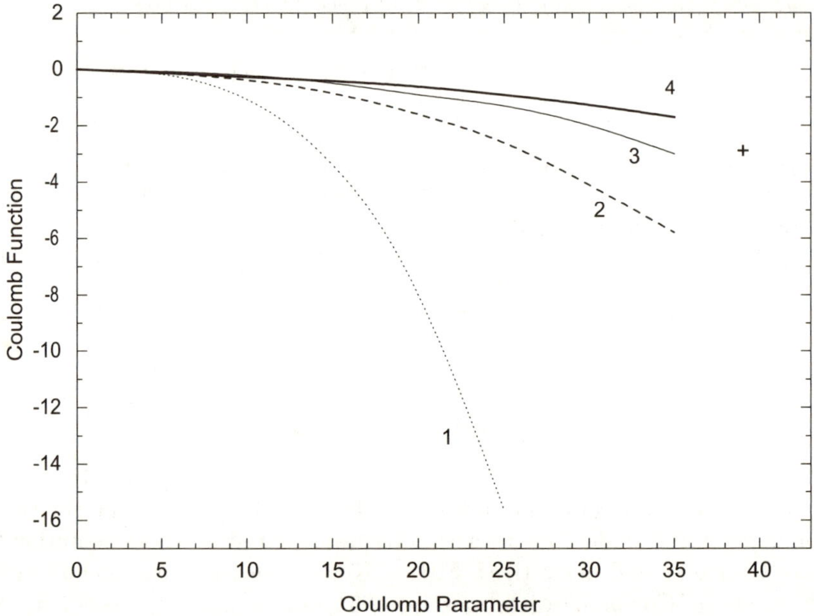

Fig. 3.1. Coulomb contribution function $\varphi(\Gamma_D)$ in different approximations: 1 - *DHLL*; 2 - *MSAV*; 3 - *MSAE*; 4 - *DHSE*; The point at $\Gamma_D = 39$ corresponds to the liquid uranium dioxide near the melting point.

between all ions present in the mixture; for example, via the reactions: $2U^{4+}{=}U^{5+}{+}U^{3+}$, $O^{2-}{+}U^{4+}{=}O^{-}{+}U^{3+}$, *etc.* The equilibrium concentrations of all the considered ions (U^{3+}, U^{4+}, U^{5+}, O^{2-} and O^{-}) in liquid $UO_{2\pm x}$ of arbitrary stoichiometry are calculated with the chemical model formalism of ionisation equilibrium in a *non-ideal* system, by using the Saha equations [58]. The same approach is applied to describe the chemical and ionisation equilibrium in the vapour phase. The composition of the equilibrium vapour over $UO_{2\pm x}$ includes both molecules: UO, UO_2, UO_3, O, O_2, and molecular ions: UO_2^+ and UO_3^-. The description of *non-congruent* phase equilibrium (evaporation) is obtained in terms of equilibrium conditions:

$$\mu_{\mathrm{O}}^{(vap)} = \mu_{\mathrm{O}}^{(liq)} = \{\mu_{\mathrm{O}^{2-}} - 2\mu_{e^-}\}^{(liq)} = \{\mu_{\mathrm{O}^{2-}} + \mu_{\mathrm{U}^{5+}} - \mu_{\mathrm{U}^{3+}}\}^{(liq)};$$

$$\mu_{\mathrm{UO}_2}^{(vap)} = \mu_{\mathrm{UO}_2}^{(liq)} = \{2\mu_{\mathrm{O}^{2-}} + \mu_{U^{4+}}\}^{(liq)}. \tag{3.10}$$

The calibration procedure of the *IM-II* ionic model is similar to that described above. From the known values of density, ρ_m, entropy, S_m, Gibbs

Table 3.3. The Ionic Diameters Used in the Extended Ionic Model *IM-II.*

Ion	U^{3+}	U^{4+}	U^{5+}	U^{6+}	O^{2-}	O^{-}	
$\sigma_i/\sigma^{0)}$	1.13	1.00	0.87	0.84			Ref.[73][(a)]
$\sigma_i/\sigma^{(0)}$	1.11	1.00	0.88		1.00	0.91	Ionic[(b)] Model

[(a)] R[U^{3+} /U^{4+} /U^{5+} /U^{6+}] = [1.165 /1.03 /0.90 /0.87] Å.
[(b)] The diameter of U^{4+}, $\sigma^{(0)}$ = 1.99 Å, is taken as a reference.

free energy, G_m, and oxygen potential, μ_O, [74] of the stoichiometric liquid, $UO_{2.00}$, at the melting temperature, the following parameters are fitted:

- reference ionic diameters, $\{\sigma^{(0)}\}[\sigma(U^{4+}) = \sigma(O^{2-}) = \sigma^{(0)}]$;
- strength of the Coulomb correction at $\Gamma_D \gg 1$: the parameter α in Eq.(3.9);
- dispersion of the ionic diameters $\{\sigma_i/\sigma^{(0)}\}$;
- energy of formation of the O^{2-} ion[3].

The values of the fitted parameters and a comparison of the predicted thermodynamic properties for liquid and vapour with existing experimental data are presented in Tables 3.2 and 3.4.

The calculations of chemical, ionisation-, and phase-equilibrium were performed with the modified computer code of the *SAHA* family [58]. Thermochemical data and excitation partition functions from the *IVTAN* database [63] have been used as input for all the atomic, molecular and ionic species. This model, which corresponds to highly ionic representation of the equilibrium composition of liquid $UO_{2\pm x}$, successfully describes the non-congruent phase equilibrium. It also incorporates a quasi-molecular representation of the vapour composition over boiling uranium dioxide. The *IM-II* model correctly predicts selected thermodynamic properties of UO_2 at the melting temperature. It provides a satisfactory qualitative description of the properties of UO_2 extrapolated to very high temperatures, and, furthermore, it predicts the decrease of the ionisation degree with increasing temperature, in accordance with the results obtained from numerical simulations of simpler ionic systems (*e.g.*, Na^{+}+Cl^{-}, [96]).

On the other hand, however, in contrast to the experiment [119], it predicts a monotonous increase of the heat capacity with increasing temperature.

[3] The experimental value is uncertain.

Table 3.4. Extended Ionic Model *IM-II*. Comparison of the Calculated and Experimental Data

Temperature, K		3120 Exp. [119]	3120 MD[125]		3120	4000	5000	6000
Density, ρ,	$10^3 kg\ m^{-3}$	8.87	7.20	Ionic model $^{(a)}$	8.88 $^{(b)}$	7.99	7.16	6.42
				INSC Database [36]	8.86	8.04	7.11	6.19
Entropy, S,	$10^{-3} J\ kg^{-1} K^{-1}$	1.183 $^{(c)}$		Ionic model	1.183 $^{(b)}$	1.285	1.381	1.466
Gibbs potential, G,	$10^4 J\ kg^{-1}$	1003 $^{(c)}$		Ionic model	-1003 $^{(b)}$	-1112	-1246	-1388
Heat capacity, C_P,	$J\ kg^{-1} K^{-1}$	440	360	Ionic model	410	416	446	486
				Experiment [119]	*450*±60	*320*±70	*320*±80	*360*±*90*
Expansivity, α_T,	$10^{-4} K^{-1}$	1.05	0.2	Ionic model	1.27	1.13	1.09	1.11
				INSC Database [36]	1.05	1.15	1.30	1.50
Compressibility, β_T,	$10^{-11} Pa^{-1}$	4.10	8.13	Ionic model	8.13	8.82	9.81	11.2
Liquid composition:								
$n_{U^{3+}}/n_{U^{4+}}$				Ionic model	0.00129	0.00920	0.0378	0.102
$n_{O^-}/n_{O^{2-}}$				Ionic model	0.00069	0.00460	0.0191	0.050
Vapour$^{(d)}$:								
$(O/U)^{(vap)}$				Ionic model	2.40	2.43	2.71	3.65
Pressure, $P_{total}^{(vap)}$	$10^3 Pa$	3.90±50%		Ionic model	3.81	124	1580	10420
				INSC Database [36]	4.69	189	2280	10910

$^{(a)}$ Reference ionic diameters, $\sigma^{(0)} = 1.99$ Å;
$^{(b)}$ Fitted from experimental and tabulated $\rho(T_m), S(T_m)$, and $H(T_m)$ values;
$^{(c)}$ Tabulated values from the updated *IVTAN* database [77], [63];
$^{(d)}$ Boiling conditions: $(\mathrm{O/U})^{(liq)} = 2$; $(\mathrm{O/U})^{(vap)} \neq 2$

3.3 Local Equations of State for the Liquid Phase: General Requirements

In general, a local (single-phase) *EOS* must satisfy the following essential requirements:

- give the correct value of the liquid Gibbs potential to ensure phase equilibrium with the vapour:

$$G(T,V^{(liq)}) = G^{(vap)}_{Exp}\left(T,P_s(T)\right);$$

- reproduce the internal energy of liquid:

$$U(T,V^{(liq)}) = U^{(liq)}_{Exp}; \tag{3.11}$$

- predict the correct $P-V$ relation in the liquid at a given vapour pressure:

$$P\left(T,V^{(liq)}\right) = P_s\left(T\right); \tag{3.12}$$

- ensure an accordance with the measured value of the isothermal compressibility:

$$\left(\frac{\partial P}{\partial V}\right)_T = -\frac{1}{\beta_T V^{(liq)}}; \tag{3.13}$$

- provide a correct prediction of the thermal expansion coefficient in the liquid phase:

$$\left(\frac{\partial P}{\partial T}\right)_V = \frac{\alpha_T}{\beta_T}; \tag{3.14}$$

- guarantee that the isobaric heat capacity of liquid be in reasonable accordance with the measurements:

$$C_P(T,V^{(liq)}) = C_P(T)_{Exp} \tag{3.15}$$

3.4 Improved Restricted Primitive Ionic Model

The restricted primitive ionic model (*IM-I*) of the liquid can be improved by adding short-range attractive contributions in the van der Waals form:

$$F = F^{(id)} + \Delta F^{(R)}_{HS} + \Delta F^{(A)}_{VdW} + \Delta F^{(C)}. \tag{3.16}$$

The ideal-gas Helmholtz free energy is given by Eq.(2.15), whilst the repulsion contribution is expressed as:

$$\frac{\Delta F_{HS}}{RT_m} = \frac{4\eta - 3\eta^2}{(1-\eta)^2}, \quad \eta = {}^1\!/_6\pi n d^3, \tag{3.17}$$

where d is a common effective hard-sphere diameter of the ions, and the additional van der Waals' term is:

$$\Delta F^{(A)}_{VdW} = -a\,n. \tag{3.18}$$

In the *IM-III*, all ions have the same potential parameters and, hence, the same size and intensity of the short-range van der Waals' attraction. Therefore, this model, apart from the Coulomb term, has two adjustable parameters controlling the short-range interionic interaction: d and a. The Coulomb excess free energy is adopted in the following form:

$$\Delta F^{(C)} = \Delta F^{(C)}_{DHLL}\,\varphi\left(\Gamma_D\right), \tag{3.19}$$

where $\Delta F^{(C)}_{DHLL}$ is the term of the Debye-Hückel limiting law ($DHLL$) Eq.(2.50) and $\varphi\left(\Gamma_D\right)$ is the non-ideality correction. This one-parametric form of non-ideality correction is quite general although not rigorous. At least all the useful approximations like *DHSE, MSAE, MSAV, etc.*, (Eqs.(2.54) to (2.56)) comply with this form. By applying the standard thermodynamic relations Eqs.(2.2) and (2.3) together with the above requirements to the *EOS*, one gets the following set of equations representing:

- phase equilibrium Eq.(3.10):

$$\varphi(\Gamma_D) = 24\eta\left(\frac{F^{(0)}_{UO_2}(T)}{RT_m} + 1 + \ln\left(P_s(T)/P_0\right)\right)$$
$$-72\eta\left(\frac{F^{(0)}_{ion}(T)}{RT_m} + \ln\left(\frac{P_0 V^{(liq)}}{RT_m}\right) - \frac{\Delta F_{HS}}{RT_m} + \frac{a}{V^{(liq)}RT_m}\right); \tag{3.20}$$

- internal energy Eq.(3.11):

$$\Gamma_D\varphi'(\Gamma_D) = 144\eta\left(\frac{\Delta U^{(liq)}}{RT_m} + \frac{a}{V^{(liq)}RT_m}\right); \tag{3.21}$$

- pressure – volume relation Eq.(3.12):

$$\varphi(\Gamma_D) - \frac{\Gamma_D}{2}\varphi'(\Gamma_D) = 72\eta\left(z_{HS}(\eta) - \frac{a}{V^{(liq)}RT_m}\right); \tag{3.22}$$

- isothermal compressibility Eq.(3.13):

$$\frac{\Gamma_D{}^2}{4}\varphi''(\Gamma_D) - \frac{3\Gamma_D}{4}\varphi'(\Gamma_D) + \varphi(\Gamma_D) = 72\eta\left(\frac{V^{(liq)}}{RT_m\beta_T} - \eta\frac{\partial z_{HS}}{\partial\eta} + \frac{a}{V^{(liq)}RT_m}\right); \tag{3.23}$$

- thermal expansion Eq.(3.14):

$$\Gamma_D\varphi'(\Gamma_D) - \Gamma_D{}^2\varphi''(\Gamma_D) = 288\eta\left(\frac{V^{(liq)}\alpha_T}{R\beta_T} - \frac{a}{V^{(liq)}RT_m}\right); \tag{3.24}$$

- heat capacity Eq.(3.15):

$$\varGamma_D \varphi'(\varGamma_D) - \varGamma_D{}^2 \varphi''(\varGamma_D) = 288\eta \left(\frac{\Delta C_V}{R}\right). \tag{3.25}$$

Input experimental data are: $V^{(liq)}, G^{(vap)}$, α_T, β_T, $C_P^{(liq)}$ and $\Delta U^{(liq)} = I(U^{4+}) + 2A(O^{2-}) + \Sigma \varGamma_{Di} D_i - \Delta H^{(vap)}$ at $T = T_m$ and P_0 =1 *atm*. In turn, $I(U^{4+})$ is the sum of the four first ionisation potentials of the uranium atom, $A(O^{2-})$ is the sum of the first two electron affinities of the oxygen atom, $\varGamma_D$ is the Coulomb non-ideality parameter in liquid phase, D_i, and $\Delta H^{(vap)}$ are, respectively, the dissociation energies in the gas phase and the heat of vaporisation. Non-ideality effects in the gas phase are neglected. These equations can be used for the determination of the *EOS* parameters or as constraints to check the validity of the model.

3.5 Compatibility Conditions

To satisfy the above equations, when considered as independent, at last six parameters are needed. Due to the linear dependence, however, not all of them can be self-consistently obtained by a parameter fitting, although the following *compatibility conditions* can be derived from Eqs.(3.20) to (3.25):

1. Identity of the left-hand sides of Eqs.(3.24) and (3.25) for thermal expansion and heat capacity leads to:

$$\frac{V^{(liq)}\alpha_T}{R\beta_T} - \frac{a}{VRT_m} = \frac{\Delta C_V}{R}. \tag{3.26}$$

   Thus the van der Waals attraction parameter, a, may be calculated directly from the measured excess heat capacity ΔC_V, the liquid molar volume at a given temperature, $V^{(liq)}$, and the thermal coefficients α_T and β_T, independently of all other parameters of the *EOS*:

$$\frac{a}{V^{(liq)}RT_m} = \frac{V^{(liq)}\alpha_T}{R\beta_T} - \frac{\Delta C_V}{R}. \tag{3.27}$$

   From existing experimental data at the melting temperature one obtains:

$$\frac{a}{V^{(liq)}RT_m} = 1.46\ . \tag{3.28}$$

2. It is easy to check that the left sides of Eqs.(3.22),(3.23) and (3.24) are linearly dependent. Therefore, they can be satisfied only when the following relation holds:

$$z_{HS}(\eta) + \eta\frac{\partial z_{HS}}{\partial\eta} = \frac{\Delta C_V}{R} + \frac{2a}{V^{(liq)}RT_m} + \frac{V^{(liq)}}{RT_m\beta_T}. \tag{3.29}$$

By using the Carnahan-Starling formula Eq.(2.30), one can immediately calculate the packing fraction, η, at the melting temperature obtaining: $\eta = 0.3235$. In addition, one can also calculate, independently of the form of $\varphi(\Gamma_D)$, the following values of the effective ion diameter, and the Coulomb parameter, Γ_D, at the melting temperature: $d = 2.189$ Å; $\Gamma_D = 39$. Combining Eqs.(3.28) and (3.18) leads to an estimation of the effective intensity of the inter-ionic short-range attraction: $\varepsilon/k = 3973\ K$.

3. There is another relation, which holds independently of the form of $\varphi(\Gamma_D)$. This follows from combining the equations for the pressure, the energy and the Gibbs potential Eqs.(3.25) to (3.27):

$$\ln P_s(T) = 3\left[\ln\left(\frac{V^{(liq)}P_0}{RT_m}\right) - \frac{a}{VRT_m} + \frac{\Delta F_{HS}}{RT_m} + \frac{F^{(0)}_{ion}(T)}{RT_m} + z_{HS}\right]$$
$$- \frac{\Delta H^{(vap)}}{RT_m} - \frac{D(UO_2)}{RT_m} - \frac{\Delta F^{(0)}_{UO_2}}{RT_m} - 1\,, \qquad (3.30)$$

This relation (which is completely independent of the Coulomb contribution) links the saturated vapour pressure $P_s(T)$, the heat of vaporisation, $\Delta H^{(vap)}$, the atomisation energy $D(UO_2)$, and the van der Waals constant, a, (already fixed), with other quantities. It cannot be regarded as an equation for the determination of the various parameters, but it can be used as the constraint for examining of the consistency of the different molecular and ionic constants. Such a check is straightforward; note, however, that the most uncertain quantities are the ionisation potentials of uranium, and the electron affinities of oxygen.

The discrepancy of up to 7.7 eV between two sets of uranium ionisation potentials, mentioned above, is high, and the estimated value [18] of $A(O^{2-}) = 6.5\ eV$ is also inaccurate. The possible error on $A(O^{2-})$ may be of the same order of magnitude as the value itself. Only the second set (*Set 2*) of ionisation potentials Eq.(3.6) proved to be consistent with the measured vapour pressure, the molecular and ionic constants, and the van der Waals constants. The value of the heat of vaporisation $\Delta H^{(vap)} = 530\ kJ\ mol^{-1}$ was obtained, which is in good agreement with the recommended experimental value of 533 $\pm 5\ kJmol^{-1}$ [77].

3.6 Determination of the Coulomb Contribution

Once the effective ion diameter of ions, d, is determined from Eq.(3.33), the value of the dimensionless Coulomb parameter, Γ_D, at any density is fixed. The value $\Gamma_D(T_m) \approx 39$ corresponding to $V^{(liq)}$ at the melting point is very large. For such high Coulomb non-ideality the predictions of different approximations, plotted in Fig. 3.1, are drastically deviant. On the other hand, from Eqs.(3.21) to (3.25) it is possible to determine the values of $\varphi(\Gamma_D)$, and its first and second derivatives $\varphi'(\Gamma_D)$ and $\varphi''(\Gamma_D)$, at least at the melting point. By solving these equations one obtains:

$$\varphi(\Gamma_D) = 72\eta\left(\frac{G^{(vap)}(T_m)}{RT_m} - \frac{F^{(0)}_{ion}(T_m)}{RT_m} - \frac{\Delta F_{HS}}{RT_m}\right); \tag{3.31}$$

$$\Gamma_D\varphi'(\Gamma_D) = 144\eta\left(\frac{\Delta U^{(liq)}(T_m)}{RT_m} + \frac{a}{V^{(liq)}RT_m}\right); \tag{3.32}$$

$$\Gamma_D^{\,2}\varphi''(\Gamma_D) = 144\eta\left(\frac{\Delta U^{(liq)}(T_m)}{RT_m} + \frac{a}{V^{(liq)}RT_m} + 2\frac{\Delta C_V(T_m)}{R}\right). \tag{3.33}$$

The values of $\varphi(\Gamma_D)$, $\varphi'(\Gamma_D)$, and $\varphi''(\Gamma_D)$ calculated at T_m are compared with predictions of the different approximations in Table 3.5. It is clear that none of them is capable to describe the experimental data. From comparison of $\varphi(\Gamma_D)$, $\varphi'(\Gamma_D)$, and $\varphi''(\Gamma_D)$ with the corresponding experimental values one can see that, whilst the *MSAE* and *DHSE* formulae provide the best approximations, they still differ significantly from the experiment. To improve the situation it is necessary to introduce a correction term to, *e.g.*, the *MSAE* formula:

$$\varphi(\Gamma_D) = \varphi_{MSAE}(\Gamma_D) + \Delta\varphi(\Gamma_D). \tag{3.34}$$

According to Table 3.5, whilst we expect that the magnitude of $\Delta\varphi(\Gamma_D)$ be lower than that of $\varphi_{MSAE}(\Gamma_D)$, its functional form is unknown. It is, however, possible to formulate the following requirements, which must be generally satisfied:

- Eq.(3.34) should reproduce the values of $\varphi(\Gamma_D)$, $\varphi'(\Gamma_D)$, and $\varphi''(\Gamma_D)$ deduced from the experiment at T_m;
- $\Delta\varphi(\Gamma_D) \to 0$ for $\Gamma_D \to 0$;
- the expansion of the correction term on powers of Γ_D must begin with the $\Gamma_D^{\,4}$-term.

The simplest form of $\Delta\varphi(\Gamma_D)$ is, therefore, the polynomial:

$$\Delta\varphi(\Gamma_D) = a\Gamma_D^{\,4} + b\Gamma_D^{\,5} + c\Gamma_D^{\,6}. \tag{3.35}$$

Another possible functional form is the Padè approximant:

$$\Delta\varphi(\Gamma_D) = \frac{A\,\Gamma_D^{\,4}}{1 + B\Gamma_D + C\Gamma_D^{\,2}}. \tag{3.36}$$

When the coefficients a, b, c (or A, B, C) are properly chosen, both expressions satisfy the above requirements and reproduce the values given in the last line of Table 3.5. The following coefficients are found:

$$a = 5.103\ 10^{-4};\ b = -1.227\ 10^{-5};\ c = 9.28\ 10^{-8};$$
$$A = 5.173\ 10^{-4};\ B = 0.01464;\ C = 9.2\ 10^{-4}.$$

Table 3.5. The Coulomb Non-Ideality Function $\varphi(\Gamma_D)$ and its Two Derivatives at $\Gamma_D(T_m) = 39$. Comparison of Different Approximations with Values Estimated from Experimental Data at the Melting Point

	$\varphi(\Gamma_D)$	$\Gamma_D \varphi'(\Gamma_D)$	$\Gamma_D{}^2 \varphi''(\Gamma_D)$
DHLL	-59319	$-1.780 \cdot 10^5$	$-3.559 \cdot 10^5$
MSAV	-7300	-17020	-22147
MSAE	-3395	-7280	-8099
DHSE	-2176	-4449	-4560
Exp.	*-2990*	*-6120*	*-6260*

The temperature dependence of the liquid heat capacity shown in Fig. 3.2 obtained on the basis of Eqs.(3.35) and (3.36) differ in curvature but both decrease with raising temperature in reasonable agreement with the experiment [119].

3.7 Conclusions

The thermodynamic properties of liquid uranium dioxide were calculated using *pure ionic models* in a restricted temperature range from the melting temperature up to $T \approx 6000\ K$. This restriction is determined by the range of validity of the physical hypotheses, which require that within this tempera-

Table 3.6. Improved Restricted Primitive Ionic Model *IM-III*. Calculated Thermodynamic Properties of the Liquid $UO_{2.00}$ at Selected Temperatures

Temperature, K		3130 $^{(a)}$	4120	4863
Volume, V,	$10^{-6}\ m^3\ mol^{-1}$	30.49	33.30	34.8
$(\partial P/\partial T)_V$	$10^3\ Pa\ K^{-1}$	2.40	1.47	1.20
Compressibility, β_T,	$10^{-11} Pa^{-1}$	4.9	4.6	4.4
Expansivity, α_T,	$10^{-4}\ K^{-1}$	1.18	0.68	0.53
Speed of Sound, V_S,	$m\ s^{-1}$	1744	1779	1837
Heat Capacity, C_V	$J\ kg^{-1}\ K^{-1}$	313	267	256
Heat Capacity, C_P	$J\ kg^{-1}\ K^{-1}$	413	318	295

$^{(a)}$ the parameters ($\varepsilon/k = 3973\ K,\ \ \sigma = 2.189$ Å) are evaluated at this temperature

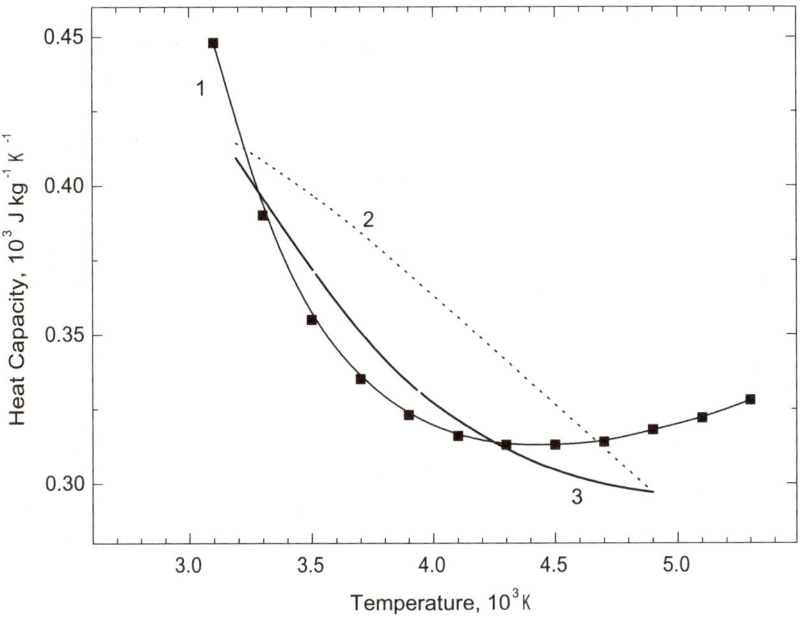

Fig. 3.2. Comparison of predicted and observed heat capacities of liquid uranium dioxide: 1 - Measurements of Ronchi *et al* [119]; 2 - Calculations based on Eq.(3.35); 3 - Calculations based on Eq.(3.36)

ture interval it is possible to neglect the effects of the saturated vapour on the properties of the liquid non-ideality. It should be stressed that *EOS*' based on the restricted ionic models (*IM-I* and *IM-III*) can be applied only to the *liquid* phase of fixed stoichiometry (U^{4+}+$2O^{2-}$), and, therefore, cannot be used for predicting the critical point and other properties related to the gas phase. At the same time, it is known [43] that at least for the 1:1 restricted primitive ionic model, a satisfactory agreement is obtained with the results of numerical simulations for the high-temperature part of the coexistence curve. The critical point can be reconstructed by taking explicitly into account the existence of effectively interacting neutral bonded complexes of ions (or molecules). For description of high-temperature ionic liquids this is essentially in line with the chemical approach.

In contrast to the *pure molecular model*, the opposite limiting case, all energies calculated in *pure ionic models* are referred to the energies of isolated ions. Their values correspond to the ionisation process $UO_2 \rightarrow U+2O \rightarrow U^{4+} + 2O^{2-}$, and differ from the molecular energy levels by the sum of the dissociation energy, the sum of four ionisation potentials of uranium, and the double electron affinity of two oxygen atoms. This sum is not very accurately known, but according to various estimations should be around 9 10^6 $J\ mol^{-1}$. This value is more than 300 times higher than the mean thermal energy at the melting temperature. Therefore, the application of the ionic model requires precise calculations of all contributions to the liquid energy (and especially the Coulomb share). Some results are presented in Fig. 3.2 and in Tables 3.1–3.6. Comparing the calculated properties with the existing experimental data, which have *not* already been used in fitting the parameters, leads to the following conclusions:

- The ionic model may be considered as a reasonable basis for the development of a *local EOS* for stoichiometric and non-stoichiometric liquid $UO_{2\pm x}$.
- The ionic model may also be used for calculations of the non-congruent evaporation of the liquid $UO_{2\pm x}$ in a restricted interval of temperatures.
- To obtain a realistic description of the thermodynamic properties of the UO_2 liquid, it is necessary to take into account both Coulomb *and* van der Waals forces. The *IM-III* model provides a simplest way to reproduce the observed decrease of the liquid heat capacity with increasing temperature, a feature, which is directly related to the Coulomb contribution.

The main advantage of the *IM-III* model surely is its correct prediction of the behaviour of C_p near the melting point. The main defect of *all* ionic models, however, is their extreme sensitivity to the fine details of the ion–ion interaction, due to the large Coulomb contribution to the free energy; additional efforts are thus needed to find a better expression for the Coulomb contribution. The main limitation of the restricted ionic models (*IM-I* and *IM-III*) is their inability to account for any variation in liquid stoichiometry, and to predict the oxygen chemical potential. Thus, this kind of models cannot be used to describe non-congruent evaporation of liquid UO_2. A further, essential limitation is the absence of electron exchange between ions, which restricts the application of the model to low temperatures. It is known that the concept of ionisation degree is meaningful not only for the vapour but also for the liquid.

The general ionic model, called *IM-II*, offers the possibility to include more than two ions, *e.g.*, U^{5+}, U^{4+}, U^{3+}, *etc.*, as well as O^{2-} and O^{-}. It should be noted that electrons, are present as a virtual component in this model. This allows to introduce a realistic oxygen potential. Hence, the extended *IM-II* model can be useful for calculation of properties of the liquid uranium dioxide at *arbitrary* values of the O/U ratio, in the presence of the non-congruent evaporation, when this is combined with a separate model for the gas phase [80]. *IM-II* fails to reproduce the correct values of the isothermal compressibility and the temperature dependence of the heat capacity; nevertheless, the realistic behaviour of the density, thermal expansion and oxygen potential of liquid $UO_{2\pm x}$, as well as the total vapour pressure, are, at least qualitatively, predicted by this model.

For many molten salts the ionisation degree decreases with increasing temperature, as we move towards the critical point[61]. When starting from a cold and dense mixture of *highly ionised* uranium and oxygen, along the coexistence curve from liquid to vapour, one has to observe the decreasing of the ionisation degree and eventually the continuous transformation of the ionic liquid to a cold *neutral* vapour. The type and location of this transformation in the uranium dioxide remains, in this context, an open question, and this represens the most important problem, irresolvable within the context of ionic models.

4

Gas-Liquid Coexistence in Uranium Dioxide

4.1 General Conditions of the Phase Equilibrium

The general condition of equilibrium, *i.e.*, the requirement of the minimum of the corresponding thermodynamic potential, (*e.g.*, of the Gibbs free energy with respect to N, P, and T, or of the Helmholtz free energy with respect to $N, \mathcal{V}$, and T taken as independent variables) includes also the phase equilibrium. If the system considered is or can be separated into two or more phases, and if the surface inter-phase contribution is negligible, the corresponding thermodynamic potential is a linear combination of the thermodynamic potentials of the single phases. For example, if only two phases exist, *e.g.*, liquid and vapour (for definiteness under isothermal – isochoric conditions), the thermodynamic potential (Helmholtz free energy) can be expressed as:

$$\mathcal{F}(T, \mathcal{V}, N_1, \ldots, N_M) = \mathcal{F}^{(liq)}(T, \mathcal{V}^{(liq)}, N_1^{(liq)}, \ldots, N_M^{(liq)}) + \mathcal{F}^{(vap)}(T, \mathcal{V}^{(vap)}, N_1^{(vap)}, \ldots, N_M^{(vap)}). \qquad (4.1)$$

Here $\mathcal{V}^{(liq)}$ and $\mathcal{V}^{(vap)}$ are the volumes of the two coexisting phases. As they occupy the whole volume of the system, it holds:

$$\mathcal{V} = \mathcal{V}^{(liq)} + \mathcal{V}^{(vap)}. \qquad (4.2)$$

Since all the particles are distributed between two phases, it follows that:

$$N_i = N_i^{(liq)} + N_i^{(vap)}; \; i = 1, \ldots, M. \qquad (4.3)$$

The *EOS* of a two-phase system, described by Eq.(4.1), depends on $M + 1$ additional parameters, *e.g.*, $y = \mathcal{V}^{(vap)}/\mathcal{V}$, and $k_i = N_i{}^{(vap)}/N_i, \; i = 1, \ldots M$, which can be defined by using the phase equilibrium condition, *i.e.*, by the requirement of the minimum of the free energy Eq.(4.1) with respect to these additional variables.

By applying the substitutions:
$\mathcal{V}^{(vap)} = y\mathcal{V}$, $\mathcal{V}^{(liq)} = (1-y)\mathcal{V}$ and $N_i^{(vap)} = k_i N_i$, $N_i^{(liq)} = (1-k_i)\, N_i$ into Eq.(4.1) and differentiating the characteristic thermodynamic potential with respect to y and k_i one obtains:

$$\frac{\partial \mathcal{F}}{\partial y} = \mathcal{V}\left[\left(\frac{\partial \mathcal{F}^{(liq)}}{\partial \mathcal{V}^{(liq)}}\right)_{\mathcal{V}^{(liq)},T,N_i^{(liq)}} - \left(\frac{\partial \mathcal{F}^{(vap)}}{\partial \mathcal{V}^{(vap)}}\right)_{\mathcal{V}^{(vap)},T,N_i^{(vap)}}\right] = 0, \tag{4.4}$$

and

$$\frac{\partial \mathcal{F}}{\partial k_i} = N_i\left[\left(\frac{\partial \mathcal{F}^{(liq)}}{\partial N_i^{(liq)}}\right)_{\mathcal{V}^{(liq)},T,N_j^{(liq)}} - \left(\frac{\partial \mathcal{F}^{(vap)}}{\partial N_i^{(vap)}}\right)_{\mathcal{V}^{(vap)},T,N_j^{(vap)}}\right] = 0. \tag{4.5}$$

Bearing in mind the definitions of pressure Eq.(2.2) and of the chemical potential Eq.(2.4), one arrives to the conditions for the free energy minimum:

$$P^{(liq)}\left(T, V^{(liq)}, N_1^{(liq)}, \ldots, N_M^{(liq)}\right) = P^{(vap)}\left(T, V^{(vap)}, N_1^{((vap))}, \ldots, N_M^{(vap)}\right), \tag{4.6}$$

$$\mu_i^{(liq)}\left(T, V^{(liq)}, N_1^{(liq)}, \ldots, N_M^{(liq)}\right) = \mu_i^{(vap)}\left(T, V^{(vap)}, N_1^{(vap)}, \ldots, N_M^{(vap)}\right). \tag{4.7}$$

The conditions formulated by Eq.(4.6)(mechanical equilibrium) and Eq.(4.7) for $i = 1, \ldots, M$ (chemical equilibrium (CCE)), together with the evident $T^{(liq)} = T^{(vap)}$ (thermal equilibrium) will be referred to as *conditions of the phase equilibrium* (CPE). This form of CPE can be applied along with different types of *EOS*. In the case of two single-phase *EOS'*, each one constructed for description of one phase only (like the *EOS* for ideal or imperfect gas, or the *EOS* of the liquid phase), the additional requirement for zero energy (the same for both *EOS'*) must be applied. Furthermore, the neighbourhood of the critical point (*e.g.*, for liquid-vapour) must be normally excluded since it cannot be correctly treated unless an additional conciliation of the two *EOS'* in this domain is provided. The second type of *EOS*, so-called unified *EOS* (*e.g.*, of van der Waals' type), which allows for the description of two or more different phases by one analytic equation, is not bounded to these constraints because of the intrinsic consistence of the properties of both phases. In the case of the unified *EOS* for the Helmholtz free energy Eq.(2.1), the *CPE* do automatically lead to the intrinsic consistence of all phase boundary parameters of the system - binodals, triple point(s), extreme points[1], *etc.*.

[1] On the other hand, one should always keep in mind that most of the unified *EOS*'s (including those discussed here) cannot describe the behaviour of two-

4.1.1 Single-Component Fluids

If a fluid is composed of only one chemical species, the general *CPE* conditions expressed by Eq.(4.6) and Eq.(4.7) are reduced to the equality of pressure and temperature in both phases to the pressure and temperature of the thermostat:

$$T^{(liq)} = T^{(vap)} = T \, , \tag{4.8}$$

$$P^{(liq)} = P^{(vap)} = P \tag{4.9}$$

along with the equality of the specific Gibbs free energies (equal to the value of the chemical potential):

$$G^{(liq)}(T,P) = G^{(vap)}(T,P) \; . \tag{4.10}$$

Sometimes the equivalent Maxwell rule, the well known loop construction on the $P(V)$ isotherm [134, 60], is used instead of Eq.(4.10) in thermodynamic calculations of phase coexistence parameters, when a *unified* thermal *EOS* , $P(T,V)$, is available for both phases. However, it should be noted that the *double-tangent* construction [134] has a more general character, and can be used for two, possibly separate branches of the Helmholtz free energy, $F(V,T)$, for coexisting phases in the $F-V$ plane. With an analogous procedure is carried out the search of the intersection point of the two branches of the Gibbs free energies of the coexisting phases in the $G-P$ plane [134].

4.1.2 Chemically Reactive Systems without Ionisation

The cardinal feature of the phase equilibrium in this kind of systems is the possibility of coexistence of two (or more) phases with different stoichiometries. This is indeed the case for all the phase transitions in the U-O system. The non-congruent phase equilibrium leads to an ensemble of phase boundaries. Due to the presence of additional degrees of freedom in the system, the phase boundaries qualitatively differ from those in a single-component fluid. The chemical potential, μ, governing the chemical and phase equilibrium in a multi-component system can be defined as an M-component vector, composed of the chemical potentials of all species: $\vec{\mu} = \{\mu_1, \ldots, \mu_M\}$, and depending, in general, on temperature and on the composition vector $\mathbf{n} = \{n_1, \ldots, n_M\}$:

$$\vec{\mu}^{(liq)}\left(T, \mathbf{n}^{(liq)}\right) = \vec{\mu}^{(vap)}\left(T, \mathbf{n}^{(vap)}\right) \; . \tag{4.11}$$

phase systems in the close vicinity of critical points, because they normally neglect the crucial correlations between density fluctuations [86]. The critical points predicted by all the mean-field (*van der Waals type*) *EOS*', including the 'true' critical points discussed in this book, differ from the real critical points both in the position on the thermodynamic surfaces and in the behaviour of the differential thermodynamic properties in their vicinity. It is believed that such critical points can be re-established in fast processes [149].

Note that the full set of equations for all species, Eq.(4.7), is restricted to the number of *basic* species, say the atomic ones or an independent set of elementary components when the global *CCE* are met in the system. For the multi-component U-O system one can write:

$$\mu_O^{(liq)}\left(T, n_O^{(liq)}\right) = \mu_O^{(vap)}\left(T, n_O^{(vap)}\right), \tag{4.12}$$

$$\mu_U^{(liq)}\left(T, n_U^{(liq)}\right) = \mu_U^{(vap)}\left(T, n_U^{(vap)}\right) . \tag{4.13}$$

In practice, for phase diagram calculations the following equivalent version of *CPE* results to be more useful:

$$\mu_{O_2}^{(liq)}\left(T, n_{O_2}^{(liq)}\right) = \mu_{O_2}^{(vap)}\left(T, n_{O_2}^{(vap)}\right), \tag{4.14}$$

$$\mu_{UO_2}^{(liq)}\left(T, n_{UO_2}^{(liq)}\right) = \mu_{UO_2}^{(vap)}\left(T, n_{UO_2}^{(vap)}\right). \tag{4.15}$$

Note that the double-tangent construction is still valid and usable for phase coexistence calculations, but in a generalised form, extended on the isothermal surface of the Helmholtz free energy in the *full-component space*, a subspace of which is defined by equal stoichiometries of both coexisting phases.

4.1.3 Congruently Coexisting (Azeotropic) Compositions

Two kinds of azeotropic compositions are important in the theory of non-congruent transitions:

- *CMC*: Congruently Melting Compositions;
- *CVC*: Congruently Vaporising Compositions (for boiling and sublimation).

In every *CMC* or *CVC* point the equilibrium conditions Eq.(4.7) or Eq.(4.10) must be satisfied, *i.e.*, the chemical potentials of all components, in our case uranium and oxygen, are equal in both coexisting phases. In particular, the values of the oxygen potential of the solid and the liquid must be equal at the *CMC* point; the same holds for the liquid and the vapour at the *CVC* point. The ensemble of the *CVC* (or *CMC*) points constitutes a set of *azeotropic curves*. Two branches may exist in each azeotropic curve: a stable and a metastable one. The general structure of all *CVC* and *CMC* curves in the U-O system is almost unknown, but is expected to be rather complex. Note that azeotropic curves cannot have self-intersections, closed loops and end-points. The intrinsic end point of a stable part of an azeotropic curve is its intersection with a phase boundary.

4.1.4 Extremal Properties of the Thermodynamic Functions in the Azeotropic Points

The characteristic property of any azeotropic (*CVC* and *CMC*) point is to be an extremum:

$$\left(\frac{\partial P}{\partial x}\right)_T = 0, \tag{4.16}$$

$$\left(\frac{\partial T}{\partial x}\right)_P = 0\,. \tag{4.17}$$

This extremum may be a *minimum,* as expected in the case of the isothermal vapour pressure minimum over boiling and saturated slightly hypostoichiometric urania. Equivalently, it may be a *maximum,* as expected in the case of the isobaric *CMC*–point of melting and freezing urania. The extremum property is a revealing attribute of thermodynamic self-consistency of any tested model of non-congruent phase equilibrium in chemically reacting and/or partially ionised systems. The capacity of an *EOS* to reproduce this extremal property is not guaranteed in advance. In fact, not all of the known theoretical *EOS* models of liquid $UO_{2\pm x}$ are successful when submitted to this check.

4.1.5 Systems of Charged Species

A remarkable feature of any Coulomb system is the existence of two versions of the chemical potential [60]: the ordinary chemical potential, $\mu\,(T,\mathbf{n})$, that is a local parameter depending on density and temperature, and the electrochemical potential, $\tilde{\mu}\,(T,\mathbf{n},\mathbf{r})$, that is the sum of the ordinary chemical potential, $\overrightarrow{\mu}\,(T,\mathbf{n})$ and the external electrostatic potential, $\psi\,(\mathbf{r})$:

$$\tilde{\mu}\,(T,\mathbf{n}) = \overrightarrow{\mu}\,(T,\mathbf{n}) + \mathbf{Z}e\psi\,(\mathbf{r})\,, \tag{4.18}$$

where $\mathbf{Z} = \{Z_1,\ldots,Z_M\}$ is the vector of the charge of all species in elemental charge units. Hence the electro-chemical potential, $\tilde{\mu}\,(T,\mathbf{n},\mathbf{r})$, is not a local parameter. For each charged species in a Coulomb system at equilibrium, the conventional chemical potentials can assume different values in the coexisting phases under *CPE*, whilst this is not the case for $\tilde{\mu}\,(T,\mathbf{n},\mathbf{r})$, whose value must be equal in the coexisting phases:

$$\tilde{\mu}^{(liq)}\left(T,\mathbf{n}^{(liq)},\mathbf{r}\right) = \tilde{\mu}^{(vap)}\left(T,\mathbf{n}^{(vap)},\mathbf{r}\right)\,. \tag{4.19}$$

This equality, when combined with the electro-neutrality condition in both phases, leads to the existence of a finite drop in the average electrostatic potential across the phase interface, $\Delta\psi$ [79]:

$$\Delta\psi \equiv \psi\,(+\infty) - \psi\,(-\infty) = \left(eZ_\alpha\right)^{-1}\left(\mu_\alpha^{(vap)} - \mu_\alpha^{(liq)}\right)\,, \tag{4.20}$$

where α is the index of any arbitrary species present in both the vapour and liquid phases. Note that the finite potential drop, $\Delta\psi = \Delta\psi\,(T)$, is an indicator of the asymmetry in the equilibrium properties of the charged species in the condensed and gaseous phases, and is identically equal to zero in symmetrical systems like, for instance, the electron-positron plasma, the primitive ionic model of electrolyte solution, *etc.*. The potential dropis a thermodynamic quantity that depends on temperature only and not on surface properties. The zero-temperature limit of this drop (along the coexistence curve) is a thermo-electro-physical property of the substance. Its value is a supplement to the set of properties that are used to characterise a real material, such as sublimation energy, ionisation potential, *etc.*. $\Delta\psi\,(T)$ tends to zero at the critical point of the gas-liquid phase transition. A special critical exponent can be defined to describe its behaviour in the vicinity of the critical point:

$$\Delta\psi\,(T) \sim |T - T_C|^{\phi}.$$

The value of the potential drop Eq.(4.20) can be directly calculated by numerical Montecarlo or molecular dynamics simulation of phase transitions in Coulomb systems when both the coexisting phases are explicitly simulated and combined. In multi-component, chemically reacting and partially ionised systems like U-O the potential drop depends also on stoichiometry:

$$\Delta\psi = \Delta\psi\,(T, x) \ . \tag{4.21}$$

Therefore, it is closely related to the chemical non-congruency in all mentioned phase transitions. Details of this dependence are presently completely unknown. The discussed relation of the *thermo-chemical* and *electro-physical* aspects of non-congruency, and the inter-phase potential drop, are very important in the case of a highly ionic representation of condensed urania, both in the solid and the liquid phase [56].

4.1.6 Chemically Reacting Fluids with Ionisation

The general *CPE* for total equilibrium of two chemically reacting (neutral) systems may be written in the form of equalities of chemical potentials for all neutral species defined by Eq.(4.6) and Eq.(4.7). In a chemical model of phase equilibrium, if ionisation takes place, the equilibrium condition Eq.(4.7) cannot be straightforwardly applied if the chemical potentials are defined by Eq.(2.4). Actually, in this case, for all charged species we have:

$$\mu_i^{(liq)}\left(T, \mathcal{V}^{(liq)}, N_1^{(liq)}, \ldots, N_M^{(liq)}\right) \neq \mu_i^{(vap)}\left(T, \mathcal{V}^{(vap)}, N_1^{(vap)}, \ldots, N_M^{(vap)}\right) , \tag{4.22}$$

so that for the electro-chemical potentials of charged species Eq.(4.19) must be applied instead of Eq.(4.7). When total *CCE* are established in both coexisting phases, for all complex particles, say AB, we have:

$$\mu_{\mathrm{AB}} = \mu_{\mathrm{A}} + \mu_{\mathrm{B}} \ , \tag{4.23}$$

and there is no need to equalise the chemical (for neutral components), or electro-chemical (for charged components) potentials; only the chemical potentials of the basic species must be equated, whereas all the others are automatically equalised[2] due to the *CCE* of Eq.(4.23).

Finally, the *CCE* between two phases may be written as Eq.(4.14) and Eq.(4.15). The presence of charged particles does not change these equilibrium conditions if ionisation equilibrium in both phases is maintained, *e.g.*, for a $\mathrm{AB} \rightleftarrows \mathrm{A}^+ + \mathrm{B}^-$ ionisation reaction we have:

$$\mu^{(liq)}_{\mathrm{AB}} = \mu^{(liq)}_{\mathrm{A+}} + \mu^{(liq)}_{\mathrm{B-}} \ , \tag{4.24}$$

$$\mu^{(vap)}_{\mathrm{AB}} = \mu^{(vap)}_{\mathrm{A+}} + \mu^{(vap)}_{\mathrm{B-}} \tag{4.25}$$

whereas

$$\mu^{(liq)}_{A+} \neq \mu^{(vap)}_{A+} \tag{4.26}$$

and

$$\mu^{(liq)}_{B-} \neq \mu^{(vap)}_{B-} \ . \tag{4.27}$$

An important point in the description of the liquid/vapour phase transition in the chemical model is its complete thermodynamic self-consistency, which ensures the validity of all appropriate thermodynamic equalities. In particular, phase coexistence generally implies that: "*If the congruent vaporisation conditions (CVC) are met at a given temperature and pressure, then the above mentioned extremum conditions with respect to composition must be satisfied both in the vapour and in the liquid phase*".

4.2 Calculation of the Equilibrium Composition and Thermodynamic Functions

The calculations of the equilibrium composition and of the thermodynamic functions in the chemical model start from the *EOS* for the Helmholtz free energy Eq.(2.1). The equilibrium composition corresponds to the minimum of the Helmholtz free energy with respect to the variation of N_i at constant temperature and volume, and with the constraints of mass, chemical proportion and charge balance, *i.e.* :
1) conservation of mass:

$$\rho \ = \sum_{i=1}^{M} n_i m_i \ = \ const \ ; \tag{4.28}$$

[2] Note that, in the particular case of uranium dioxide, it is convenient to use the chemical potentials of the compound itself and that of oxygen as the basic ones.

2) chemical proportions:

$$\frac{\sum_{\alpha} n_i \nu_i^{(\alpha)}}{\sum_{\alpha} n_j \nu_j^{(\alpha)}} = const \; ; \tag{4.29}$$

3) and electro-neutrality:

$$\sum_{i=1}^{M} n_i Z_i = 0 \; . \tag{4.30}$$

Generally, the chemical potential Eq.(2.4) depends on the densities of all species, through the interaction corrections. When the explicit expressions for all contributions to the Helmholtz free energy in Eq.(2.58) are fixed, the minimisation of the free energy becomes a mere numerical problem. Two approaches are possible:

1. direct minimisation of Eq.(4.1), or
2. solution of the equation system for chemical and ionisation equilibrium Eq.(4.6)- Eq.(4.7) under the balance constraints Eq.(4.28) - Eq.(4.30).

Sometimes a procedure consisting of a combination of these two approaches is more efficiently used in numerical applications. When the equilibrium composition, **n**, is evaluated, all thermodynamic functions can also be calculated from the thermal and/or caloric equation of state, being self-consistently defined as first derivatives (Eqs.(2.2) and (2.3)), of the free energy expressed by Eq.(2.1). In spite of the complex dependence of the free energy Eq.(2.48) on T and V via the variable equilibrium composition, its first derivatives Eqs.(2.2)–(2.5), *i.e.*, pressure, entropy and chemical potentials, are equal to the corresponding partial derivatives of the free energy at fixed composition at the minimum of $\mathcal{F}(T, \mathcal{V}, N_j)$. As far as the second derivatives (heat capacities, compressibility, *etc.*) are concerned, the additional dependence of the free energy $\mathcal{F}(T, \mathcal{V}, N_j)$ on the equilibrium composition must be taken into account explicitly. In practice, these values are calculated via a direct numerical derivation of Eq.(2.2) or Eq.(2.3). When the equilibrium composition, **n**, and the thermodynamic functions are calculated for a given temperature, density and chemical composition, then the parameters of the relevant thermodynamic processes - like phase transitions lines, Hugoniot curves, or parameters of any iso-process (isobaric, isentropic, *etc.*) - can be also calculated with the same iterative procedure.

The advantage of the *chemical* versus the *physical* model is the reliability of the re-normalised effective interaction of the free (unbounded) particles, as well as the maximal exploitation of any empirical information for the description of contributions of internal degrees of freedom of complex particles, like energies of ionisation and dissociation, excitation energy levels, *etc.*. On the other hand, the main drawback is the conventional sense of all interaction potentials, defined as effective functions. It is well known that in this case, in order to avoid double counting for every pair of *free* and *bonded* particles, one must account for the inaccessibility of some regions of the phase space. Therefore, the main problem of the chemical model is the self-consistency of the definition of the re-normalised interaction potentials of free pairs (or triplets, *etc.*), and the restriction of the accessible excited states of the bound pairs. This gives rise to the well-known problem of the partition function cut-off and to the directly connected problem of re-normalisation of the effective depth of the short-range attraction between free particles [70, 58, 87]. The calculations performed in the context of the *EOS'* described in the next Chapters have the following, common limitations:

- The excitation partition functions of all complex particles are assumed to be equal to their ideal-gas counterparts [10.1] and to be in accord with the recommendations, contained in Ref. [77], whereby the reduced thermodynamic potential $\Phi^0(T) \equiv G(T)/T$ is used for this extraction.
- No cut-off procedure and no perturbation of the internally bound excited states of complex particles due to inter-particle interaction in the dense liquid are taken into account.
- Only single-stage ionisation is included in the *MIX-I* model. The main charged species are the molecular ions UO_3^-, UO_2^+ and UO^+.

On the other hand, the chemical model makes it possible to compose an arbitrary stage of ionisation and an arbitrary structure of complex particles in the frame of a unified calculation procedure. In principle, once a proper definition of the effective interaction of free particles is given, and the bound states are properly restricted, the *MIX-II* model can be applied. This model has no restrictions like those mentioned above, and can be used for the description of a wide variety of states ranging from the presumably highly ionised liquid at very high temperatures, to a 'cold' vapour represented by a neutral molecular mixture.

4.3 General Structure of the Liquid-Vapour Phase Boundaries in $UO_{2\pm x}$

The degree of non-congruency in all phase transitions (for example, in evaporation) depends on the relation between phase parameters and the chemical and ionisation equilibrium. In some cases, this phenomenon is negligible; however, in the liquid/vapour phase transition non-congruency may be very pronounced. In the U-O system, the structure of the phase boundaries is rather complicated. Therefore, for the sake of clarity, the following discussion will start from the phase transition in a simple reference case, the single substance approximation (*SSA*). In this context, the term *FCE* mode will be used [77]. In the *FCE* mode one assumes the equality of the total Gibbs free energy in the liquid and the vapour phases having equal stoichiometries:

$$(\mathrm{O/U})^{(liq)} = (\mathrm{O/U})^{(vap)} = const\ .$$

This corresponds to the well-known Maxwell construction on the $P(V)$ isotherm. It is important that, in many aspects, the phase transition in the *FCE* mode provides lower or upper bounds for some properties of the real, non-congruent phase transition. In the *FCE* mode the variable $P_S\,(T)$ is a single-valued function and the two-phase region in the P-T plane is represented by a single line. The general features of this phase transition in uranium dioxide are known from the thermodynamic theory of mixtures [89]. The two-phase region in the $P-T$ plane is no longer a single line - like in *FCE* mode - but a two-dimensional domain, whose boundary parameters depend on the O/U-ratio. Its form may be complex, but in the case of $UO_{2.00}$ we presume that this domain has the characteristic *banana-like* shape [115, 134]. In the general case, the single *FCE*-coexistence curve splits into two boundaries:

Boiling curve (BC) - It corresponds to the start of boiling when the liquid is *expanded* at constant temperature and stoichiometry. The boiling curve describes the total equilibrium of a stoichiometric liquid with the non-stoichiometric vapour:

$$(\mathrm{O/U})^{(liq)} \neq (\mathrm{O/U})^{(vap)} = 2.00\ . \tag{4.31}$$

Saturation curve (SC) - This characterises the start of the saturation of the vapour when this is compressed at $T, x = const$, or $P, x = const$. The saturation curve corresponds to the total equilibrium of the stoichiometric vapour with the non-stoichiometric liquid:

$$(O/U)^{(liq)} \neq (O/U)^{(vap)} = 2.00\ . \tag{4.32}$$

The saturation curve corresponds to a *local* equilibrium of the $UO_{2.00}$ vapour with the surface of the condensed phase, which is oxygen-depleted.

The two above mentioned curves cannot intersect in the $P-T$ plane (like, for instance in the *FCM* evaporation) but can osculate. In uranium dioxide, at each value of (O/U), both curves are expected to touch in one point. A set of these points for different values of the O/U-ratio forms the azeotropic line, corresponding to the congruent vaporisation conditions. It has been proved [77] that the *CVC*-line exists for $T < T_m$, but the existence of this line at $(\mathrm{O/U}) = 2.00$ and $T > T_m$ is still uncertain[3]. The results of the numerical calculations performed with the described models do not predict *CVC*-points for $\mathrm{O/U} = 2.00$ at any temperature.

The boundaries of the UO_2 two-phase region in the $P-T$, $V-T$, $P-V$, $H-T$, $H-P$ planes are no longer a set of coupled pairs of coexisting states. All these pairs are to be represented on $P-x$, $T-x$, *etc.*, diagrams.

Consequently, the two-phase part of the isotherm in the $P-V$ plane between the boiling and the saturation curve is no longer a horizontal line

[3] Melting point is meant here at ambient pressure.

as for the *FCE* mode. Hence, the picture of the two-phase region, as it is presented in Fischer's work (Figs. 2, 3, 7, 8 in Ref. [39]) is probably incorrect. Obviously, the general thermodynamic stability conditions $(\partial P/\partial V)_{T,x} \geq 0$ must be satisfied everywhere.

The mass and number densities of the coexisting phases are no longer the parameters of order (for instance, the density of the liquid may, in principle, be less than that of the gaseous phase), though in the case of the gas-liquid coexistence in $UO_{2.00}$, the calculations always predict the usual order in the values of liquid and vapour densities and enthalpies, at least at temperatures far from the critical point.

Within the two-phase region of the non-congruent phase transition in the $P-V$ and $P-T$ planes lies the *FCE* mode, including the *FCE*-critical point. The following relations must hold:

$$P(T,x)^{(BC)} \geq P(T,x)^{(FCE)} \geq P(T,x)^{(SC)} , \tag{4.33}$$

and

$$V^{(vap)}(T,x)^{(FCM)} \geq V^{(vap)}(T,x)^{(FCE)} \geq V^{(liq)}(T,x)^{(FCE)} \geq V^{(liq)}(T,x)^{(BC)} . \tag{4.34}$$

The equalities in Eqs.(4.33) and (4.34) correspond to the congruent vaporisation condition.

The heat of vaporisation at constant pressure (temperature) is no longer a single-valued function of temperature (or pressure) as it is in the *FCE* mode. Consequently, the Clausius-Clapeyron equation in its simplest form is no longer valid. Nevertheless, the heat of equilibrium vaporisation $L(T) = H^{(vap)}(T, P_{SC}(T)) - H^{(liq)}(T, P_{BC}(T))$ has as its upper limit the corresponding value of the heat of vaporisation under *FCE*conditions at the same temperature or pressure, $L^{(FCE)}(T)$. When these are equal, the vaporisation becomes congruent. In summary, we have:

$$L(T) \geq L^{(FCE)}(T) , \tag{4.35}$$

$$L(P) \geq L^{(FCE)}(P), \tag{4.36}$$

$$L^{(CVC)}(T) = L^{(FCE)}(T), \tag{4.37}$$

$$L(P)^{(CVC)} = L^{(FCE)}(P). \tag{4.38}$$

Here is:

$$L^{(FCE)}(T) = H^{(vap)}(T)^{(FCE)} - H^{(liq)}(T)^{(FCE)} , \tag{4.39}$$

$$L^{(FCE)}(P) = H^{(vap)}(P)^{(FCE)} - H^{(liq)}(P)^{(FCE)} . \tag{4.40}$$

There is no point on the boundary of the global equilibrium two-phase region where the conditions of the ordinary critical point are fulfilled, *i.e.*:

$$(\partial P/\partial V)_{T,x} = 0, \tag{4.41}$$

$$(\partial^2 P/\partial V^2)_{T,x} = 0. \tag{4.42}$$

Instead of the single critical point as in the *FCE* mode, there exists a finite critical region encompassing several extrema. The critical point of uranium dioxide obtained by Fischer in 1987 [39] and 1992 [41], by solving Eqs.(4.39) and (4.40) for the sum of the partial pressures of the uranium-bearing species cannot be a real critical point, since it does not correspond to any of the above mentioned extrema.

Three points are usually matter of discussion in the theory of solutions:

- the point of maximal pressure on the *BC* - cricondenbar, or *CCP*-point,
- the point of maximal temperature on the *SC* - cricondentherm, or *CCT*–point,
- the true critical point where:

$$(\mathrm{O/U})^{(liq)} = (\mathrm{O/U})^{(vap)}\ . \tag{4.43}$$

Two additional extrema are also useful in thermodynamic analyses:

- the point of maximal O/U-ratio on the *BC* - or *CCX*-point;
- the point of minimal O/U-ratio on the *SC* - or *CMINX*-point.

There may be several of such *CCX*- and *CMINX*–points on the boiling and saturation curves, when at least one *CVC*–point exists. It should be noted that, for all the above mentioned extrema, the slope of the isotherm $P(V)$ is negative :

$$(\partial P/\partial V)_{T,x} < 0. \tag{4.44}$$

4.4 Equilibrium Properties and Composition of $UO_{2\pm x}$

4.4.1 Liquid Phase at Low Temperatures

Over a wide range of pressures, and up to moderately high temperatures, including the region of the gas-liquid coexistence in $UO_{2\pm x}$, the equilibrium composition of the U-O system in the quasi-chemical representation exhibits some remarkable features concerning the relation between the congruency of evaporation and the parameters of the underlying particle interactions in the condensed phase.

The equilibrium composition of the vapour and the liquid phases at low temperatures is almost independent of the presence of the oxygen species: O, O_2, O^+, O^-, and O_2^-. Moreover, in many cases the set of dominant species includes only three constituents. This peculiarity simplifies the thermodynamic analysis of the composition under *CPE*, and entails some consequences, which are very important for the description of the liquid-vapour phase transition. In the relieved state ($P \approx 0$) of the condensed phase the dependence of the thermodynamic properties on pressure is very weak. Therefore, for the calculation of the phase equilibrium between any number of condensed phases at $P \approx 0$ one can consider their properties on the zero isobars only. So is, for example, for the solid - liquid phase equilibrium in the ordinary melting of urania, and also for the liquid-liquid transitions in the low-temperature hypo- and hyper-stoichiometric U–O system including the predicted triple point at $T \approx 2700\ K$ (liquid–liquid–solid) at O/U< 2.0, and the another, more hypothetical, triple point at $T \approx 2400\ K$ (solid–liquid–liquid) at O/U> 2.0.

4.4.2 Vapour Phase without Ionisation

A number of thermodynamic quantities are functions only of temperature and of the O/U-ratio, *i.e.*, they do not depend on the total vapour pressure. In particular, the partial pressures of atomic and molecular oxygen, and hence their chemical potentials, obey this rule. This is due to the equilibrium conditions Eq.(4.11), whose left-hand side does not depend on pressure. This concept can be more formally expressed as follows:

(i) If the equilibrium vapour obeys ideal-gas law, and
(ii) the equilibrium composition is neutral, *i.e.*, the vapour consists only of molecules UO_n , $n = 1, 2, \ldots$

then

a) the partial pressures and the chemical potentials of atomic and molecular oxygen, and
b) the molar volumes of all UO_n species
are independent of the total vapour pressure, and depend only on temperature and on the O/U-ratio. In the particular case of stoichiometric vapour (O/U $= 2.00$), all the above mentioned quantities depend only on temperature.

This implies that:

- All functions mentioned in a) and b) can be calculated from the chemical equilibrium constants provided by standard databases (*e.g.*, [64] or [62]), and, at the same time, all these functions can be measured. A comparison of the model predictions with the experiment can, therefore, be directly carried out for test and validation purposes.
- For every temperature, the *EOS* model correlates all thermodynamic variables in a) and b).
  Using these dependencies one can convert the measurements of any of the above mentioned variables (say $P_{O_2}, P_{UO_3}/P_{UO}$, *etc.*) to obtain others (*e.g.*, the value of the O/U-ratio).

This is also valid in view of converting calculated quantities into experimentally verifiable variables (for instance, the O/U-ratio in the equilibrium vapour). We shall return to this point below, when experimental data will be invoked to calibrate the model.

4.4.3 Partially Ionised Vapour Phase

The features described above can be generalised in models including charged species. In these cases, points (i) and (ii) of the previous section must be extended boy add the requirement of electroneutrality, namely:

(iii) If the equilibrium vapour obeys an ideal-gas laws, and

(iv) the equilibrium composition includes only the following types of molecules: UO_n, $n = 1, 2, \ldots$ and ions[4] $UO_n^{(Z_n)}$, $n = 1, 2, \ldots$; $Z_n = \pm 1, \pm 2, \ldots$, and

(v) the electroneutrality condition is fulfilled, then

c) all thermodynamic variables, and, in particular, the molar (volume) fractions of all charged species: $UO_n^{(Z_n)}$ (including electrons, if they are present), as well as other dimensionless parameters, such as the degree of ionisation, *etc.*, depend only on temperature and on the O/U-ratio, but not on the total vapour pressure.

Conditions (iii, iv) are violated in the neutral vapour if the volume fractions of the O, O_2, O_3, U_2O_2, U_2O_3, U_2O_4, *etc.*, species are not negligible. This can be important when the fraction of atomic oxygen becomes significant. In vapours containing charged species, conditions (iii, iv) are also violated if the concentration of such charged species as e^-,O^+, O^-, O_2^-, $U_2O_2^{\pm}$, $U_2O_3^{\pm}$, *etc.*, are not small compared with the total charge density. In practice, condition c) is violated when the electron density is of the same order of magnitude as that of the dominant molecular ions UO_2^+ and UO_3^-.

The real situation is illustrated below by the partial pressure of molecular oxygen $P_{O_2}(T)$ calculated as a function of temperature for $O/U = 2.00$ from Breitung and Reil's experimental curve [16]:

$$\log\left(P_S^{(BR)}(T) \big/ MPa\right) = 23.7989 - 29605.5/T - 4.75783 \log(T/K) \quad (4.45)$$

where three different $P_S(T)-$ dependencies are considered corresponding to:

1) $P_S^{(BR)}$, Eq.(4.45),
2) $P_S^{(BR)}$, multiplied by a factor of two,
3) $P_S^{(BR)}$, multiplied by a factor of 1/2.

All calculations were performed using the thermodynamic data of Ref.[64]. The results are presented in Table 4.1.

As one can see from this Table, the values of the O_2 partial pressure at relatively low temperatures ($T < 4500\ K$) are are in the three cases very close; their differences are proportional to the partial fraction of atomic oxygen, whose presence violates condition (iv) above. One can also see that at a constant total vapour pressure the partial pressures of oxygen P_O becomes higher when the temperature and the O/U-ratio increase.

---

[4] The presence of the U^{2+}, U^{3+}, *etc.*, ions complies with this condition, whilst the presence of the O^- and O^{2-} ions violates it.

Table 4.1. The Variation of the Partial Pressure of the Molecular Oxygen with the Variation the Total Pressure of the $UO_{2.00}$ Vapour for Three $P(T)-$ Curves: Exact, Doubled, and Half-Pressure, Corresponding to Eq.(4.45)

Temperature, K	3120	3817	4000	5000	6000
$P_S = P_S^{(BR)}(T), bar$	0.0485	1.0	1.82	19.00	77.42
$x_O(P_S^{(BR)})$	0.0016	0.0093	0.0135	0.067	0.181
$P_{O_2}(P_S^{(BR)}), bar$	$2.231\ 10^{-7}$	$8.761\ 10^{-5}$	$2.938\ 10^{-4}$	$3.911\ 10^{-2}$	0.604
$P_{O_2}(P_S^{(BR)})/P_{O_2}(2P_S^{(BR)})$	0.977	0.947	0.949	1.087	1.541
$P_{O_2}(P_S^{(BR)})/P_{O_2}(P_S^{(BR)}/2)$	1.009	1.002	0.987	0.771	0.517

The calculations of the vapour composition from the *IVTAN* database [64] and from the Bober and Singer's [54] values of $P_S\,(T)$:

$$\log\left(P_S^{(BS)}\,(T)\Big/MPa\right) = 5.049 - 23042/(T/K)$$

give the 1% - and 5% - boundaries for the partial volume fraction of atomic oxygen, x_O reported in Table 4.2.

This estimation shows that beyond these boundaries the accuracy of the quantities mentioned in a) and b) is limited by the values of x_O indicated in Table 4.2.

Some consequences concerning the predicted equilibrium composition of the $UO_{2\pm x}$ vapour can now be drawn:

- The pressure-independence may be useful for calculations of evaporation in the *FCM* mode [20]. In this case, the value $(O/U)^{(vap)}$ is presumed to be the same as in the bulk of the condensed phase (bcp): $(O/U)^{(bcp)} = (O/U)^{(bcp)}$. Thus, if $(O/U)^{(bcp)}$ is unknown, the temperature dependence of any quantity described by conditions a) and b) provides information on the stoichiometry of the condensed phase - yet, a sufficiently accurate thermodynamic database must be utilised. On the other hand,

Table 4.2. The 1% - and 5% -Boundaries of the Atomic Oxygen Volume Fraction, x_O in the $UO_{2\pm x}$ Vapour

Temperature, K	2000	2500	3000	3120	3500	4000	4500
(O/U) $[x_O = 0.01]$	2.90	2.52	2.22	2.16	2.00	1.77	
(O/U) $[x_O = 0.05]$		2.90	2.65	2.58	2.40	2.16	1.93

the exact knowledge of $(\mathrm{O/U})^{(bcp)}$ provides the possibility of an additional consistency check of the thermodynamic calculations via the measurement of any quantity entering in a) and b).

- In the case of equilibrium vaporisation, on the boiling curve (BC), the known dependence of one of the quantities mentioned in a) and b) as a function of O/U (for example, the data of Chapman et al. [24]) may be converted by the model into the dependence of $(\mathrm{O/U})^{(vap)}$ *vs.* $(\mathrm{O/U})^{(bcp)}$, with the consequent possibility of deducing the O/U values corresponding to *FCM* conditions and congruent vaporisation (if this really exists).

The values of the calculated functions mentioned in a) and b) are affected by uncertainties in the input data of the partition functions and in the energies of formation of the species. Particularly important is the value of the electron affinity $A(UO_3)$. For example, the same P_{total}/P_{UO_2} at T_m :

$$P_{total}[16]/P_{UO_2}[64] \cong (4.8\ kPa/2.3\ kPa) \approx 2.00 \tag{4.46}$$

corresponds to markedly different values of $(\mathrm{O/U})^{(vap)}$:

$$(\mathrm{O/U})^{(vap)} = 2.535\{A(UO_3) = 200\ kJ/mol[64]\}\ ; \tag{4.47}$$

$$(\mathrm{O/U})^{(vap)} = 2.271\{A(UO_3) = 500\ kJ/mol[63]\}\ . \tag{4.48}$$

Finally, three aspects must be further investigated:

- The behaviour of the quantities mentioned in a) and b) at low temperatures, and the possible existence of discontinuities.
- The range of validity of conditions a) and b) for O/U far from stoichiometry.
- The influence of complex molecules such as U_2O_2, U_2O_3, U_2O_4, whose properties are known ([64]), and others such as U_2O_5, *etc.*, which can be considered as bound states of interacting 'basic' neutral oxide molecules $UO_i{-}UO_j$, or of charged ones, such as $UO_2^+{-}UO_3^-$, produced by reactions of type:

$$\mathrm{UO} + \mathrm{UO} \leftrightarrow \mathrm{U_2O_2}; \mathrm{UO_2}^+ + \mathrm{UO_3}^+ \leftrightarrow \mathrm{UO_2} + \mathrm{UO_3} \leftrightarrow \mathrm{U_2O_5}; \tag{4.49}$$

5

Application of the Chemical Model within the van der Waals Approximation

5.1 Non-Congruent Evaporation over $UO_{2\pm x}$

Almost all previous calculations of the thermodynamic properties of liquid UO_2 were carried out in the *single substance approximation (SSA)*, where vapour and liquid phases have the same elemental and molecular composition (*FCE* mode). In this Chapter, a preliminary account for non-congruent evaporation is presented, based on a relatively simple and easy manageable van der Waals-type *EOS* . The main problem is here the dependence of non-congruency of vaporisation on the particle interaction parameters, in particular, on the their relative sizes. This model can be first used to describe the gas-liquid phase transition in congruently evaporating uranium dioxide, and then extended, through a more complicated numerical procedure, to evaluate the non-congruent coexistence of vapour and liquid phases. The results of these calculations are described in the following sections.

5.1.1 Assumptions

Composition

The *MIX* model is used here as a starting point, where $UO_{2\pm x}$ is treated as a mixture of neutral and charged particles, whereby only molecular ions and single-ionised positive and negative atomic ions are taken into account.

Inter-particle forces

In this simplified model the particle interactions can be summarised as follows:

Short-range repulsion

The hard-spheres mixture approximation (*HSM*, see Eq.(2.32)) [92, 13] is taken, since, when the particle sizes are significantly different, it is more accurate than the one-fluid generalisation Eqs.(2.29),(2.30) of the one-component Carnahan-Starling formula. The difference in size between a given species and its ionic counterpart is neglected, as in earlier calculations [76]. Furthermore, the approximation of additive diameters, instead of additive volumes, is used here:

$$\sigma(UO_j) = A + Bj \quad (A, B = const;\ j = 0, 1, 2, 3).$$

Short-range attraction

The van der Waals *EOS* form Eq.(2.43) is adopted. Considering the great uncertainty in the definition of the short-range interaction in a dense liquid, one can assume that the *VdW* attraction parameter, a_{ij}, is equal for all uranium-bearing species:

$$a_{_{VdW}}(UO_i{}^{(Z1)} - UO_j{}^{(Z2)}) = const \ (i, j = 0, 1, 2, 3;\ Z_1, Z_2 = +1, 0, -1). \tag{5.1}$$

The mutual attraction of the oxygen species (O and O_2) and the attraction between oxygen and the uranium-bearing species is assumed to be *one tenth* of the mutual attraction of these latter:

$$a_{_{VdW}}(O_m - O_n) = a_{_{VdW}}(O_m - UO_j) = 0.1\ a_{_{VdW}}(UO_i - UO_j) = a_{_{VdW}} \quad (m, n = 1, 2;\ \ i, j = 0, 1, 2, 3).$$

The intensity of attraction, $a_{_{VdW}}$, is eventually used as an adjustable parameter. All differences in intensity of attraction between given species and their associated ions are neglected.

Coulomb interaction

The charged species (atomic and molecular ions) are assumed to behave like *charged hard spheres* , whose sizes and short-range attraction are the same as for their neutral counterparts. From a test of the different models for the Coulomb correction (Eq.(2.54)-(2.56)) it can be found that only two of them are consistent with the fundamental assumptions of the *chemical model* (Section 2.4). All attraction corrections can be associated with an effective interaction of species in the *free* state, while the interactions in the *bound* states are taken into account as internal excitations of the complex particles. The final calculations are made by using the Debye-Hückel (*DH*) approximation for charged hard spheres, the *DHSE* model expressed by Eq.(2.54) and Eq.(2.56). In fact, it was shown by Fisher *et al.* [43] that, in spite of its simplicity, this approximation is sufficiently accurate to describe the liquid/vapour phase transition in the context of the *restricted primitive model* of charged hard spheres (discussed in Section 2.3.7), if these are combined with an explicit account for *bound* pairs, as separate species subjected to effective mutual interactions.

5.1.2 Calibration

Though the model contains several adjustable parameters, only few of them are actually relevant for calibration purposes. The adopted calibration procedure is described in detail in Section 5.3. The following thermodynamic quantities are used as input values:

- Melting density $=8870\ kg\ m^{-3}$.
- Partial vapour pressure of UO_2 (at $T = T_m$) $= 2.3\ kPa$.
- O/U ratio in the gas phase at $T = 3500\ K$. From experimental measurements of the ratio $P(\mathrm{UO})/P(\mathrm{UO_2})$ [108] one calculates the value[1] O/U = 2.29.

The applied calibration procedure leads to the definition of the interaction parameters listed in Table 5.1.

Table 5.1. Interaction Parameters of Species Adopted Within the *VdW-DH-HSM* Approximation

<table>
<tr><th>Species</th><th>O</th><th>OO_2</th><th>U</th><th>UO</th><th>UO_2</th><th>UO_3</th></tr>
<tr><td>Diameters, Å</td><td>2.40</td><td>3.53</td><td>3.36</td><td>3.56</td><td>3.74</td><td>3.90</td></tr>
<tr><td>b_{VdW}, m^3</td><td colspan="2">$2.76\ 10^{-25}$</td><td colspan="4">$2.484\ 10^{-24}$</td></tr>
<tr><td>a_{VdW}, $J\ mol^{-1}$</td><td colspan="2">$1.382\ 10^{6}$</td><td colspan="4">$1.244\ 10^{7}$</td></tr>
</table>

The primary adjusted parameters are the diameters and the UO_2-UO_2 attraction parameter. On the other hand, the model ability to predict the correct O/U ratio is mainly determined by the two species UO_3 and UO. Particularly, the key parameter for fitting the oxygen chemical potential in the liquid is the difference between the intrinsic volumes of these two species. It should be recalled (see Section 4.4) that the oxygen chemical potential, μ_O, of liquid uranium dioxide just above T_m is effectively independent of the interaction parameters of the oxygen species. In the model computation via *VdW-DH-HSM*, the approximation of additive volumes can be used for the set of radii of uranium oxides UO_n, like in Section 5.1. This approximation, when combined with the equal attraction constants of UO_n, is equivalent to an *unaltered* ideal-gas-like dissociation equilibrium of the UO_2 molecule in the liquid. Comparison tests with values of the radii dispersion beyond the additive-volumes approximation show that in all cases the change of intrinsic volume upon UO_2 dissociation remains the most important parameter. By varying this value one can fit the difference of the *dissociation* conditions in the liquid and in the gas phase. As already mentioned in Section 4.4, the dissociation degree is related to the observable value of the rate of change of the oxygen chemical potential with the O/U ratio. This value was measured by Chapman *et al.* [24] and proved to be very large. However, the analysis of Chapman's experiment (Section 4.4) reveals large uncertainties which do not enable these data to be used for fitting purposes.

[1] Note that if this experimental value is correct, evaporation in the reported experiments did not take place in the forced-congruent mode.

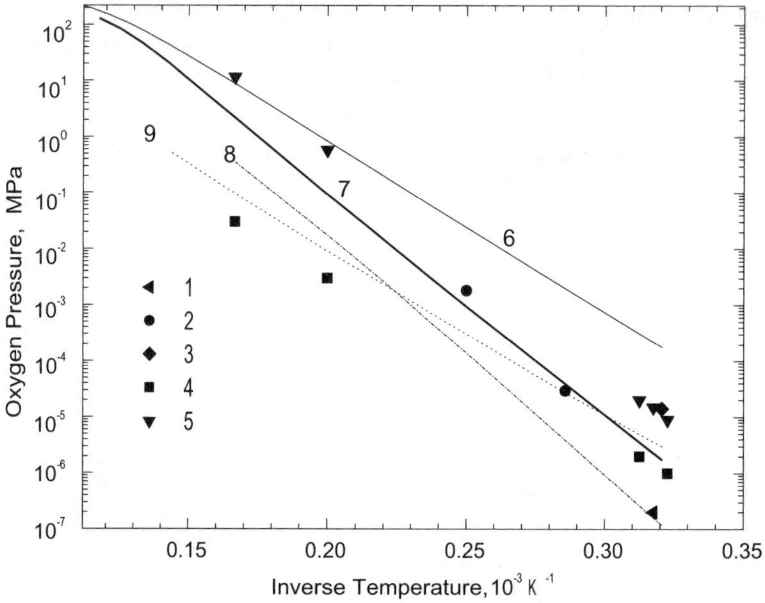

Fig. 5.1. Comparison of the predicted partial pressure of oxygen (P_{O_2}) over boiling uranium dioxide with experiment and other calculations:
1 – experimental point of Chapman [24] for $(O/U)^{(liq)} = 2.0$;
2 – values of oxygen pressure, reconstructed from the *ITU*-1985 data of UO_n-composition measurement [104];
3 – oxygen pressure, reconstructed from the measured ratio of total and partial UO_2 pressures;
4 – calculation results corresponding to Eq.(22) in Ref. [74];
5 – calculation results of Green and Leibowitz [54].
6 to 8 – Present calculations for:
 6 – non-symmetrical (non-additive) radii dispersion rule
 $d(U/UO/\ UO_2\ /\ UO_3 = 3.34\ /3.74/3.74/4.34$ Å;
 7 – *VdW-DH-HSM*- approximation (Section 5.2);
 8 – gas-phase oxygen pressure at *FCE* coexistence;
9 – pressure of molecular oxygen reconstructed from Fischer's calculations of atomic oxygen pressure over $UO_{2.00}$ [39];

5.1.3 Results

Equilibrium composition, thermodynamic functions and gas-liquid coexistence parameters, calculated within the *VdW-DH-HSM* approximation, are presented in Figs. 5.4 to 5.3 and in Tables 5.1 to 5.6. They correspond to *boiling conditions, i.e.*, to the *global* equilibrium of stoichiometric liquid – non-stoichiometric vapour with:

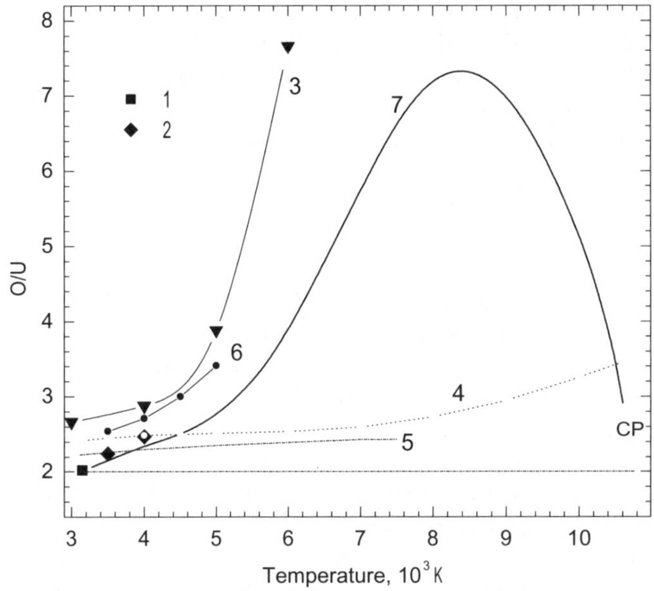

Fig. 5.2. Comparison of calculated O/U ratio in the equilibrium gas phase over boiling $UO_{2.00}$ under non-congruent vaporisation:
1 – experimental point measured by Chapman [24] in liquid stoichiometric UO_2;
2 – values of oxygen pressure reconstructed from *ITU*-1985 data on vapour composition;
3 – calculation results of Green and Leibowitz [54];
4– O/U ratio reconstructed from the Fischer's [39] calculations: $O/U = (O/U)_S + P_O/P_S$
5– calculation with a very weak dispersion of the UO_n radii and a high value of the affinity of UO_3($500\ kJ\ mol^{-1}$);
6– O/U ratio from the calculated composition of Ohse *et al* [104];
7– *VdW-DH-HSM*- approximation

$$(O/U)^{(vap)} \neq (O/U)^{(liq)} = 2.00\ . \tag{5.2}$$

The phase coexistence with the same interaction parameters can also be calculated in the *FCE* mode for comparison. The structure of the non-congruent boiling curve is more complicated than in the *FCE* mode. Note, that the calculations at $T > T_m$ on all tested versions do never satisfy the (possible) congruent vaporisation conditions:

$$(O/U)^{(vap)} = (O/U)^{(liq)}\ . \tag{5.3}$$

As already mentioned, three extremal points can appear on the boiling curve, *BC*: the *CCX*-point (maximum of O/U ratio in coexisting vapour), the *CCP*-

point (cricondenbar), and the critical point (CP). Two other extreme points may appear on the saturation curve, SC: the *minx*-point (minimum of O/U ratio in the coexisting liquid), and the CCT-point (cricondentherm).

At the critical point, the boiling curve approaches continuously the saturation curve. Though this is not proved for a general case, in uranium dioxide, under conditions of non-congruent boiling, these points are ordered as follows:

$$T_{CCX} < T_{CCP} < T_C < T_{CCT} \; ; \tag{5.4}$$

$$P_{CCX} < P_{CCT} < P_C < P_{CCP} \; . \tag{5.5}$$

The results of the calculations are presented in Fig. 5.1 to Fig. 5.4.

Fig. 5.1 presents the value of partial pressure of molecular oxygen over molten $UO_{2.00}$. The experimental point of Chapman [24] for O/U = 2.0, and the values of oxygen pressure reconstructed from the ITU-measurement data of UO_n-composition [108], are also shown. Some other calculation results are presented for comparison: those published by Green and Leibowitz [54] and Fisher [39], as well as theoretical values reported by Hyland [76]. The results proved to be surprisingly close to the values provided by both model versions (models I and II) reported in Ref. [108] (not presented in Fig. 5.1). Two other curves display calculated values. The first corresponds to the *FCE* mode in the same *VdW-DH-HSM* approximation, where the vapour is forced to be stoichiometric (O/U =2.0). The curve 5 corresponds to a pure neutral composition with the set of comparatively large diameters, shifted toward that of UO_3 (here the ratio $d(UO_3)/d(UO_2)$ is of primary importance):

$$d\{O/O_2/U/UO/UO_2/UO_3\} = \{2.40/3.53/3.36/3.75/3.75/4.23(\text{Å})\}. \tag{5.6}$$

Fig. 5.2 presents the results of the O/U calculation in the equilibrium gas phase over boiling $UO_{2.00}$ under non-congruent vaporisation conditions. The available experimental data and the results of other calculations are also plotted.

Fig. 5.4 represents the density of coexisting phases *vs.* temperature, in the *FCE* mode and under *global* equilibrium, calculated within the *VdW-DH-HSM*- approximation. The experimental density data and their high-temperature extrapolation are shown for comparison. Two numerical simulation points of the density of liquid in the pure *ionic* model [125] are also plotted.

Contrary to the case of the *FCE* evaporation regime, where a large number of other calculations exist, in the case of *global* equilibrium the *only* work available for comparison is that of Fischer [39]. In fact, though other studies are dealing with non-congruent vaporisation of uranium dioxide, they operate with a *separate* description of the liquid and vapour phases, and are confined to the relatively low temperature region ($T < 6000\ K$). Therefore, the following discussion has to be restricted to an internal comparison of different model versions within the same approach.

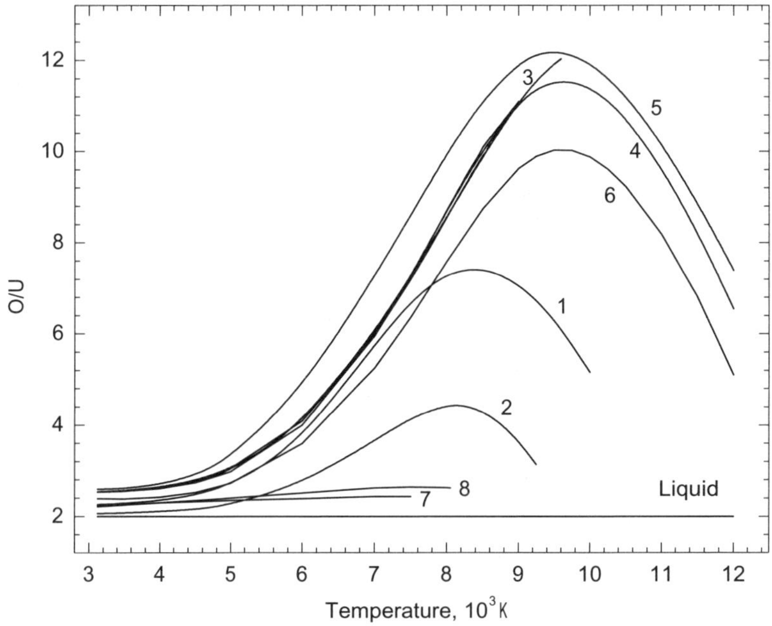

Fig. 5.3. Calculated O/U vapour ratio over boiling $UO_{2.00}$ under non-congruent vaporisation conditions:
1 - *MIX-I* model within the *VdW+HSM+DHS* approximation, assuming an 'intermediate' level of UO_n-radii dispersion. For the initial $O/U^{(vap)} = 2.26$ and oxygen potential ;
2 - neutral model (*NM-I*, Section 2.4) with the *weak dispersion* of UO_n diameters: *low* O/U in the vapour at $T = T_m$- (O/U = 2.06), and *low* values of *oxygen pressure* ($P_{O_2} \cong 0.26\ Pa$) [76];
3 - neutral model (*NM-1*) with the *large dispersion* of UO_n diameters: *high* $(O/U)_m^{(vap)} = 2.54$ and, consequently, *high oxygen pressure* and *high oxygen potential* in the vapour ($P_{O_2} \cong 14\ Pa$);
4 - *MIX-I* model with charged species included (new *IVTAN*-96 data) $A(UO_3) = 500\ kJ\ mol^{-1}$, and with the same dispersion of UO_n diameters as in 3. The 'soft' form of the Coulomb correction is assumed here;
5 - The same as in 4, but with the *DHS*-form of the Coulomb correction. (O/U = 2.54);
6 - The same as in 4, but with the lower UO_3 electron affinity $A(UO_3) \cong 350\ kJ\ mol^{-1}$;$(O/U)_m^{(vap)} = 2.38$;
7 - *MIX-I* model with charged species and weak dispersion of UO_n-radii (as in 2), and a *high* value of the UO_3 electron affinity $A(UO_3) \cong 500\ kJ\ mol^{-1}$; $(O/U)_m{}^{(vap)} = 2.23$;
8 - The same as 7, but with the *IVTAN*-82 molecular partition functions, $(O/U)_m{}^{(vap)} = 2.23$

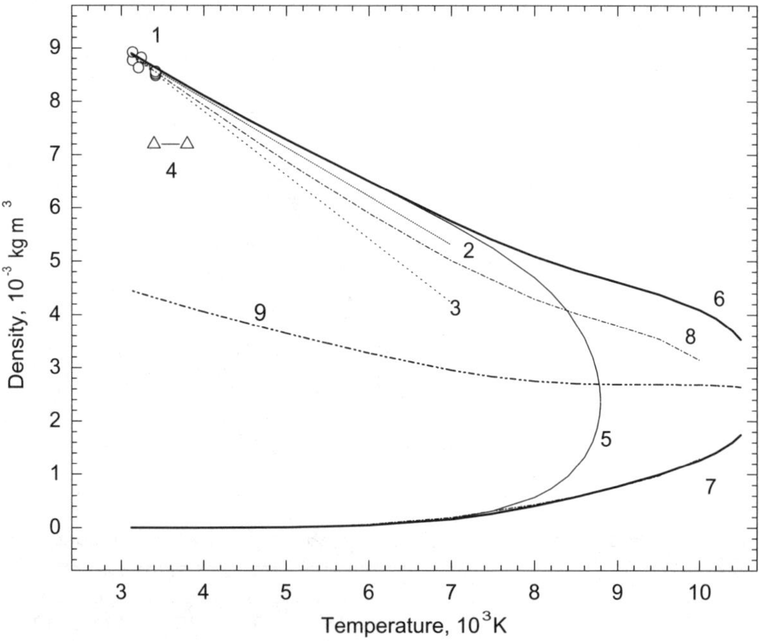

Fig. 5.4. Density of coexisting gas and liquid phases for non-congruently boiling uranium dioxide. Comparison with experimental and other calculated data:
1 - measurements of Drotning and Christensen [26];
2 - original Drotning's fit of the measurements data from Ref. [31];
3 - best common fit of the Drotning and Christensen experimental data [26];
4 - results of direct numerical simulation of Sindzigre and Gillan [125] (pure ionic model for liquid uranium dioxide at $P = 0$);
5 - partially-congruent mode of the gas-liquid equilibrium (*FCE*) within the Van-der-Waals + Debye-Huckel + Hard-Spheres-Mixture (*VdW-DH-HSM*) approximation;
6,7 - coexisting liquid (6) and gas (7) densities of non-congruent boiling uranium dioxide within the (*VdW-DH-HSM*) approximation and with the approximation of additive volumes for the radii dispersion of UO_n $(n = 0, 1, 2, 3)$;
8 - coexisting liquid (the same as in 6) with a non-additive radii dispersion (effective diameters: $U/UO/UO_2/UO_3 = 3.35/3.74/3.74/4.08$ Å) equivalent to suppressed dissociation in the liquid phase;
9 - the mean value of the liquid and vapour densities (diameter)

Density

Up to temperatures about 7000 K the passage from the *FCE* mode to the global equilibrium *does not* noticeably change the properties of the *liquid* phase. This is the key feature of the non-congruent vaporisation regime. All properties of liquid remain close to those of the stoichiometric liquid along the zero-pressure isobar. This feature was used in the previous estimations of non-congruent vaporisation of $UO_{2.00}$ [76]. Though more sophisticated, the present calculations confirm all the previous estimates. The curves plotted in Fig. 5.4 represent the densities of the coexisting phases under global equilibrium. The density of the gas phase corresponds to a variable stoichiometry. Note that this curve *is not the boundary* of the two-phase region of $UO_{2.00}$.

All approximations within the *chemical model* give markedly higher values of the critical density in the *FCE* mode. This effect is further enhanced at the global equilibrium conditions. The sensitivity of the results to the variation of the model parameters is discussed below (Section 5.3). Note also that the empirical rule of rectilinear diameters evidently fails in the case of non-congruent equilibrium. Consequently, estimates of the critical parameters based on this rule are very likely spurious.

Vapour pressure

The main feature of the transition from the *FCE* mode to the global equilibrium is the significant increase of the vapour pressure over boiling uranium dioxide. Preliminary estimations of this increase were reported in Ref. [76] for low temperatures. The pressure increase (in comparison with the *FCE*) is mainly due to the increase of the partial pressure of UO_3 (at high temperatures oxygen becomes dominant in this sense). As it was noticed in Section 5.1, the *FCE* vapour pressure in all versions of the chemical model considerably exceeds the pressure recalculated from the results of the in-reactor vaporisation experiment [21] at $T > 6000\ K$. This discrepancy in the evaluation of the boiling pressure increases even more in the case of global equilibrium. As expected, the equilibrium boiling pressure decreases after reaching its maximum near the critical point. One can locate this maximum using the *VdW-DH-HSM*- approximation at $T_{CCP} \cong 10400\ K$. Its magnitude, $P_{max} \cong 1.8\ GPa$, is very sensitive both to the details of the adopted approximation for the description of non-ideality effects, and to the individual partition functions of the uranium-bearing species. The magnitude of P_{max} is surprisingly high in all *EOS* versions, and this is quite a general feature of the non-congruent evaporation.

Other properties of the liquid phase

The properties of liquid uranium dioxide up to $T = 7000\ K$ are almost independent of the kind of equilibrium, and are very near to those of the liquid at the zero-pressure isobar. Thus, the composition of the liquid and its differential quantities under global equilibrium are, to a first approximation, the same as in the *FCE* mode.

Heat of vaporisation

The heat of phase transition at constant pressure (temperature) under non-congruent evaporation conditions is not a function of temperature (pressure)

Table 5.2. The Thermodynamic Parameters at the *FCE* Critical Point, Calculated with the *VdW* Model in Comparison with the Results of Some Other Calculations

	T_C K	P_C MPa	ρ_C $kg\,m^{-3}$	$z_C^{(a)}$	L/RT_C	ρ_m/ρ_C	x_O	$K_d{}^{(c)}$
VdW+DH	8789	383	2330	0.606	$6.4^{(b)}$	3.80	0.19	0.84
Fischer [41]	10600	388	1560	0.761	$5.0^{(d)}$	5.68	0.31	
ibid.$^{(e)}$		158		0.310				
Mistura *et al*[105]	7563	141	1730	0.350	$7.0^{(b)}$	5.13	0.0	0.0
Simple VdW$^{(f)}$	9367	222	2140	0.359		4.15	0.0	0.0

$^{(a)}$ $z_C = P_C m_{UO_2}/\rho RT_C$.
$^{(b)}$ L is the heat of vaporisation at the melting point of $UO_{2.00}$.
$^{(c)}$ $K_d \equiv \{n_{UO}\ n_{UO_3}\}^{1/2}/n_{UO_2}$.
$^{(d)}$ $L = 440\ kJ\ mol^{-1}$.
$^{(e)}$ Uranium-bearing species only.
$^{(f)}$ Ref. [76].

only, as in the case of the *FCE* mode. The value of heat of vaporisation can always be calculated if one knows the thermodynamic parameters along both two-phase boundaries (*i.e.*, for boiling and saturation, respectively). The vaporisation heat in the *FCE* mode is the *lower* bound for the heat of vaporisation under global equilibrium:

$$L(T)_{\text{total}} = H(T, P_{SC}(T))^{(vap)} - H(T, P_{BC}(T))^{(liq)} > L(T)_{FCE}\,. \qquad (5.7)$$

Comparison of the boiling liquid enthalpy in the *FCE* mode, with that of non-congruent boiling liquid shows that the increase in the vaporisation heat is negligible up to the close vicinity of the critical point. The main contribution to this increase is caused by the difference in enthalpy of the vapour phase.

Critical point at global equilibrium

The value of the critical temperature and density in the *VdW-DH-HSM* approximation can be estimated by extrapolation of the calculated O/U ratio to its value at critical point $(O/U)_C = 2.00$. This gives[2]:

$$T_C \cong 10650\ K;\quad \rho_C \cong 2600\ kg\ m^{-3};\ V_C \cong 0.104\ m^3\ kmol^{-1};\ z_C \cong 1.6\ ;$$

[2] Here $z_C = P_C m_{UO_2}/\rho RT_C$. The high value of z_C is not due to predominance of repulsive forces at the critical point of the non-congruent phase transition, but mainly due to dissociation. It should be also noted that criticality in such a system corresponds to the loss of positive definiteness of the matrix $|\ T\partial\mu_k/\partial N_j\ |$, where μ_k is the chemical potential.

$$(T_C)_{\text{total}}/(T_C)_{FCE} \cong 1.2; \quad (V_C)_{\text{total}}/(V_C)_{FCE} \cong 1.1; \quad (V_C/V_m) \cong 3.4 \,. \quad (5.8)$$

The parameters at the *FCE* critical point are compared with results of some other calculations in Table 5.2.

Oxygen partial pressure over molten $UO_{2.00}$

The *VdW-DH-HSM* equation of state is calibrated by using data set (labelled *Set 2* in Fig. 5.1) reconstructed from the *ITU* vapour composition measurements [108]. The oxygen pressure values calculated on this basis differ from the experimental data: for instance, at T_m the present results are much greater than those of Chapman [24], and much lower than the oxygen pressure (curve 3 in Fig. 5.1) reconstructed from the ratio of the experimental total to partial pressures ($P_{O_2} = 14\ Pa$ at T_m). As for the other *calculations*, they are in good agreement with the data calculated by both models I and II in Ref.[104], and are in satisfactory agreement with the (far) high-temperature extrapolation of Hyland's model of oxygen potential [74]. The results obtained are described at $T < 7000\ K$ by the simple linear Arrhenius fit:

$$\ln\{P_{O_2}\} = \Delta G_{O_2}/kT = A + B/T; (A, B = const) \,. \quad (5.9)$$

At low temperatures the results are close to those of Fischer [39], but exceed them drastically at high-temperatures. However, Fischer's values lie below the oxygen pressure corresponding to the equilibrium stoichiometric vapour (O/U = 2.00), calculated with the *IVTAN* database. Curve 5 in Fig. 5.1 illustrates the influence of the dispersion of the $UO/UO_2/UO_3$ radii on the oxygen pressure under global equilibrium. A shift of this dispersion towards the larger diameter of UO_3 gives rise to non-realistic values of the oxygen potential.

O/U *ratio in the vapour phase over molten* $UO_{2.00}$

Fig. 5.2 shows the behaviour of the O/U ratio in the vapour phase over molten $UO_{2.00}$ under non-congruent vaporisation conditions. The experimental data are represented by the values reconstructed from the following sources:

- the oxygen pressure measurements (Chapman *et al.* [24]);
- the ratio P_{UO}/ P_{UO_2} - measured by (Ohse *et al* [104]), O/U = 2.535;
- the ratio $P_{\text{total}}(T_m)/P_{UO_2}$ (P_{total} [16], P_{UO_2} [64]).

If one compares the calculated data with the results of other models, namely of Fischer [39], Green and Leibowitz [54], and Ohse *et al* [104], the present results are characterised by a high peak of O/U at $T \sim 8000\ K$. Its magnitude is very sensitive to the details of the *EOS* model adopted. The total O/U ratio, accounting also for the oxygen pressure in the vapour phase, displays the expected qualitative trend: after reaching a maximum, it definitely tends to the value 2.00, corresponding to the critical point. This can be compared with

Fischer's total O/U values deduced from his concentrations of the uranium-bearing species with addition of the contributions of atomic and diatomic oxygen, according to the following two procedures:

$$1)\ (O/U)_{\text{total}} = (O/U)_S + (P_{\text{O}} + P_{\text{O}_2})/P_S;$$

$$2)\ (O/U)_{\text{total}} = (O/U)_S + P_{\text{O}}/(n_{UO_2} kT)\ . \tag{5.10}$$

Here n_{UO_2} is the total number density of the uranium-bearing species. The first $(\text{O/U})_{\text{total}}$ value in Eq.(5.10) is shown in Fig. 5.2. It differs from 2.00, and does not tend to this value when $T \to T_C$. The second value of the $(\text{O/U})_{\text{total}}$ ratio can be recalculated only at the critical point, where it gives the equally incorrect value $(\text{O/U})_c = 2.45$.

The slope of O/U as a function of temperature tends to zero when T tends to T_m. Thus, one can suppose that the approximation adopted *does not* indicate existence of congruent vaporisation at temperatures lower than the melting point (*i.e.*, in the metastable liquid). Calculations of the O/U ratio of the vapour over the metastable super-cooled liquid $UO_{2.00}$ at temperatures from 3000 down to 2000 K are presented in Table 5.3.

The *VdW-DH-HSM* approximation predicts quite high values of the $(\text{O/U})^{(vap)}$ ratio at high temperatures. At the same time, the predicted oxygen chemical potential of the liquid is in good agreement with the linear extrapolation of the oxygen potential model of Ref.[74], which is valid for *solid* uranium dioxide. On the other hand, the *SST* model of Fischer gives a rather low value of O/U in the vapour, and, at the same time, the calculated oxygen chemical potential is inconsistent with the oxygen pressure of the *stoichiometric* vapour (*cf.* Fig. 5.1).

5.2 Oxygen Potential

5.2.1 Simplified Calibration Procedure

A simple procedure for adjusting selected particle interaction parameters in the chemical model of liquid $UO_{2\pm x}$ consists of the following steps:

1) the liquid density the melting point is imposed:

Table 5.3. Stoichiometry of the Vapour over Molten $UO_{2.00}$ within the *VdW-DH-HSM* Approximation

Temperature, K	2250	2500	2750	3000	3120
$(\text{O/U})^{(vap)}$	2.207	2.221	2.235	2.250	2.258

$$\rho_m^{(liq)}(T = T_m; P = 4.85\ kPa) = 8870\ kg\ m^{-3}. \quad (5.11)$$

In practice, this is equivalent to the zero-pressure liquid property:

$$P^{(liq)}(T_m, \rho_m) \approx 0.$$

2) One fixes the vapour pressure over molten $UO_{2.00}$ at the experimental value of the total pressure $P \cong 5\ kPa$. In fact, when non-stoichiometry of the equilibrium vapour is ignored, the second condition is equivalent to assigning to the Gibbs free energy in the stoichiometric liquid that of the stoichiometric vapour at $T = T_m$ and $P = 4.85\ kPa$:

$$G^{(liq)}_{UO_{2.00}}(T_m;\ P = 0) \ = \ G^{(vap)}_{UO_{2.00}}(T_m;\ P = 4.85\ kPa).$$

If the real vapour is non-stoichiometric, the partial pressure of $UO_2^{(vap)}$ at $T \approx T_m$ should be used when fitting the Gibbs free energy of the ideal gas. This value can be obtained from two literature sources:

1. experimental measurements performed in *ITU* in 1985 (P_{UO_2} at $T = T_m$ [108]).
2. extrapolated values of P_{UO_2} (T_m) over molten $UO_{2.00}$ available in existing databases.

For instance, one can take the values of the Gibbs free energy $G^{(liq)}_{UO_{2.00}}$ of liquid UO_2 at $T = T_m$ from Refs. [64] or [62]; In fact, all values are found to be close enough.

The value of P_{UO_2} from the *IVTAN* database is more accurate (~15%). The precision of the experimental value of *ITU*-1985 is still under discussion, but one can assume that this pressure is within the uncertainty interval of the *IVTAN* database: P_{UO_2} =2.3 ± 0.35 kPa

5.2.2 Calibration in the Case of Non-Congruent Evaporation

The simplified procedure for calibration of the inter-particle force parameter involves fitting of two quantities only: (1) the liquid density at the melting

Table 5.4. Partial Pressure of UO_2 at $T = T_m$ over Molten $UO_{2.00}$

Source	P_{UO_2}, kPa
IVTAN-82 [62]	2.33
IVTAN-96 [64]	2.26
Experiment *ITU*-1985	2.50

Table 5.5. Calculated and Measured [108] $K_d^{(vap)} \equiv \{n_{UO_3} n_{UO}\}^{1/2}/n_{UO_2}$

T, K	*IVTAN* (1996)	*IVTAN* (1982)	Ohse *et al*-I	Ohse *et al*-II	Fischer (1987)	ANL	$ITU^{(a)}$
3120	0.092	0.098			0.066	0.051	
3150	0.129		0.0966	0.681	0.112	0.067	0.186

$^{(a)}$ Fischer's extrapolation [39] of *ITU*-1985 experimental data [108].

point and (2) the partial vapour pressure of $UO_{2.0}$. These two reference data are sufficient to define the model in the *FCE* (*forced congruent equilibrium*) approximation. If the non-congruency of evaporation is taken into account, the oxygen potential in the liquid phase becomes the third, and most important fitting parameter. Quantities such as the vapour-phase oxygen potential or the partial pressure P_{O_2} over stoichiometric liquid $UO_{2.00}$ and its derivative with respect to $(O/U)^{(liq)}$ are also of primary importance. Any of the values of P_O, P_{O_2}, μ_O and μ_{O_2} measured at low pressures and at temperatures near the melting point can be used for fitting, because each of them can be converted into any other quantity, including the oxygen chemical potential and the O/U ratio in the equilibrium vapour.

Three calibration procedures were examined, involving the following sets of experimental input:

- Results of the direct partial pressure measurements of O_2 over molten $UO_{2\pm x}$ from the experiment of Chapman *et al.*[24].

  The partial pressure of O_2 over a sample of $UO_{2\pm x}$ with controlled O/U ratio at the bulk temperature of the molten $UO_{2\pm x}$ (3150 K) was measured in this experiment, when, however, the temperature of the (solidified) surface layer was only 2200 K. The pressure P_{O_2} was determined in the interval $(O/U)_{cond} \approx$ 1.90-2.10. The vapour is supposed to be in equilibrium with the molten bulk, and the measured pressure is approximated by the equation:

  $$log(P_{O_2}/atm) = 41.1\ (O/U)^{(liq)} - 87.9\ .$$

  Thus, if one assumes that the Chapman's results are sufficiently accurate and that the concept of pure molecular liquid is adequate, one must conclude that the dissociation degree of UO_2 in the liquid is much lower than in the coexisting vapour.

  Furthermore, an important question is whether it is meaningful to use P_{O_2} data over the solidus to calibrate the parameters of the *EOS* for liquid. The answer is affirmative, since the chemical model is able to reproduce important features of the oxygen properties of solid urania, in spite of the absence of any reference to a crystalline structure. Actually, this is one of the most convincing and attractive aspects of this model.

- The ratio Eq.(4.46) of the measured total pressure (Bober and Singer [16]) to the calculated partial pressures P_{UO_2}[64] of the vapour at the melting point.

  The calculation of the equilibrium composition of the vapour over $UO_{2\pm x}$ shows that the ratio of the total pressure to the partial pressure of UO_2 has a minimum at O/U = 2.00 in the region of low temperatures (close to T_m), and increases with increasing x. In particular, by assuming the measured total pressure of Bober and Singer [16], and the partial pressure of UO_2 calculated from the Gibbs free energy of condensed $UO_{2.00}$, recommended in the *IVTAN*-96 database [64], one obtains:

$$P_{\text{total}}(Bober - Singer)/P_{UO_2}(IVTAN) \cong 4.8\ kPa/2.3\ kPa \approx 2.0\ . \tag{5.12}$$

  After calculation of the vapour composition that reproduces this ratio with the use of the same database, one obtains for the vapour over liquid $UO_{2.00}$:

$$(O/U)^{(vap)} = 2.535; \quad P_{O_2} = 14\ Pa; \quad \Delta G_{O_2} = -227.8\ kJ\ mol^{-1}. \tag{5.13}$$

  These values are very sensitive to the input thermodynamic data. For instance, the same ratio Eq.(5.12) when combined with the value of the electron affinity (500 $kJ\ mol^{-1}$) leads to following values:

$$(O/U)^{(vap)} = 2.27; \quad P_{O_2} = 0.477\ Pa; \quad \Delta G_{O_2} = -317.2\ kJ\ mol^{-1}\ . \tag{5.14}$$

  Furthermore, the accuracy of these values is rather low because of the poor precision of P_{total} (Bober and Singer) and P_{UO_2} (*IVTAN*) that leads to a considerable uncertainty of the ratio Eq.(5.12). This is also true for the dependence of Eq.(5.13) and Eq.(5.14) on $(O/U)^{(vap)}$.

- The results of measurements of the partial pressures of the neutral species over molten UO_2 in the temperature interval 3270-4500 K (Ohse *et al*[108]), Table 5.5.

5.3 Composition of the Liquid Phase

The concept of composition is under several aspects merely conventional in liquids, especially at high temperatures. Thus, one must distinguish features that are observable in some way, from those that are merely internal attributes of the *chemical model.* The most important features of the equilibrium composition of liquid uranium dioxide [3] can be summarised as follows:

[3] Note that two versions are considered here: i) the Neutral Molecular Model (Section 2.4.2), and ii) the ionic *MIX-I* model, with neutrals + molecular ions and single-ionised atoms + electrons (Section 2.4.4)

The equilibrium fractions of all oxygen species $[O/O_2\ /O^-/O_2^-]$ in the liquid phase are effectively irrelevant within a realistic assumption of non-ideality involving suitable interatomic potentials of the oxygen species. This property entails that the chemical potentials of atomic and molecular oxygen, as well as all their derivatives with respect to O/U, do not appreciably depend on the oxygen interaction parameters. The situation may be illustrated by the two following examples:

Example A : - Molecular Model of neutral liquid $UO_{2\pm x}(x \ll 1)$
$(T \sim T_m;\ \rho \sim \rho_m)$
Composition: $[O/O_2/U/UO/UO_2/UO_3]$, with estimated dominant species: $UO/UO_2/UO_3$.
$x_{UO_i} \gg x_O + x_{O_2}$ (i=1,2,3)
Input parameters: T; ρ; O/U.
Interaction parameters: $\{\sigma_{ij}\}; \{\varepsilon_{ij}\}$.
The system of three equations for
n_{UO}, n_{UO_2} and n_{UO_3} to solve is:
1) $(O/U) \equiv 2 \pm x = 2 + (n_{UO_3} - n_{UO})/(n_{UO} + n_{UO_2} + n_{UO_3})$;
2) $\rho(T, n_{UO}, n_{UO_2}, n_{UO_3}) = \rho_m$;
3) $(n_{UO}\ n_{UO_3})/(n_{UO_2})^2 \equiv K_d^2 = K(T, \rho, \{n_i\}, \{\sigma_{ij}\}, \{\varepsilon_{ij}\})$.
After solving these equations one obtains:
1) the equilibrium composition $[n_{UO}, n_{UO_2}, n_{UO_3}]$;
2) the oxygen chemical potential,

$$\mu_O = \mu_{UO_3} - \mu_{UO_2} = \mu_{UO_2} - \mu_{UO} = (\mu_{UO_3} - \mu_{UO})/2\ ; \tag{5.15}$$

3) all derivatives $(\partial^k \mu_O/\partial x^k)_{T,P}$.

Example B: - Ionic *MIX* Model of liquid $UO_{2\pm x}(x \ll 1; T \sim T_m;\ \rho \sim \rho_m)$
Composition: $[O_2/O/U/UO/UO_2/UO_3\ /\ U_iO_j^+/\ U_iO_j^-]$ $(i, j = 0, 1, 2, 3)$.
Dominant species (calculation results):

$$\{UO/UO_2/UO_3/UO^+/UO_2^+/UO_3{}^-\}$$
$$x_{UO_i} \gg x_O + x_{O_2} \quad (i = 1, 2, 3)\ .$$

Input parameters: T; ρ; O/U.
Interaction parameters: $\{\sigma_{ij}\}; \{\varepsilon_{ij}\}$.
We have here a system of six equations for six variables:
$n_{UO}, n_{UO_2}, n_{UO_3}, n_{UO_+}, n_{UO_2{}^+}, n_{UO_3{}^-}$:
1) $O/U \equiv 2 + x = 2 + \frac{n_{UO_3} + n_{UO_3}^- - n_{UO} - n_{UO+}}{n_{UO} + n_{UO^+} + n_{UO_2} + n_{UO_2+} + n_{UO_3} + n_{UO_3}^-}$;
2) $\rho(n_{UO}, n_{UO_2}, n_{UO_3}, n_{UO+}, n_{UO_2{}^+}, n_{UO_3{}^-}) = \rho_m$,
3) $(n_{UO}\ n_{UO_2})/(n_{UO_2})^2 \equiv K_d^2 = K(T, \rho, \{n_i\}, \{\sigma_{ij}\}, \{\varepsilon_{ij}\}, \{Z_i\})$
(Chemical Equilibrium);
4) $n_{UO+} + n_{UO_2{}^+} = n_{UO_3{}^-}$
(Electroneutrality);

5)$(n^{+}_{\mathrm{UO_2}}\ n^{-}_{\mathrm{UO_3}})/(n_{\mathrm{UO_2}} n_{\mathrm{UO_3}}) = K_{23}(T,\rho,\{n_i\},\{\sigma_{ij}\},\{\varepsilon_{ij}\},\{Z_i\})$
(Ionisation Equilibrium);
6)$(n_{\mathrm{UO^+}} n_{\mathrm{UO_3^-}})/(n_{\mathrm{UO}} n_{\mathrm{UO_3}}) = K_{13}(T,\rho,\{n_i\},\{\sigma_{ij}\},\{\varepsilon_{ij}\},\{Z_i\})$.

By solving the system of equations one obtains:
1) the equilibrium composition $[n_{\mathrm{UO}}, n_{\mathrm{UO_2}}, n_{\mathrm{UO_3}}, n_{\mathrm{UO^+}}, n_{\mathrm{UO_2^+}}, n_{\mathrm{UO_3^-}}]$,
2) the oxygen chemical potential

$$\mu_{\mathrm{O}} = \mu_{\mathrm{UO_3}} - \mu_{\mathrm{UO_2}} = \mu_{\mathrm{UO_2}} - \mu_{\mathrm{UO}} = (\mu_{\mathrm{UO_2}} - \mu_{\mathrm{UO}})/2 = (\mu_{\mathrm{UO_2}} + \mu_{\mathrm{UO^+}})\ ; \quad (5.16)$$

3) all derivatives $(\partial^k \mu_{\mathrm{O}}/\partial x^k)_{T,P}$.

Analysing these results, one concludes that liquid uranium dioxide at low temperature is analogue to a thermostat that determines the oxygen chemical potential from the account of the actual presence of atomic or molecular oxygen in equilibrium composition. This is in turn defined by the reactions:

$UO_2(c) \leftrightarrow UO(c) + O(g)$ or $UO_3(c) \leftrightarrow UO(c) + O_2(g)$

The same is valid for any other species, for example UO_2, in a purely ionic representation of the liquid, where:

$U^{+4}(c) + 2O^{-2}(c) \leftrightarrow UO_2(g)$

In fact, the problem of the oxygen potential in this context is equivalent to the general non-ideality problem of condensed $UO_{2\pm x}$.

When the non-ideality model and the interaction parameters of the dominant species are adopted, only the oxygen chemical potential and its derivatives $(\partial\mu_{\mathrm{O}}/\partial x)_{T,P}$ may be unambiguously determined. In the inverse procedure, experimental information about μ_{O} and $(\partial\mu_{\mathrm{O}}/\partial x)_{T,P}$ can be used for testing the non-ideality models and the assumed interaction parameters (Section 5.1). The relation between the non-ideality problem and the equilibrium composition in liquid uranium dioxide is illustrated by the example of a restricted neutral (molecular) model (*RNM*), reported in Section 2.3. Precisely, the effect of non-ideality on the equilibrium composition is governed by the corrections introduced in the chemical potentials of the species: $\Delta\mu_{UO}$, $\Delta\mu_{UO_2}$, and $\Delta\mu_{UO_3}$, so that the equilibrium constant may be written in the form:

$$(n_{\mathrm{UO}}\ n_{\mathrm{UO_3}})/(n_{\mathrm{UO_2}})^2 \equiv K_d{}^2(T,\{n_i\}) \equiv K(T,\{n_i\}) = K_0(T)\exp\{-\Delta E_D/kT\}\ . \quad (5.17)$$

Here ΔE_D is the effective shift of the dissociation energy E_D of the reaction $2UO_2 \leftrightarrow UO + UO_3$ in the liquid, *i.e.*:

$$\Delta E_D \equiv 2\ \Delta\mu_{\mathrm{UO_2}} - \Delta\mu_{\mathrm{UO}} + \Delta\mu_{\mathrm{UO_3}}\ . \quad (5.18)$$

The oxygen chemical potential is:

$$\mu_{\mathrm{O}} = \mu_{\mathrm{UO_3}} - \mu_{\mathrm{UO_2}} = \mu_{\mathrm{UO_2}} - \mu_{\mathrm{UO}} = (\mu_{\mathrm{UO_3}} - \mu_{\mathrm{UO}})/2\ . \quad (5.19)$$

It is here convenient to use the *primitive* version of the *RNM* model, with equal interactions for all three species, as a reference system. In this case $\Delta E_D \equiv 0$, so that the equilibrium dissociation constant of Eq.(5.17) and the oxygen chemical potential in liquid of Eq.(5.19) are equal to the ideal-gas values of $K^0(T)$ and $\mu_{\mathrm{O}}(T)$ - note that $\mu_{\mathrm{O}}(T)$ does not depend on pressure. When the interaction parameters of the various species are different, but O/U = 2.00, and the densities of UO_3 and UO are forced to be equal, the following simple relation holds:

$$\mu_O \equiv (\mu_{UO_3} - \mu_{UO})/2 = \mu_O(T) + (\Delta\mu_{UO_3} - \Delta\mu_{UO})/2 \; . \tag{5.20}$$

According to the *RNM*-model, at low temperatures and within a realistic assumption on non-ideality effects, the density of UO_2 prevails in the composition of liquid $n_{UO_2} \gg n_{UO} + n_{UO_3}$ Thus, when $(\mathrm{O/U})^{(liq)} = 2.00$, the oxygen chemical potential is determined by the non-ideality corrections of the minor species only. Furthermore, it is well-known that the ideal-gas dissociation constant $K^0(T)$ and the dissociation degree ${K_d}^0(T) \equiv \{K^0(T)\}^{1/2}$ decrease exponentially with $1/T$ when $T \to 0$, and the slope of the oxygen partial pressure over solid $UO_{2\pm x}$ as a function of x tends to infinity for $T \to 0$. Eq.(5.20) makes it possible to correlate these functions with the interaction corrections in the liquid. The following relation holds:

$$\left(\frac{\partial \ln P_{O_2}}{\partial x}\right)_{P,T} = \frac{2}{kT}\left(\frac{\partial \mu_O}{\partial x}\right) \cong 2 + K_d^{-1} \to K_d^{-1} \quad (K_d \ll 1) \; . \tag{5.21}$$

Eq.(5.21) relates the observable value of the slope of the oxygen pressure *vs.* stoichiometry of condensed uranium dioxide with the non-ideal equilibrium constant of dissociation. Even though Eq.(5.21) is not strictly exact in complicated liquid models where charged species are included (Section 4.1), it remains qualitatively valid in the general case. Finally, it can be realised that the relations corresponding to Eq.(5.21), represent general, key issues in the calibration procedure of the oxygen potential.

5.4 Discussion

Like in ordinary liquids, the value of the critical temperature of UO_2 in the *FCE* mode is strongly correlated with the vapour pressure at the melting point, and, therefore, with the Gibbs free energy and the heat of vaporisation[4] of liquid UO_2. Within this main trend, a weak dependence of the *FCE* critical temperature on the UO_j radii dispersion is found. Widening this dispersion lowers T_c by about 1000 *K*, in both *TPT* and *VdW* approximations. These features are illustrated in Table 5.2 and Figs. 5.1 to 5.4.

[4] The *FCE* critical temperature increases with the heat of vaporisation, whereby the ratio between critical temperature and heat of vaporisation slightly increases when the *TPT* approximation is used instead of the simplified *VdW-DH-HSM*.

As for the real critical temperature in the regime of non-congruent evaporation, its correlation with the input parameters is more complicated. Besides the dependence on the heat of vaporisation, a strong correlation with the oxygen chemical potential of the liquid is found. This determines most of the parameters of the non-congruent phase transition, including the critical temperature: first the real critical temperature and the two other extrema, the maximal pressure and the maximal O/U ratio on the boiling curve.

For the case of a neutral-composition model, the results of two calculation versions with a differing initial vapour O/U ratios are shown in Figs. 5.2 and 5.3. The first one (curve 2 in Fig. 5.3) corresponding to a weak dispersion of the UO_n diameters, exhibits at T_m a low O/U ratio in the vapour, $(O/U)^{(vap)} = 2.06$, as well as too low values of the oxygen pressure and oxygen potential ($P_{O_2} \cong 0.26\ Pa$).

The other curve (curve 3 in Fig. 5.3) corresponds to a larger dispersion of radii and to a high resulting O/U vapour ratio, and, consequently, to a high resulting oxygen pressure and oxygen potential of the vapour ($P_{O_2} \cong 14\ Pa$, at T_m). It is noteworthy that such a large degree of dispersion is necessary to fit the ratio of two empirical pressures: the total vapour pressure over molten $UO_{2.00}$ of Bober and Singer [16] ($P_{\text{total}} \cong 4.85\ kPa$) and the tabulated partial pressure of UO_2 from the *IVTAN* database that corresponds to the result of the *ITU* experiment ($P_{\text{total}} \cong 2.3\ kPa$, Table 5.4). The resulting O/U ratio in the vapour at the melting point is $(O/U)_m = 2.53$.

The value of the oxygen potential in both versions 3) and 4) in Fig. 5.3, giving very close results for the O/U ratio up to T $\sim$ 9000 K, is near to the linear extrapolation (of $\ln(P_{O_2})$ *vs.* $1/T$) of the adopted low-temperature oxygen potential model of solid $UO_{2\pm x}$ (*cf.* Fig. 5.1). Both cases 3) and 4) can be compared with one of the best versions, *i.e.*, the *MIX* model within the *VdW+HS+DHS* approximation (Section 5.2, Figs. 5.1 – 5.4) and an intermediate radii dispersion; this produces intermediate values of the O/U ratio, oxygen pressure and oxygen potential of the vapour ($P_{O_2} \cong 1.77\ Pa,\quad T = T_m$). All three versions 2), 3), and 4) exhibit the same qualitative form of the O/U *vs.* T dependence, with the varying maxima of the O/U ratio, evidently correlated with the oxygen potential of liquid UO_2.

The presence of charged species makes the picture more complicated, but two effects can immediately be seen. First, one observes the influence of new species in the equilibrium composition (typically, the molecule of UO_3 with its high electron affinity), that can affect the results, even if the Coulomb corrections are not taken into account. Another effect is due to the parameters of short-range attraction and repulsion of ions, and to their coupling with the corresponding parameters of the relevant neutral molecules. This problem has not yet been studied.

A further effect is the sensitivity of the results to the choice of the form of the non-ideality Coulomb corrections. It has been pointed out that the effect (perturbation) of these corrections may be very strong, and depends on the analytical approximant assumed. The range of presumed variation of the

Coulomb non-ideality corrections are so large that use of any 'strong' form them, like, *e.g.*, the well-known *DHLL* approximation, leads to total breakdown of the thermodynamic stability conditions. This problem must be further studied. At the present stage, the influence of the Coulomb corrections may be illustrated by two examples of application of a rather 'soft' correction. Curve 4) in Fig. 5.3 demonstrates the effect of the charged species (*MIX* model) in combination with the *DHGA* approximation for Coulomb corrections, and with the presently recommended moderate value of the UO_3 electron affinity, $A(UO_3) = 200\ kJ\ mol^{-1}$. Both curves 3 and 4 show a very similar increase of the vapour O/U ratio up to $T \sim 9000\ K$, whilst at higher temperatures a marked divergence is predicted. This divergence is, however, associated not with the Coulomb corrections, but rather with the growth of the thermal ionisation. The O/U ratio of the vapour further increases (curve 5) when the *DHSE*-form of the Coulomb correction is used: $(O/U)^{(vap)} = 2.54$. The influence of the dispersion of UO_n diameters - within the same composition and Coulomb non-ideality model - is shown in Fig. 5.3 (curves 5 and 1). The latter corresponds to the intermediate level of the UO_n diameters dispersion with the consequent initial $(O/U)^{(vap)} = 2.26$; the corresponding oxygen potentials are shown in Fig. 5.1.

The *VdW-HS-DHS* version of the equation of state exhibits a great sensitivity of the non-congruency of the phase transition in $UO_{2\pm x}$ at high temperatures to the assumptions regarding the intensity of interaction between oxygen and the molecules UO_n. In the calculations presented in Section 5.2 this interaction is represented by the *VdW* attraction constant, assessed to approximately one tenth of the corresponding constants for the UO_n - UO_i attraction (see also Section 5.2). In this case, the value of the maximal O/U ratio decreases by about 20%.

Thus, one can conclude that, contrary to the case of low temperatures, where the properties of liquid $UO_{2\pm x}$ depend only weakly on the oxygen interaction parameters (see Section 4.4), at higher temperatures these parameters are of increasing importance. The non-congruency of the phase transition in UO_2 is strongly influenced by the value of the UO_3 electron affinity. This is illustrated by the curves 6, 7, and 8 in Fig. 5.3, where results of calculations with $A(UO_3) = 200\ kJ\ mol^{-1}$ and $A(UO_3) = 350\ kJ\ mol^{-1}$ (curves 4 and 6, respectively) are compared. The latter case corresponds to $(O/U)^{(vap)} = 2.38$. Furthermore, a combination of a high value of the UO_3 electron affinity ($A(UO_3) \cong 500\ kJ\ mol^{-1}$) with a weak dispersion of UO_n diameters (curve 2 in Fig. 5.3) produces extremely low values of the O/U ratio ($(O/U)^{(vap)} = 2.23$) and of the oxygen pressure ($P_{O_2} = 0.264\ MPa$; $T = T_m$). The last curve (curve 8) illustrates the small effect of the updated molecular partition functions in the *IVTAN*-96 with respect to those of the *IVTAN*-82 database. Curves 7 and 8 in Fig. 5.3 correspond to the same value of the UO_3 electron affinity ($A(UO_3) \cong 500\ kJ\ mol^{-1}$) but to differing databases.

The sensitivity of the boiling pressure in the *FCE* mode and under global equilibrium is qualitatively similar to that of the critical temperature and the

Table 5.6. The Maximal Boiling Pressure at the Boiling Curves

Version[a]	1	2	3	4	5	6	7	8
P_{CCP}, MPa	1800	> 800	> 2000	> 4000	5070	> 2900	> 300	> 550

[a] see Fig.5.3, p.81.

maximal O/U ratio discussed above. The *FCE* critical pressure is correlated with the critical temperature and the heat of vaporisation, as in the case of simple substances. The main difference is a remarkable enhancement of the *FCE* critical pressure in the real, chemically reacting UO_2, even in the simplified *single-substance approximation (SSA)*. This effect is illustrated in Fig. 5.3 and presented in Table 5.6 and discussed in Sections 5.1, 5.2.

Under non-congruent vaporisation conditions the most interesting aspect is the value of maximal pressure on the boiling curve (P_{CCP}, in the notation adopted in Section 4.3). This value may be much higher than the pressure at the real critical point, P_C, and even higher than the *FCE* critical pressure, $P_C^{(FCE)}$. In conjunction with the maximal O/U ratio, the value of P_{CCP} increases with the increase of the dispersion of the UO_n diameters, and of the consequent initial vapour O/U ratio and oxygen potential over molten $UO_{2\pm x}$. At the same time, increasing of the UO_3 electron affinity leads to a decrease of the maximal pressure P_{CCP}. A general view of the pressure *vs.* temperature dependence on the boiling curve is similar to the O/U *vs.* temperature dependence, observable in Fig. 5.3. The values of the maximal pressure on the curves of Fig. 5.3 are presented in Table 5.6.

5.5 Justification of the MIX Models

Before closing this Chapter, a word is in order on the arguments in favour of the examined approach and on its physical justification.

The main difficulty in the construction of a physically convincing equation of state of UO_2 is the complexity of both liquid and vapour structures, *i.e.*, a presumably dominant ionic character of the condensed state, and a slightly ionised reacting molecular mixture in the gaseous state. Since solid UO_2 may be successfully described as a pure mixture of U^{4+} and O^{2-} ions, and the vapour as a neutral molecular mixture, the structure of the liquid is probably somewhere in-between. Now, the *MIX* model (particularly *MIX-II*) makes it possible to construct a unified and correlated description of both phases - and hence of the critical point - by accounting for both ionic and molecular features.

Furthermore, it allows, in a natural way, for the presence of an arbitrary set of constituent particles of differing structures and arbitrary state of ionisation, whose composition is self-consistently determined in a unified general calculation procedure. Therefore, if one can define the effective interaction potential of the free particles, and properly restrict the bound states, then it is possible, at least in principle, to provide a continuous description of all possible states of the system, from the presumably highly ionised liquid state to the relatively cold neutral molecular mixture of the vapour.

Though the dilemma of a *physical* or *chemical* representation necessarily arises in any attempt of theoretical description of strongly coupled chemically reacting Coulomb systems, the essential advantage of the *MIX* model is furnished by the (normally weak) re-normalised effective interaction of the free (unbounded) particles. Moreover, a maximum amount of empirical information can be invoked and exploited in describing the contributions of the internal degrees of freedom of complex (bounded) particles, *e.g.*, energies of ionisation and dissociation, excitation energy levels *etc.*. Furthermore, such an approach is widely and successfully used in the theory of *strongly coupled plasma* (*SCP*), and has found an important technological application in the design of the Gas-Core Nuclear Reactor [58].

The essential equivalence of the *physical* (pure ionic) and *chemical* (quasi-molecular) description of a cold, dense liquid is proved by the Stillinger-Reiss theorem [128, 117], which demonstrates the formal correspondence of the partition function of ionic liquids to that of liquids consisting from virtual molecular clusters. The equivalence of these two approaches in the description of high-temperature gaseous and liquid uranium dioxide in the vicinity of the critical point, constitutes the main argument in favour of the *DHSE* approximation for the Coulomb contribution. The argument is corroborated by the results of Fisher and Levin [43]. These authors have shown that, at least for the 1:1 *restricted primitive ionic model*, a satisfactory agreement with the results of numerical simulations in the evaluation of the critical point and of the high-temperature part of the coexistence curve can be obtained, even when the simplest Debye-Hückel (*DHSE*) approximation is combined with the explicit assumption of interacting neutral bounded ionic pairs (molecules), both in the vapour and in the liquid phase. This is indeed the key issue of the *chemical approach* of the *MIX* model.

Both the physical (ionic) and the chemical (*MIX*) representations correspond to different, but not incompatible, views on the structure of materials. The *MIX* model represents the disordered gas and liquid as an ensemble of conventionally distinguishable clusters, mainly consisting of one uranium atoms surrounded by several oxygen atoms, freely exchanging oxygen and electrons. In addition to the clusters, *free* atoms and molecules of uranium and oxygen (U, U_2, O, O_2) are also included in the picture, as separate species with their own intrinsic volumes.

The use of the *hard spheres mixture* approximation for taking into account the complicated *configurational* effects remains the corner stone of the *MIX* model. The second important aspect is the self-consistent definition of the numerous effective potentials corresponding to the various interactions between clusters.

The assumption of liquid UO_2 being an ensemble of molecular-like virtual clusters $UO_i^{(0,\pm)}$ naturally raises further considerations on the possible role of more complicated clusters, like U_2O_2, U_2O_3, U_2O_4, U_2O_5, *etc.*. The next step in this direction might be the development of models for chain-structured friable liquids build of U_iO_j $(i, j = 2, 3, ...\infty)$ molecules. In Ref.[123] the hypothesis of the possible coexistence (probably involving a first-order phase transition discontinuity) of two different kinds of ionic liquids - 'normal' compact one at low temperatures and high densities, and a chain-structured one in the hot expanded state close to the critical point - might open an ampler perspective in the description of the critical region, since the formation of chains and clusters U_iO_j with i and $j \gg 1$ is quite convincing in high-temperature dense vapours and in expanded liquids.

The last, not least advantage of the *MIX* model in the description of the liquid/vapour phase transition resides in its complete thermodynamic self-consistency, which ensures the validity of all relevant thermodynamic relationships. In particular, the phase coexistence, combined with the chemical and ionisation equilibrium, implies the following property: *"If the chemical model predicts that congruent vaporisation conditions exist at a given temperature and pressure, then the well-known extremum conditions are necessarily satisfied:* $(\partial P^{(vap)}/\partial x)_T = 0$ *and* $(\partial T^{(vap)}/\partial x)_P = 0$". This statement is also valid for an equilibrium system containing charged particles, in spite of the non-equalities of the chemical potentials of the charges in the two phases.

6

New Equation of State for Fluid Uranium Dioxide Based on Thermodynamic Perturbation Theory

Among the many possible types of *VdW*-like *EOS*', only one has the essential advantage of a solid theoretical background, namely that based on *thermodynamic perturbation theory* (*TPT*) [89, 150]. Since the first successful theories of simple fluids developed by Barker and Henderson [10] and by Chandler, Andersen and Weeks [6], this model, in conjunction with computer simulation experiments and integral equation theory, has become one of the most useful predicative tools for fluid and fluid mixtures.

The first application of *TPT* to fluids with variable composition (dissociating liquid nitrogen dioxide) was reported in Ref.[144]. In the case of chemically reacting fluids, the *TPT* approach is in many aspects very similar to that developed for non–reacting fluid mixtures, with, however, two additional features: first, the evaluation of the equilibrium composition at a given V and T, second, the computation of composition-sensitive properties.

An important point, however, is that the predicted composition can become ambiguous at high densities. Actually, it is possible to construct different schemes to specify the chemical composition of the liquid phase, from a pure physical description of a mixture of elemental particles (*e.g.*, for liquid UO_2, the atoms U and O or the ions U^{4+} and O^{2-}, U^{2+} and O^-, *etc.*), to the limiting case where the liquid is composed only of UO_2 molecules (*SSA*). The origin of this ambiguity is clear: in the dense, condensed phase the intermolecular distances are so short that it is hard to distinguish the neutral molecule UO_2 from three atoms O+U+O (or ions $O^{2-}+U^{4+}+O^{2-}$, $O^-+U^{2+}+O^-$) in close proximity. For neutral reacting gases this question has often been discussed, whilst, for ionic liquids the solution of the problem has been addressed in terms of the Stillinger-Reiss theorem [128, 117] (see the more detailed discussion of this problem in Section 2.4).

Being closely related to the classic van der Waals *EOS* mentioned above, the equations of state based on *TPT*, can be applied in a wide range of states, from densities near the solid state density until to the critical point. To date, there are numerous applications of the *TPT* to different kinds of fluids, including those with non-spherical, multipolar molecules, to liquid metals and mixtures of simple, as well as complex molecular fluids, and to metallic alloys, where different reference systems are used (hard spheres, soft spheres, rigid dumbbells, non-spherical convex bodies and mixtures of them) [19].

TPT-models can be implemented on the groundwork of a well-established formalism, that is briefly summarised in Section 6.1. A short review of the *TPT* basic principles is presented in Section 6.1.1. Equations of state based on *TPT* can be extended to multi-component mixtures in an approximate way. Details of this one-fluid approach are given in Section 6.1.2.

Finally, using of a centre-symmetrical Lennard-Jones (12-6) potential to describe the interactions between pseudo-molecules, as well as other components of the liquid mixture, is too drastic a simplification. In reality, these molecules (or ion clusters) may be non-spherical in shape, entailing a significant anisotropy in their mutual interactions. That is why the parameters of the Lennard-Jones (12-6) potential must be considered as effective quantities averaged over all orientations. This needs, however, a more detailed verification. In Section 6.2, we present an attempt to account for the effects of non-central intermolecular interactions, and their influence on the parameters of the effective Lennard-Jones (12-6) potential.

6.1 Thermodynamic Perturbation Theory

Peierls [111] first proposed the thermodynamic perturbation theory (*TPT*) starting from the assumption that the total energy of a macroscopic system can be split into two parts: $U = U^{(0)} + w$, where $U^{(0)}$ is the total energy of an unperturbed *reference system*, and w a correction, which can be treated as small in a certain sense. The total free energy, $\mathcal{F}$, of the system can be expressed in terms of the free energy of the reference system, $\mathcal{F}^{(0)}$, and the averaged value of the perturbation, w:

$$\mathcal{F} = \mathcal{F}^{(0)} + \overline{w} - \frac{1}{2kT}\overline{(w - \overline{w})^2} + \ldots \tag{6.1}$$

Note that whilst the *first-order* correction is simply equal to the mean value of the perturbation energy, w, averaged over the ensemble of the unperturbed system (and hence may have either signs), the *second-order* correction is always negative, so that in the particular case of $\bar{w} = 0$ the perturbed free energy is always *lowered* by the perturbation. This important feature of the *TPT* has become the basis of many classical variational approaches.

Obviously, the structure and the thermodynamic properties of the unperturbed system must be completely characterised. In fact, the great success of *TPT* in the statistical theory of liquids was achieved only when both perturbed and reference systems could be fully realised. Many computer simulation experiments have established that the radial distribution functions of a realistically described fluid (*e.g.*, one whose molecules interact via the Lennard-Jones (12-6) potential) are very similar to those of a model fluid having purely repulsive intermolecular interactions. This similarity is mainly due to the insensitivity of the structure of a real fluid to the weak attractive forces acting between non-charged, non-polar molecules of simple fluids. This fact enables the free energy (and hence all thermodynamic functions) to be expanded in a power series involving only the attractive part of the interaction energy, $w = U_N - U_N^{(0)}$.

According to Eq.(6.1) one can write:

$$\frac{F}{kT} = \frac{F^{(0)}}{kT} + \frac{n}{kT}\langle w \rangle - \frac{1}{2}\left(\frac{n}{kT}\right)^2 \left(\langle w^2 \rangle - \langle w \rangle^2\right) + \dots . \tag{6.2}$$

Here F is the Helmholtz free energy per particle of the real fluid, $U_N = \sum_{i<j} \Phi(r_{ij})$ is its total potential energy; $F^{(0)}$ and $U_N^{(0)} = \sum_{i<j} \Phi^{(0)}(r_{ij})$ are the same functions for the model *reference* fluid, with purely repulsive interactions; $n = N/V$ is the particle number density; $\langle w \rangle = 2\pi \int g^{(0)}(r) w(r) r^2 dr$ is the value of the *perturbation* potential averaged over the ensemble of the reference system; $w(r)$ is a pair contribution to w, *i.e.*:

$$w = \sum_{i<j} w(r_{ij}), \qquad w(r) = \Phi(r) - \Phi^{(0)}(r).$$

We must note that in any approach based on *TPT* the subdivision of the interactions into a strong and steep repulsive core contribution, and a relatively small, weakly varying attractive part is essential. It is not required, however, that the second term in Eq.(6.1) be necessarily small; the only constraint is the rapid convergence of the series Eq.(6.2). The perturbation corrections may, for example, be due to the short-range non-valence attractive part of the interaction of charged or reacting particles [143].

6.1.1 Simple Fluids

In *TPT*-based applications, one should distinguish between the strict, rigorous approach of the mathematical method itself and some methods where simplified expressions of $P(V, T, N_i)$ are adopted, which are in turn deduced from predictions of rigorous *TPT* for simplest molecular models, and eventually fed by empirically fitted parameters. In fact, there are several sources of uncertainties in this type of *EOS*':

- those related to the approximate nature of the *first-order TPT* approach;
- those due to the application of *TPT* relations, validated only for simpler molecular models, to more complex non-spherical, polar molecules, *etc.*;
- those caused by the absence of accurate molecular constants and intermolecular potential parameters;
- finally, one should also consider the errors due to simplifications of one-component *TPT* formulae and their approximate extension to multi-component mixtures.

The results obtained with the simplest form of *TPT*-based *EOS* [143] show that use of such an approach may be helpful to predict thermodynamic properties at high temperatures. When *TPT* is applied to a particular fluid, the main problem is how to choose the reference system and ascertain the relevant parameters. If one supposes that the free energy of the reference system can be considered within the simplest *excluded volume theory* and the attractive forces are sufficiently weak and long-ranged for ignoring the nearest order structure, then the well known van der Waals *EOS* Eq.(2.39) is obtained:

$$P = \frac{RT}{V - b} - \frac{a}{V^2}\,, \tag{6.3}$$

where $a = -2\pi N_a \int w(r) r^2 dr$ incorporates the effects of attraction between molecules, and $b = 4\ V_0$, where $V_0 = {}^1\!/_6 \pi d^3$ is the molecular volume and d is he effective molecular diameter.

A more coherent account of the repulsive forces leads to modifications of the first term of this equation. For example, if instead of $V/(V - b)$ the Carnahan–Starling expression for the hard-sphere compressibility factor Eq.(2.30) is adopted, one obtains an *EOS* which is more accurate [65], at high temperatures, than the corresponding *VdW-EOS.*

The approach proposed by Mistura, Magill and Ohse [105] for UO_2 was already mentioned as *PHC-EOS* in Chapter 1, Table 1.1, and in Chapter 2, Eq.(2.43). These authors actually replaced the hard-sphere repulsive term of the van der Waals *EOS* by the expression for hard convex molecules proposed by Boublik Eq.(2.31).

All the equations of this kind are usually called *VdW*-like *EOS*'s. They all have many features in common. First, they predict $\Delta C_V = C_V - C_V^{(id)} = 0$. Secondly, they produce a weak temperature dependence of the second virial coefficient, $B(T) \sim T^{-1}$, at low temperatures, in contrast to the rigorous statistical mechanical theory [150], which predicts a dependence of type $\ln|B\,(T)\,| \sim T^{-1}$. Furthermore, the compressibility factor at the critical point is unrealistically high for non-reacting fluids (~0.375). This is due to assumptions with no rigorous grounds.

Any approach restricted to a *first-order TPT* neglects fluctuations, which become more important the closer is the considered state to the critical point,ad hence does not permit a detailed description of the critical region. Up to now, valid *EOS* expressions have been obtained only for the simplest molecular models such as spherical particles with short-range interaction, steep repulsion at short distances together with weak attraction on a longer range (simple liquids).

Two main approaches exist in the case of simple fluids. Barker and Henderson [10] proposed an analytical approach to *TPT* of a real fluid whose molecules interact via a spherically symmetric potential, having both repulsive and attractive parts. These authors expanded the free energy using a specific parameterisation of the interaction potential $\Phi(r) \Rightarrow \Phi(\alpha, \gamma, r)$, which tends to the interaction potential for hard spheres when $\alpha \to 0$ and $\gamma \to 0$.

$$F = F^{(0)} - 2\pi n d^2 g_{HS}^{(0)}(\eta, d) \left\{ \int_0^{\sigma} \exp\left(-\Phi(r)/kT\right) dr + d - \sigma \right\} + \\ +2\pi n \int_{\sigma}^{\infty} w(r) g_{HS}^{(0)}(\eta, r)\, r^2 dr + \dots , \tag{6.4}$$

where $g_{HS}^{(0)}(\eta, r)$ is the radial distribution function of the hard-sphere fluid, and $\eta = {}^1\!/_6 \pi n d^3$ is its packing fraction. Starting from Eq.(6.4), Barker and Henderson defined a temperature dependent hard sphere diameter, $d_{BH}(T)$, in the following way:

$$d_{BH}(T) = \sigma - \int_0^{\sigma} \exp\left(-\Phi(r)/kT\right) dr \ . \tag{6.5}$$

Here σ is the smallest (short-distance) root of the equation $\Phi(\sigma) = 0$. This definition renders the second term of Eq.(6.4) equal to zero, whereby, the hard sphere diameter $d_{BH}(T)$ is a weakly decreasing function of temperature only. Levesque and Verlet [95] have proposed a simple approximation for this dependence: $d/\sigma = \mathcal{P}(T^*)$, where $\mathcal{P}(T^*)$ is a polynomial function of the reduced temperature, $T^* = kT/\varepsilon$, for the case of a one-component fluid with a Lennard-Jones (12-6) interaction potential :

$$\Phi_{LJ}(r) = 4\varepsilon \left[\left(\frac{\sigma}{r}\right)^{12} - \left(\frac{\sigma}{r}\right)^{6} \right] . \tag{6.6}$$

In a following work, Verlet and Weiss [136] presented a more accurate approximation. Substituting the potential Eq.(6.6) into Eq.(6.4), they obtained the following first-order approximate expression for free energy:

$$\frac{F}{RT} = \frac{F_{HS}^{(0)}(\eta)}{RT} - n \frac{a}{RT} \left[\left(\frac{\sigma}{d}\right)^{12} M_{12} - \left(\frac{\sigma}{d}\right)^{6} M_6 \right], \tag{6.7}$$

where

$$a = \frac{16}{9} \pi N_a \varepsilon \sigma^3 \tag{6.8}$$

is the van der Waals attraction parameter, and

$$M_k = 2\pi \int_{\sigma}^{\infty} (r/d)^{-k} g_{HS}^{(0)}(\eta, r) r^2 dr \tag{6.9}$$

is the k-th moment of the hard-sphere radial distribution function.

Andersen, Chandler and Weeks [6] used a quite different approach, based on the functional differentiation method to expand the free energy of a simple fluid:

$$F = F^{(0)} + \int \frac{\delta F}{\delta \Phi(r)} \Delta\Phi(r) d\mathbf{r} + \frac{1}{2} \iint \frac{\delta^2 F}{\delta\Phi(r)\delta\Phi(r')} \Delta\Phi(r)\Delta\Phi(r') d\mathbf{r} d\mathbf{r}' + \dots \tag{6.10}$$

where $\Delta\Phi(r)$ denotes the difference between pair interactions in the real and the reference fluids, and $\delta F/\delta\Phi(r)$ denotes the functional derivative. The subdivision of the potential into a reference part $\Phi^{(0)}(r)$ and a perturbation, $w(r) = \Phi(r) - \Phi^{(0)}(r)$, was done in Ref.[6] as follows:

$$\Phi^{(0)}(r) = \begin{cases} \Phi(r) + \varepsilon, r < r_0 \\ 0 \quad\quad , r > r_0 \end{cases}$$

and

$$w(r) = \begin{cases} -\varepsilon \ , r < r_0 \\ \Phi(r), r > r_0 \end{cases}, \tag{6.11}$$

where: $\varepsilon = -\Phi(r_0)$, and $\Phi'(r_0) = 0$.

The reference hard-sphere diameter, $d(n, T)$, is defined in Ref. [6] in such a way that it is also a weakly decreasing function of temperature and density. Andersen, Chandler and Weeks' theory seems to be more accurate, especially at low temperatures, but, due to the density-dependent diameter, it is much harder to use in practical applications.

6.1.2 One-Fluid Approximation

In this section, the basic approximate formulae used in *TPT* for multi-component mixtures are reported for the simplest case, *one-fluid approximation*, with the purpose of illustrating the nature and the role of the parameters in this kind of applications[1].

The definition of the effective one-fluid potential parameters of a mixture, σ_{MIX} and ε_{MIX} is:

$$\sigma^3_{MIX} = \sum_{ij} x_i x_j \sigma^3_{ij}; \tag{6.12}$$

$$\sigma^3_{MIX} \varepsilon_{MIX} = \sum_{ij} x_i x_j \sigma^3_{ij} \varepsilon_{ij}; \tag{6.13}$$

where σ_{ij} and ε_{ij} are the parameters of Lennard-Jones potential Eq.(6.6); $x_i = N_i/N$ is the mole fraction of i-th component, $N = \sum_i N_i$. The volume/geometrical-mean rule is applied for mixed interactions:

$$\sigma^3_{ij} = {}^1\!/_2 \left(\sigma^3_{ii} + \sigma^3_{jj}\right),$$

$$\varepsilon_{ij} = \sqrt{\varepsilon_{ii}\varepsilon_{jj}}. \tag{6.14}$$

The compressibility factor, $z{=}PV/RT$, and the excess hard-sphere free energy per mole are expressed as functions of the packing fraction according to the Carnahan-Starling formula:

---

[1] This approximation is used below for comparison purposes and referred to as the *simple TPT-EOS*.

$$z_{HS} = \frac{1+\eta+\eta^2-\eta^3}{(1-\eta)^3}; \tag{6.15}$$

$$\frac{\Delta F_{HS}}{RT} = \frac{4\eta-3\eta^2}{(1-\eta)^2}. \tag{6.16}$$

The packing fraction η is:

$$\eta = {}^1\!/_6 \pi n\, d^3_{MIX}\,, \tag{6.17}$$

where n is the number density and d_{MIX} is defined as a function of σ_{MIX}. The effective diameter is given by Eq.(2.38):

$$\frac{d_{MIX}}{\sigma_{MIX}} = \frac{0.9955+0.168\;T^*}{1+0.1958\;T^*}\,, \tag{6.18}$$

where $T^* = kT/\varepsilon_{MIX}$.
The van der Waals attraction parameter is:

$$a_{VdW} = -\frac{16}{9}\varepsilon_{MIX}\sigma^3_{MIX}. \tag{6.19}$$

The simplified expressions for the moments M_{12} and M_6 Eq.(6.9) given by Verlet and Weiss [136]:

$$M_{12} = \frac{2}{9}\pi d^3_{MIX}\frac{1-0.797\xi^2-0.48\xi^3}{(1-\xi)^2}; \tag{6.20}$$

$$M_6 = \frac{2}{3}\pi d^3_{MIX}\frac{1-0.691\xi-1.169\xi^2+0.751\xi^3}{(1-\xi)^2}; \tag{6.21}$$

$$\xi = \eta - \eta^2/16 \tag{6.22}$$

provide the following simple equation for the total excess free energy of the mixture:

$$\Delta\mathcal{F}_{MIX} = kT\sum_i N_i\left[\frac{\Delta F_{HS}}{RT}(\eta) - \frac{32}{3}\,\eta T^*\, f(\eta,T^*)\right]. \tag{6.23}$$

The *TPT* correction factor, $f(\eta,T^*)$, is defined as the ratio of the attraction term:

$$A(\eta,T^*) = \left(\frac{\sigma_{MIX}}{d_{MIX}}\right)^{12} M_{12}(\eta,T^*) - \left(\frac{\sigma_{MIX}}{d_{MIX}}\right)^{6} M_6(\eta,T^*)$$

to its limiting value for $\eta \to 0$:

$$f(\eta,T^*) = A(\eta,T^*)/A(0,T^*).$$

The excess compressibility factor $\Delta z = z - 1$ has the form:

$$\Delta z_{MIX} = \Delta z_{HS}(\eta) - \frac{32}{3}\eta T^* f_1(\sigma_{MIX}, \varepsilon_{MIX}, \eta, T^*) \ , \tag{6.24}$$

where

$$f_1(\eta, T^*) = \eta \frac{\partial}{\partial \eta} f(\eta, T^*) \, .$$

The excess chemical potential $\Delta\mu_i$ of the i-th component is:

$$\frac{\Delta\mu_i}{RT} = \frac{\Delta F_{MIX}}{RT} + \Delta z_{MIX}\left[\left(\frac{\sigma_{ii}}{\sigma_{MIX}}\right)^3 - 1\right] + \frac{\Delta U_{MIX}}{RT}\left[\frac{\varepsilon_i}{\varepsilon_{MIX}} - \left(\frac{\sigma_{ii}}{\sigma_{MIX}}\right)^3 - 1\right] , \tag{6.25}$$

where:

$$\frac{\Delta U_{MIX}}{RT} = -T\frac{\partial}{\partial T}\left(\frac{\Delta F_{MIX}}{RT}\right) \tag{6.26}$$

is the reduced excess internal energy of the mixture, and the value of ε_i is defined by the average:

$$\varepsilon_i = \frac{1}{\sigma^3_{MIX}} \sum_j x_j \sigma^3_{ij} \varepsilon_{ij} \ . \tag{6.27}$$

6.2 Fluids Composed of Molecules with Anisotropic Interaction

Most molecules and clusters in a fluid are not centre-symmetric, so that their interactions are anisotropic. On the other hand, contrary to crystals, the properties of a liquid are isotropic, and the directional dependence of the interactions is actually hidden in thermodynamics[2]. In ordinary liquids the anisotropic interactions affect the thermodynamic functions only *implicitly*, forming the short-range order structures. In fact, the anisotropic forces in such fluids are present always as an average over all possible mutual orientations of molecules. The problem is therefore to find a proper averaging procedure and to convert the effect of anisotropic forces into an effective correction of the isotropic interactions.

This problem, called sometimes the 'sphericalisation' problem [91], is completely solved only in the case of weak long-ranged multipole interactions on the basis of the *TPT*-approach of Gubbins and Gray [53]. In the general case, especially for strong non-central repulsion forces, no coherent approach has been developed so far. The method described in this Section, was proposed in Ref. [77] as an alternative to empirical sphericalisation procedures [91]. It can be treated as a generalisation of the Barker-Henderson's formulae on fluids, constructed from molecules with dominating anisotropy of repulsion forces.

[2] Excluding such systems as liquid crystals.

Let us start from the most general form of a non-central interaction: $\Phi = \Phi(r,\omega)$, where r is the distance between the molecular centres, $r = |\mathbf{q_1} - \mathbf{q_2}|$, ω is the set of all angular variables, *e.g.*, for linear molecules like UO_2 $\omega = \{\varphi_1, \varphi_2, \theta_1, \theta_2\}$ where θ and φ are the azimuthal and polar angles determining the direction of arbitrary axes passing through the centres of the molecules with respect to the line connecting these centres.

Without any restriction, this potential may be written in the form:

$$\Phi(r,\omega) = \Phi_0(r) + \Delta\Phi(r,\omega). \tag{6.28}$$

Here $\Phi_0(r) = \langle\Phi(r,\omega)\rangle$ is the central part of the potential, defined as the interaction of two molecules averaged over all orientations, $< \ldots >$ represents the operator of averaging over all angles θ_i and φ_i:

$$\langle \ldots \rangle = \frac{1}{16\pi^2}\int_0^{\pi} sin\theta_1 d\theta_1 \int_0^{\pi} sin\theta_2 d\theta_2 \int_0^{2\pi} d\varphi_1 \int_0^{2\pi} \ldots d\varphi_2. \tag{6.29}$$

It is evident, that such a definition implies that the averaged value $\langle\Delta\Phi(r,\omega)\rangle$ is equal to zero.

We shall now suppose, that the central contribution in Eq.(6.28) may be approximated by a Lennard-Jones (12-6) potential $\Phi_0(r) = \Phi_{LJ}(r, \sigma_0, \varepsilon_0)$ with given effective parameters ε_0 and σ_0. In other words, ε_0 and σ_0 may be considered as some averaged values of the depth of the potential well and the mean distance of closest approach for the potential of Eq.(6.28). Physically, it is clear that negligence of the anisotropic part $\Delta\Phi(r,\omega)$ in Eq.(6.28), as well as the approximation of Eq.(6.28), *e.g.*, by Eq.(6.6) is valid only in the limit of high temperatures, *i.e.*, when $kT \gg \varepsilon_0$. We also suppose that the anisotropic part of Eq.(6.28) may be expressed in the form:

$$\Delta\Phi(r,\omega) = \frac{\phi(\omega)}{r^m}, \tag{6.30}$$

i.e., that it is possible to separate the angular and spatial variables, and to represent the dependence on distance by an inverse power function. In this case, the potential expressed by Eqs.(6.28) - (6.30) maintains a sufficient general character. For instance, it includes as a special case the classical Stockmayer potential for rigid polar molecules:

$$\Phi(r,\theta_1,\theta_2,\varphi) = 4\varepsilon_0\left[\left(\frac{\sigma_0}{r}\right)^{12} - \left(\frac{\sigma_0}{r}\right)^{6}\right] - \frac{D^2}{r^3}[2cos\theta_1 cos\theta_2 - sin\theta_1 sin\theta_2 cos\varphi], \tag{6.31}$$

where D is the dipole moment of the molecule, and $\varphi = \varphi_1 - \varphi_2$.

6.2.1 Averaged Diameter of Non-Spherical Molecules

Let us first estimate the value of σ, *i.e.*, the root of the potential Eq.(6.28). It is important because this value corresponds to the linear parameter in Eq.(6.6) for a given molecular orientation, according to the definition:

$$\Phi(\sigma(\omega), \omega) = 0 \tag{6.32}$$

or $\Phi_0(\sigma(\omega)) + \Delta\Phi(\sigma(\omega), \omega) = 0$. We shall estimate the difference

$$\Delta\sigma(\omega) = \sigma(\omega) - \sigma_0 \tag{6.33}$$

by expanding Eq.(6.28) in a power series with respect to Eq.(6.33) and taking into account only the linear terms.

By substituting Eq.(6.33) into Eq.(6.32), remembering that $\Phi_0(\sigma_0) = 0$, one obtains:

$$\begin{aligned} \Phi_0(\sigma(\omega)) &= \Phi_0(\sigma_0) + \Phi_0'(\sigma_0)\Delta\sigma(\omega) + \dots\,, \\ \Delta\Phi(\sigma(\omega), \omega) &= \Delta\Phi(\sigma_0, \omega) + \Delta\Phi'(\sigma_0, \omega)\Delta\sigma(\omega) + \dots\,. \end{aligned} \tag{6.34}$$

Solving now the linear Eq.(6.34) with respect to $\Delta\sigma(\omega)$, we get:

$$\Delta\sigma(\omega) = -\frac{\Delta\Phi(\sigma_0, \omega)}{\Phi'(\sigma_0) + \Delta\Phi'(\sigma_0, \omega)}. \tag{6.35}$$

After evaluating $\Delta\sigma(\omega)$ by averaging both sides of Eq.(6.35) over the angular variables, we find:

$$\langle \Delta\sigma(\omega) \rangle = -\frac{\langle \Delta\Phi(\sigma_0, \omega) \rangle}{\Phi_0'(\sigma_0)} + \frac{\langle \Delta\Phi(\sigma_0, \omega)\Delta\Phi'(\sigma_0, \omega) \rangle}{\left[\Phi_0'(\sigma_0)\right]^2} + \dots\,.$$

Taking now into account the symmetry properties of the anisotropic part of the potential, we notice that the first term at the right-hand side goes to zero. Preserving only the first non-vanishing contributions, and taking into account Eq.(6.30), we find:

$$\langle \Delta\sigma(\omega) \rangle = -\frac{m \langle \phi^2(\omega) \rangle}{\sigma_0^{2m-1} \left[\Phi_0'(\sigma_0)\right]^2}. \tag{6.36}$$

Thus, the mean value $\langle \Delta\sigma(\omega) \rangle$ is always non-positive. We should notice that this result is valid for *any type* of anisotropic interaction, *i.e.*, for arbitrary $\phi(\omega)$ and any positive m.

By differentiating Eq.(6.6) and substituting the result in Eq.(6.36), we obtain:

$$\langle \sigma \rangle = \sigma_0 \left(1 - \frac{m \langle \phi^2 \rangle}{(24\varepsilon_0 \sigma_0^m)^2} \right). \tag{6.37}$$

We introduce the following simpler notations:

$$\langle \sigma \rangle = \langle \sigma(\omega) \rangle, \quad \langle \phi^2 \rangle = \langle \phi^2(\omega) \rangle,$$

and define Q as a dimensionless anisotropy parameter:

$$Q^2 = \frac{\langle \phi^2 \rangle}{(4\varepsilon_0 \sigma_0^m)^2}, \tag{6.38}$$

where Q represents the ratio of the mean squared value of the anisotropic contribution to the corresponding isotropic repulsive interaction.

Then Eq.(6.37) can be finally rewritten as:

$$\langle \sigma \rangle = \sigma_0 \left(1 - \frac{mQ^2}{36}\right). \tag{6.39}$$

6.2.2 Effective Molecular Diameter

For the calculation of the effective particle diameter we apply the following generalisation of the Barker-Henderson definition:

$$\langle d \rangle = \left\langle \int_0^{\sigma(\omega)} (1 - \exp\{-\beta \Phi(r,\omega)\})dr \right\rangle. \tag{6.40}$$

As above, the angular brackets mean averaging Eq.(6.29) over the angular variables, and the $\sigma(\omega)$ function is the root (or the zero) of the potential function for a given mutual orientation of the molecules; $\beta = 1/kT$ is the inverse statistical temperature. Let expand the integrand in Eq.(6.40) over the anisotropic contribution up to quadratic terms:

$$\langle d \rangle = \langle \sigma \rangle - \left\langle \int_0^{\sigma(\omega)} \exp\{-\beta \Phi_0(r)\} \left(1 + \frac{1}{2}\beta \left(\frac{\phi(\omega)}{r^n}\right)^2\right) dr \right\rangle,$$

where the condition $\langle \phi(\omega) \rangle = 0$ was taken into account.

Considering now that:

$$\int_0^{\sigma(\omega)} \exp\{-\beta \Phi_0(r)\}dr \approx \int_0^{\sigma_0} \exp\{-\beta \Phi_0(r)\}dr + \sigma(\omega) - \sigma_0,$$

and that:

$$\int_0^{\sigma(\omega)} \exp\{-\beta \Phi_0(r)\} \frac{dr}{r^{2m}} \approx \int_0^{\sigma_0} \exp\{-\beta \Phi_0'(\sigma_0)(r - \sigma_0)\} \frac{dr}{\sigma_0^{2m}} = \frac{kT}{24\varepsilon_0 \sigma_0^{2m-1}},$$

designating the effective diameter for isotropic interaction according to Eq.(2.37):

$$d_0 = \sigma_0 - \int_0^{\sigma_0} \exp\{-\beta \Phi_0(r)\}dr,$$

and expanding the integrand into Taylor series at $r = \sigma_0$, we come to the relationship:

$$\langle d \rangle = d_0 - \sigma_0 \frac{\langle \phi^2 \rangle}{48\varepsilon_0 kT (\sigma_0)^{2m}}. \tag{6.41}$$

The value $\langle d \rangle$ in Eq.(6.40) is always positive when Eq.(6.41) is applied to sufficiently small anisotropy corrections. Therefore, it is convenient to replace Eq.(6.41) by an equivalent positive-defined Padé-approximant, and finally represent the averaged diameter in terms of the anisotropy parameter Eq.(6.38):

$$\langle d \rangle = d_0 \left(1 + \frac{\sigma_0}{d_0} \frac{\varepsilon_0 Q^2}{3kT} \right)^{-1}. \tag{6.42}$$

6.2.3 Estimation of the Anisotropy Parameter

The uranium dioxide molecule has a symmetrical linear structure, with zero dipole moment. From this it follows, that the repulsive forces may play an important role in the anisotropic molecular interactions. Hence, we reasonably put $m = 12$, in accordance with the power dependence on r of the repulsive contribution in Eq.(6.6). However, to perform such an evaluation, at least one additional estimation is needed for the anisotropy parameter Q of Eq.(6.38).

The requirement that the anisotropy parameter be equal to the ellipticity of the molecular shape can serve here as a simplest hypothesis. We can approximately assume for the dioxide molecule that:

$$Q \approx \frac{\sigma_{\max} - \sigma_{\min}}{\sigma_0}. \tag{6.43}$$

Here $\sigma_{\max}$ and $\sigma_{\min}$ are, respectively, the maximal and minimal linear dimensions.

Let us apply such an estimate, and start from diameters of atoms and molecules recommended in Appendix A.3. We then obtain:

$$Q \approx \frac{\sigma_O + 2L_{U-O} - \sigma_U}{\sigma_{UO_2}} = \frac{1.8 + 2 \times 1.8 - 2.9}{3.8} = 0.66 \,. \tag{6.44}$$

A comparison of the values found empirically by fitting of the corresponding parameters in the equation of state at T_m, with the effective potential parameters $\langle \sigma \rangle$ and $\langle \varepsilon \rangle$, as well as with the effective diameter $\langle d \rangle$ calculated on the basis of the above equations is presented in Table 6.1.

The value of $Q \sim 0.66$ is close to that fitted from experimental data. Table 6.1 also includes an estimate of the value of the mean well depth, $\langle \varepsilon \rangle$, ensuing directly from the fact that anisotropy does *not* affect the attractive part of the potential. In this case, the condition: $\varepsilon_0 \sigma_0^6 = \langle \varepsilon \rangle \langle \sigma_0 \rangle^6$ must be satisfied. Hence:

$$\langle \varepsilon \rangle = \varepsilon_0 \left(\frac{\sigma_0}{\langle \sigma \rangle} \right)^6. \tag{6.45}$$

Table 6.1. Anisotropy Effect on the Effective Potential Parameters

Parameter	Equation	Estimation	Fitting
$\langle\sigma\rangle/\sigma_0$	Eq.(6.39)	0.85	0.79
$\langle\sigma\rangle/\langle d\rangle$	Eqs.(6.42), (6.39)	1.18	1.15
$\langle\varepsilon\rangle/\varepsilon_0$	Eq.(6.45)	2.6	2.76

6.3 Conclusions

The anisotropy corrections discussed in this section are based on rough approximations, however they give an idea of the magnitude of the anisotropy effects in the macroscopic expressions of molecular interactions. The sign of the anisotropy effect is certain, but the value of the anisotropy correction depends on the assumptions involved, on the definition of Q, and on the bond lengths and dimensions of the atoms used in the calculation. The final values of the potential parameters can be further adjusted on the basis of the available experimental data, however, it is important that anisotropy corrections be taken into account, and general formulae be applied to all the relevant molecular clusters.

Finally, the following conclusions can be drawn without any doubt:

- Any attempt to correctly describe the properties of fluids formed by anisotropic molecules in the context of pseudo-spherical models entails a remarkable *decrease* of the effective molecular diameter $\langle\sigma\rangle$, irrespective of the type of anisotropic interaction, compared with the case of direct application of an isotropic potential.
- The effective diameter, $\langle d\rangle$, used in *TPT*, must also decrease, in comparison with the case of isotropic potential, but more than $\langle\sigma\rangle$ does. Therefore, when the anisotropy of repulsive interaction is taken into account, the $\langle\sigma\rangle/\langle d\rangle$ ratio noticeably grows compared with the case of no correction.
- The effective potential well depth, $\langle\varepsilon\rangle$, being proportional to the sixth power of the $\sigma_0/\langle\sigma\rangle$, markedly increases, as compared to the isotropic one, ε_0.

7

Thermodynamic Properties of UO_2, as Predicted by the New Equation of State

This chapter contains the description of the eventually recommended equation of state for uranium oxide, called here *INTAS*-99-*EOS*, as expressed in terms of models and sub-models presented in the previous Chapters. In the frame of the *chemical model* approach, this equation probably represents the most comprehensive *EOS* of $UO_{2\pm x}$, affordable by optimal use of the presently available theoretical tools and experimental databases. The equation is of a *MIX* type, as sketched in Chapter 2; the details of its structure are explained and discussed in Chapters 3 to 6. After a recapitulation of the fundamental hypotheses and a summary of the main features, this Chapter is dedicated to the presentation of an extensive set of numerical calculations and to the comparison with predictions of alternative models.

7.1 Summary of the Model Features

We summarise here the salient points of the *INTAS*-99-*EOS* development and formulation.

The calculation of the equilibrium composition and of the thermodynamic functions starts from the definition of the Helmholtz free energy, $F(V,T,N_j)$, and a set of arbitrary numbers of particles of all species, N_j. The free energy is formulated within the composite *MIX-I* chemical model as the sum of three contributions: the ideal-gas part, $F^{(id)}$, the repulsive $\Delta F^{(R)}$, the short-range attractive interactions of neutral and charged particles, $\Delta F^{(A)}$, and the Coulomb interaction of charged particles, $\Delta F^{(C)}$,*i.e.*:

$$F(T,V,N_j) = F^{(id)} + \Delta F^{(S)} + \Delta F^{(C)}, \tag{7.1}$$

where $\Delta F^{(S)} = \Delta F^{(R)} + \Delta F^{(A)}$ results from the sum of the contributions of all short-ranged interparticle forces. The set of equilibrium concentrations, $n_i = N_i/\mathcal{V}$, corresponds to the conditional minimum of the total free energy, $\mathcal{F} = (\Sigma N_i)\, F$, with respect to the variation of N_i at constant temperature and total volume, $\mathcal{V} = \Sigma N_i V$, under the constraint of:

- conservation of mass: $\Sigma N_i m_i = const$;
- chemical proportion: $\Sigma N_i \nu_i^{(\beta)} / \Sigma N_j \nu_j^{(\alpha)} = const$;
- electro-neutrality: $\Sigma N_i Z_i = 0$.

The adopted composite model accounts for the specific properties of the system investigated, describing, in a unified formalism, a *multi-component mixture* of chemically reactive, strongly interacting neutral and charged molecules and atoms [99].

The contribution of neutral and charged-neutral particle interactions to the free energy, $\Delta F^{(S)}$, is formulated according to the modified thermodynamic perturbation theory described in the previous Chapter. The parameters of the central-symmetrical Lennard-Jones (12-6) potential for the interactions between the components of the liquid mixture are written as reported in Chapter 6 in order to account for anisotropy when this is required by the individual repulsive interactions. Consequently, the parameters of the Lennard-Jones (12-6) potential are considered as appropriate averages of their anisotropic counterparts over all mutual orientations of particle pairs.

The hard-spheres mixture approximation, Eq.(2.32), was used for the contributions of the reference repulsive interactions of all (ionic and neutral) species to the free energy.

A simplified version of the expressions Eq.(6.20)-(6.22) was then assumed for the first-order *TPT* short-range attractive contributions to $\Delta F^{(S)}$, within the one-fluid approximation for mixtures, Eq.(6.7), with an effective diameter, d_{MIX}.

Finally, the definition of the effective molecular diameter, $d(T)$, in Eq.(6.42) was applied to d_{MIX} and was expressed through the effective parameters of the isotropic part of the potential $< \sigma >, < \varepsilon >$, and a dimensionless anisotropy parameter, Q, according to Eqs.(6.18) and (6.42).

The non-ideality correction, $\Delta F^{(C)}$, due Coulomb long-range ion-ion interactions, has the form of the modified *DHSE* approximation described in Chapter 3, Eq.(3.9) [140, 44] ($\alpha = 0.75$). The effective diameters of the charged particles differ from those of their neutral counterparts. It was however assumed, that the diameters of the ionic species, $\sigma_n^{(ion)}$, obey to the same dispersion law and the same temperature dependence as the diameters of their neutral counterparts.

7.2 Calibration

The *INTAS*-99-*EOS*, like all *EOS*' based on the *MIX* approach, is a multi-parametric equation of state. This feature significantly complicates the calibration procedure. All interaction parameters were first theoretically evaluated[1], and a sensitivity study was performed to subdivide them according

[1] See Appendix A.3

to the strength of their influence on the properties of liquid $UO_{2\pm x}$. It was found that the *EOS* predictions are most sensitive to the molecular sizes and attraction parameters of UO_2, and to the anisotropy parameter, Q. These are called *dominant* parameters. A subset of the *subdominant* parameters was identified, affecting only *some* thermodynamic properties.

The calibration procedure involves consecutive fitting of the dominant and subdominant parameters. The relative values of all other parameters with respect to the dominant ones are pre-fixed.

The adjustment of the dominant and subdominant parameters is made on the basis of the properties of liquid $UO_{2.00}$, *i.e.*: density, Gibbs free energy, heat capacity, compressibility and oxygen potential at T_m. The three-stage calibration procedure is performed as follows:

1. In the first stage, the following thermal and caloric properties of liquid UO_2:
   a) liquid density (8870 $kg\ m^{-3}$),
   b) Gibbs free energy / partial vapour pressure of UO_2 (2.3 kPa),
   c) liquid heat capacity and compressibility (as additional constraints),
   are fitted in the *SSA*-representation by varying only the dominant parameters σ_{UO_2}, ε_{UO_2}, and Q. The fitted experimental measurements are sufficiently accurate, and the dominant model parameters adequately sensitive to their variation.
   This stage, in conjunction with the initial theoretical estimations, provides the first guess for the fitted variables.
2. In the second stage, the same properties and, additionally, the oxygen potential[2], *i.e.*, the partial vapour pressure of O_2 ($P_{O_2} = 6\ Pa$) over $UO_{2\pm x}$ at T_m, are fitted, starting from the results of the previous stage, in the *full-component MIX-I* composition model[3], by varying both dominant and subdominant parameters.
3. In the last stage, the ratio of ionic/neutral sizes is fixed, and the *CVC* point at $T = 3400\ K$ is reproduced, whereby all properties fitted in the first two stages are taken as strong constraints.

No additional properties are fitted at higher temperatures.

The effective interaction parameters found for neutral and ionic species are listed in Table 7.1.

If one compares the interaction parameters with a pre-averaged isotropic potential (see Table. 6.1, p.111), the anisotropy of the molecular interaction

[2] The value of the oxygen potential of liquid $UO_{2\pm x}$ at T_m was supposed to be equal to the oxygen potential of the solid $UO_{2.00}$ recommended by Hyland (Appendix A.1). This equality corresponds to the assumption that $T = T_m$ is a point of congruent melting composition (*CMC*), where two basic chemical potentials, say of uranium and oxygen, are the same in two phases.

[3] It was assumed, that the ion/neutral diameter ratio is fixed, and the additional parameter, the value of this ion/neutral ratio, is fitted to reproduce the $AB = A^+ + B^-$ dissociation energy.

Table 7.1. Interaction Parameters of the Species Adopted in the *INTAS*-99-*EOS*

Species	O	O_2	U	UO	UO_2	UO_3
$\langle\sigma\rangle$, Å	1.83	2.22	2.48	2.67	2.92	3.23
$\sigma^{(ion)}$, Å	2.25	2.73	3.05	3.28	3.59	3.97
$\langle\varepsilon\rangle/k$, K	168	337	210	20350	21830	14100

causes a remarkable *decrease* of the effective molecular diameter, $\langle\sigma\rangle$. The effective diameter, $\langle d\rangle$, decreases even more strongly than $\langle\sigma\rangle$.

For UO_2 a reasonable correspondence was found between the value of the effective parameter, ε, in the simple *TPT-EOS* ($\varepsilon/k \simeq 8200\ K$), and the value of ε_0 in the *INTAS*-99-*EOS* ($\varepsilon_0/k = 21830/2.6 \simeq 8400K$, see also Tables 6.1 and 7.1). However, the value of $\langle d\rangle$ ($\langle d\rangle = 2.92/1.18 \simeq 2.47$ Å) is appreciably lower than that of d_{BH} in the simple *TPT-EOS*. This entails a remarkably better prediction of the liquid compressibility of the *INTAS*-99-*EOS* with respect to the simpler *TPT-EOS*.

7.3 Validation of the INTAS-99-EOS

Before introducing calculation results and their comparison with experimental data, it must be pointed out that most of the predicted effects are common to all the versions of the chemical (or molecular) representation of uranium dioxide fluid. Actually, substantial differences have been only found by comparing chemical with purely ionic models.

This similarity is caused by three features in the formulation of the chemical model.

- The first one is the rigidity of the short-range inter-particle repulsion (which in the simple *TPT* as well as in the *VdW-EOS* model approaches that of hard spheres).
- The second one is the large (and unperturbed) value of the internal excitation energy of the molecules that provides an essential contribution to the heat capacity of liquid and vapour at high temperatures.
- The third (and most important) feature is the effect of chemical reactions in the gas and the liquid.

Results of *INTAS*-99-*EOS* calculations are presented and compared with available experimental data in Figs. 7.1 to 7.12 and in Tables 7.2 to 7.3. The comparison is in some cases extended to the *VdW-EOS* and to the simpler

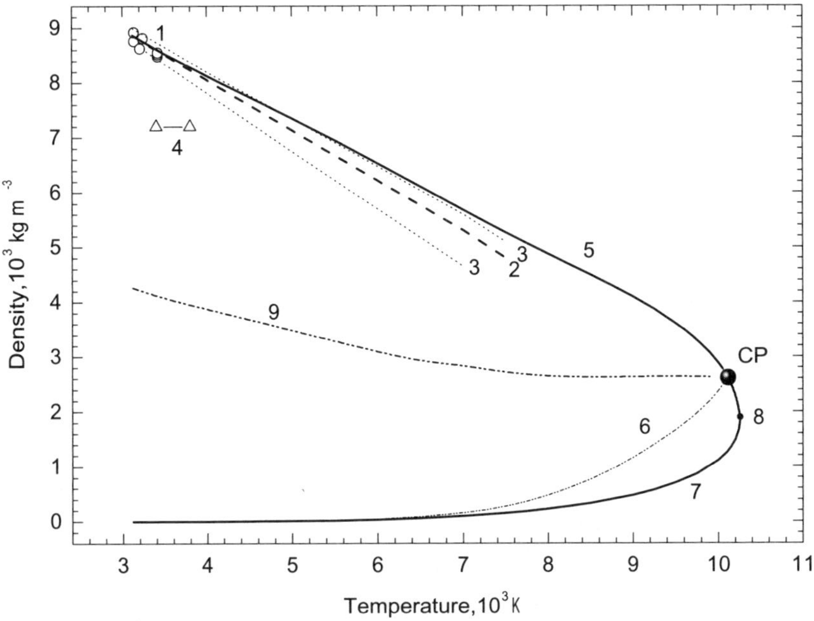

Fig. 7.1. Comparison of the calculated densities with experimental data and computer simulation:
1 – measurements of Drotning and Christensen [26];
2 – original Drotning's fit of the Ref.[31] data (recommended by *INSC*);
3 – uncertainty limits (recommended by *INSC*);
4 – *MD* simulation [125] for restricted ionic model $U^{4+}+O^{2-}$ at $P = 0$.
INTAS-99-EOS:
5 – density of the boiling $UO_{2.00}$; (*BC*)
6 – vapour density (O/U≥ 2) over non-congruently boiling $UO_{2.00}$;
7 – density of the saturated vapour $UO_{2.00}$(*SC*);
8 – *CCT* - point of maximal temperature (cricondentherm);
9 – $\frac{1}{2}(\rho^{(liq)}(BC) + \rho^{(vap)}(BC))$ (diameter);
CP – critical point of the non-congruent $UO_{2.00}$ evaporation

TPT-EOS in two modes, respectively: the *FCE* mode[4] and the total, non-congruent equilibrium mode.

[4] The *FCE* mode corresponds to the partial-equilibrium coexistence of two phases with the same stoichiometry, what physically corresponds to the *frozen diffusion approximation* [77] in the description of the phase coexistence. The *FCE* mode equalises the Gibbs free energies of the two coexisting phases. The parameters of such an equilibrium can be found by using the classic Maxwell's *double-tangent construction* on the Helmholtz free energy *vs.* volume plot at constant temperature and O/U ratio.

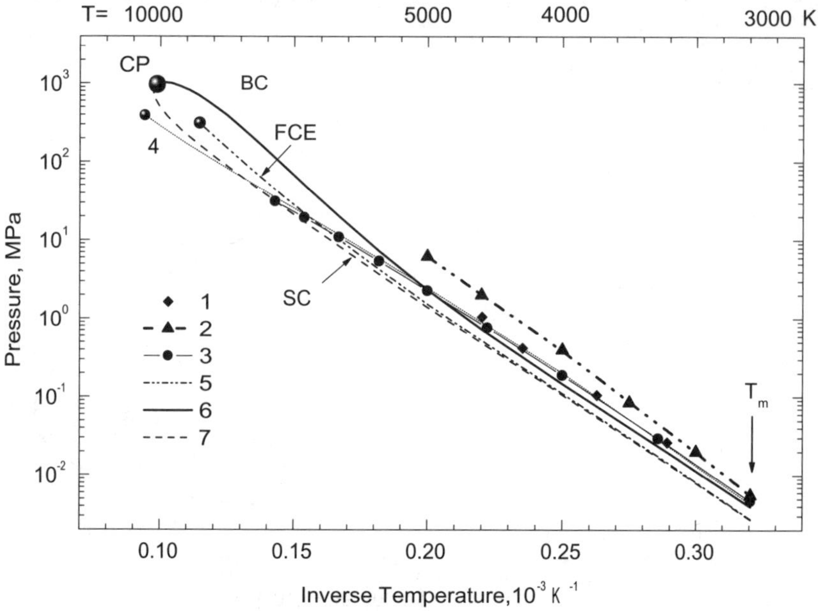

Fig. 7.2. Total vapour pressure over liquid UO_2. Comparison of the calculated non-congruent gas-liquid $P-T$ boundaries in boiling (BC) and saturation (SC) regimes with the *INSC* recommendations and other calculations: 1 – measurements of Bober and Singer [16]; 2 – *INSC* recommendation (1981) ; 3 – *INSC* recommendation (2000) [36]; 4 – non-congruent boiling curve for neutral composition; with pseudocritical point (Fischer,1992 [41]) ; 5-7–*INTAS*-99-*EOS*: 5 – *FCE* (double-tangent construction); 6 – boiling curve (BC- boiling of $UO_{2.00}$ liquid); 7 – saturation curve (SC- saturation of $UO_{2.00}$ vapour)

7.3.1 Density

Fig. 7.1 presents the densities of the liquid on the boiling curve and of the vapour on the saturation curve, (BC: $(O/U)^{(liq)}$=2.0; SC: $(O/U)^{(vap)}$=2.0), together with the density of the vapour coexisting with boiling $UO_{2.00}$ ($(O/U)^{(vap)} > 2$). Two fits of available experimental data (Appendix A.1) on liquid density, extrapolated to high temperatures, are added for comparison, *i.e.*:

- the original fit of Drotning [31] Eq.(A.3), p. 136, and
- the best fit of the united experimental data of Drotning and Christensen [26] between T_m and 3400 K : $\rho(T) = 12596 - 1.195\ (T/K)\ kg\ m^{-3}$.

In this figure are also displayed the results of computer simulations with the pure ionic model (two points) [125]. Comparison with the calculations of the *FCE* coexistence based on the *VdW-DH-HSM* approximations for neutral and ionic compositions (Fig. 5.4, p.82) reveals only little differences, mainly at high temperatures. This is likely a result of the same calibration conditions applied in both cases.

Fig. 7.1 shows significant deviations from the empirical rule of rectilinear diameters of Calliet and Mathias. The 'diameter' of the coexistence curve is markedly above the linear extrapolation of its low-temperature part. This property is a common feature not only of the *VdW-DH-HSM* and the *INTAS*-99-*EOS* but also of *most* of the *EOS'* based on the *chemical model.* The magnitude of this deviation and the density range where it occurs are remarkably large.

7.3.2 Total vapour pressure

The comparison of the predicted equilibrium vapour pressure with the literature data serves as a first validation of the model up to temperatures of the order of 5500 K. Fig. 7.2 shows the calculated total vapour pressure over liquid UO_2 under boiling and saturation regimes compared with the existing high-temperature measurements. These were obtained from complex experiments performed in a nuclear reactor, where rapid energy insertions were applied to small uranium oxide samples under conditions near to saturation. It should be remarked that the pressures under the two regimes differ by up to one order of magnitude.

Fig. 7.2 shows the vapour pressure *vs.* inverse temperature, as calculated via the simple *TPT-EOS* for a neutral composition and its comparison with experimental data of Bober and Singer [16] (*FCE* mode, double-tangent construction) between 3500 K and 4500 K. The following fit of in-pile experimental data, in the range 4700 - 8000 K [21]: $\log(P/MPa) = 23.7989 - 29605.5\ (T/K)^{-1} - 4.75783\ \log(T/K)$ was used for comparison.

The low-temperature vapour density is close to that of an ideal, chemically reacting-gas mixture, and does not depend on non-ideality effects. However, the density of the non-stoichiometric vapour in equilibrium with boiling $UO_{2.00}$ at $T > 7000\ K$ remarkably deviates from the saturated vapour density (see Fig. 7.1).

Both the simple *TPT-EOS* and the *INTAS*-99-*EOS* (as well as the *VdW-DH-HSM-EOS*, under *FCE*) predict a total vapour pressure rather close to the linear Arrhenius extrapolation up to $T = 7000\ K$. Near the critical point the slope markedly increases. This is a common feature for all *EOS*' based on the chemical model, and is in striking contradiction with the calculations of Fischer [39], and with the values reconstructed from the liquid $H - P$ measurements by Breitung and Reil [21]. It should be stressed that two constituents of the Clausius-Clapeyron equation - the heat of the vaporisation in the *FCE* mode, $L_{FCE}(T)$, and the compressibility factor difference, $\Delta z_{FCE}(T)$, are not constant, contrary to the case of an ideal gas. Both values noticeably

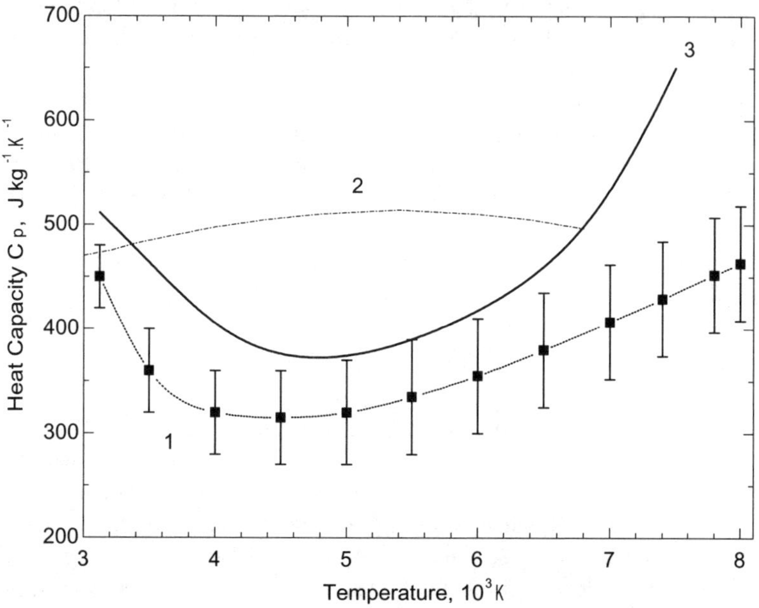

Fig. 7.3. Heat capacity of the liquid uranium dioxide. Comparison of predictions and measurements: 1 - measurements of Ronchi *et al* [119]; 2 - C_p calculated by Fischer; 3 - *INTAS*-99-*EOS*

differ from their ideal-gas counterparts, and the mentioned deviation of the $\log P$ *vs.* $1/T$ line directly results from this difference:

$$\left(\frac{d\ \ln P}{d\,(1/T)}\right)_{FCE} = L_{FCE}/\Delta z_{FCE}(T)\ ; \qquad (7.2)$$

Table 7.2. Partial Vapour Pressures of UO_2 (MPa)

Temperature, K	3120	3500	4000	4500	5000	6000	7000
IVTAN-82 [62]	0.00226	0.0135	0.0662	0.22	0.53	1.67	
IVTAN-96 [64]	0.00233	0.0131	0.0756	0.29	0.83	3.82	10.6
VdW-DH-HS	0.00233	0.0147	0.0890	0.33	0.92	3.74	9.8
INTAS-99-*EOS*	0.00231	0.0136	0.0662	0.32	0.75	3.24	9.5

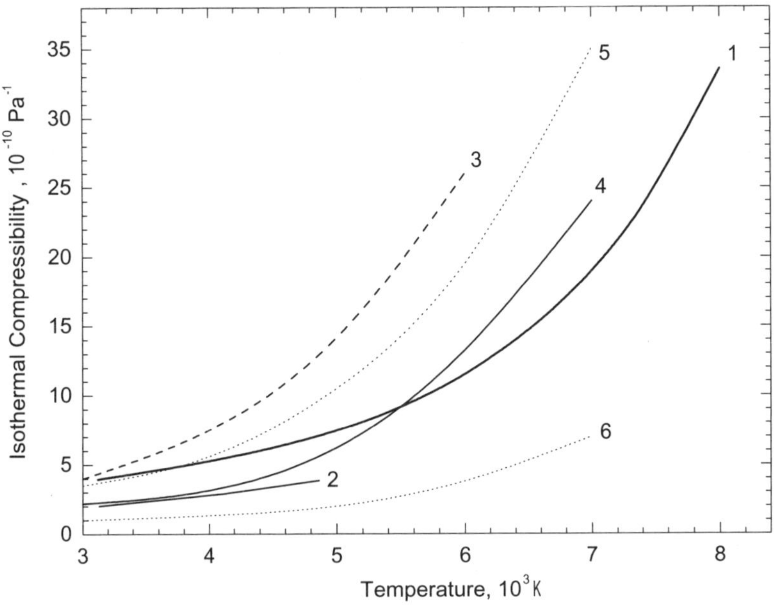

Fig. 7.4. Isothermal compressibility of the liquid UO_2: 1 – *INTAS*-99-*EOS*; 2 – simple *TPT-EOS*; 3 – Fink *et al* (1981) [78]; 4 – Fischer (1987) (*SST*) [39]; 5 – Breitung and Reil, [21], upper limit; 6 – *ibid.*, lower limit

$$L_{FCE}(T) = \left(H(T)^{(vap)} - H(T)^{(liq)}\right)_{FCE}; \; \Delta z_{FCE}(T) = \left(z^{(vap)} - z^{(liq)}\right)_{FCE}.$$

Linearity of the vapour-pressure log P *vs.* $1/T$ relationship under *FCE* up to $T \approx 7000\ K$ results from the cancellation of these deviations in the quotient of $L(T)$ and $\Delta z(T)$. The total vapour pressure depends on properties of both liquid and vapour phases. However, the partial vapour pressure of UO_2 is mainly ruled by the liquid phase - but only at temperatures not too close to the critical point ($T \leq 7000\ K$). The comparison of the *INTAS*-99-*EOS* with the extrapolated values of the *IVTAN*-96 database, and with predictions of other *EOS'* is presented in Table 7.2.

7.3.3 Heat capacity and thermal coefficients

The heat capacity and the thermal coefficients predicted by the *INTAS*-99-*EOS*, are presented in Table 7.3, and in Figs. 7.3, 7.4, and 7.5, where they are compared with selected experimental data (Appendix A.1) and *INSC* recommendations [78].

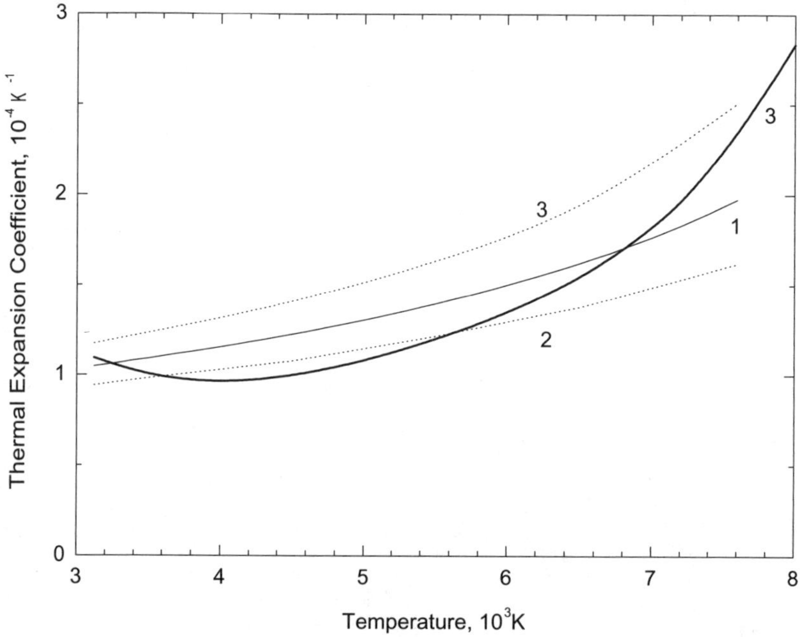

Fig. 7.5. Thermal expansion of the liquid UO_2: 1 – Breitung and Reil [21] (recommended by *INSC*); 2 – lower limit of the *INSC* recommendations; 3 – upper limit of the *INSC* recommendations; 4 – *INTAS-99-EOS*

Fig. 7.3 presents a comparison of the calculated and experimental values of the heat capacity. The results are compared with the experimental data of Ronchi *et al* [119], which extend up to 8000 K.

Table 7.3. Heat Capacity and Thermal Coefficients at the Melting Temperature

	C_P $J\,kg^{-1}K^{-1}$	C_P/C_V	β_T $10^{-10}Pa^{-1}$	β_S	α_T $10^{-5}K^{-1}$	V_S $m\ s^{-1}$
Calculated:						
VdW-DH-HSM	474	1.61	2.22	1.38	1.07	2860
Simple *SSA-TPT*	384	1.28	1.92	1.49	0.68	2760
INTAS-99-EOS	511	1.26	3.95	3.13	1.09	1900
Experimental:	440±50	1.5±0.2		3.2	1.03	1850

The model is corroborated by the fairly good agreement of the predicted heat capacity with the measurements [119]. In fact, it should be noted that so far none of the existing liquid UO_2 models was able to correctly reproduce the observed temperature dependence of C_p.

The main success of the simple *TPT* model was its ability to reproduce (in average) the experimental heat capacity data for the liquid phase up to 5000 K. The *INTAS*-99-*EOS* calculation extends this agreement up to the upper temperature limit of the experiment (8000 K). However, the simple *TPT* is not the best in reproducing the compressibility and the Gibbs free energy of liquid uranium dioxide as given in the *IVTAN*-96 database.

Comparative calculations show that the heat capacity and thermal coefficients of the liquid at the melting point are insensitive to the inclusion of atomic oxygen and atomic uranium in the liquid composition, as well as to the dispersion of the UO_n radii.

7.4 Non-Congruent Equilibrium and Critical Point

The thermodynamic functions, the equilibrium composition, and the parameters of the liquid-gas coexistence region are calculated under forced congruent equilibrium (*FCE* mode) and non-congruent (global equilibrium) conditions.

The calculated values of the thermodynamic functions for non-congruent evaporation of liquid $UO_{2\pm x}$ are presented in Tables 7.4 to 7.6 and illustrated in Figs. 7.6 to 7.8. Extensive thermodynamic tables are collected in Appendix A.4.

The unusual structure of the gas-liquid phase transition boundary, which exhibits a *essential difference* with respect to that of ordinary liquids, is an important aspect of the model predictions. This feature is due to non-congruency of the coexisting phases in $UO_{2\pm x}$. Both the total vapour pressure and the degree of oxygen enrichment/depletion in the vapour phase strongly depend on the evaporation regime. Therefore, for the two extreme cases, distinct functions $P = P(T)$ can be calculated, respectively corresponding to:

- *slow* evaporation under *global equilibrium conditions* resulting in the *boiling* curve, *BC*, and
- *fast*, non-equilibrium, forced-congruent evaporation mode (*FCM*), resulting in the *saturation* curve,*SC*.

Curve 1 in Fig. 7.6 was obtained by enforcing a partially-congruent phase equilibrium (*FCE* mode) with $(U/O)^{(liq)} = (U/O)^{(vap)}$. This can be compared with its non-congruent counterparts (*BC* and *SC* evaporation, curve 2). The predictions of Fischer [39] (curve 3) are also plotted in the same graph.

The total vapour pressure in the two-phase region is plotted as a function of density (Fig. 7.10) and enthalpy (Fig. 7.11). In the latter graph it can be seen that the isotherms cross the boiling curve with a continuous curvature, produced by the non-congruency of the vaporisation process.

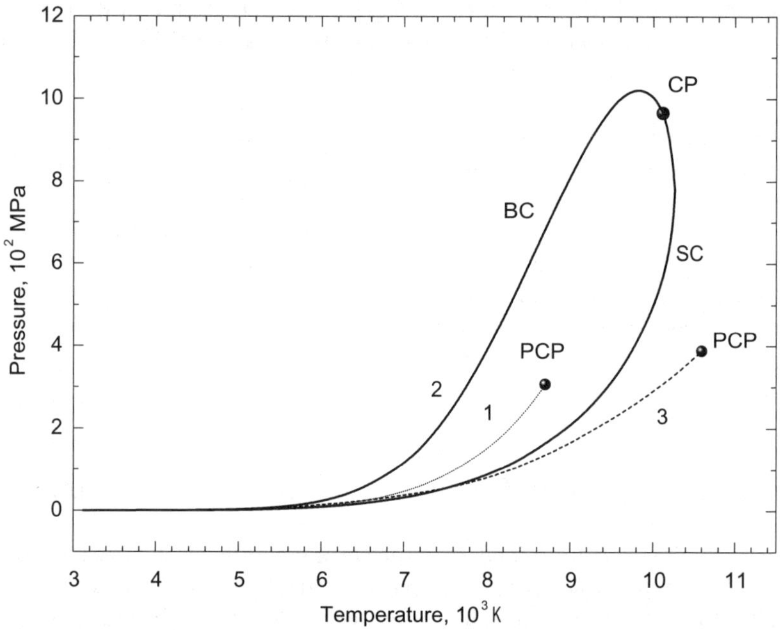

Fig. 7.6. Total and partial (*FCE*) vapour-liquid phase equilibrium in stoichiometric uranium dioxide: 1 - gas-liquid coexistence in *FCE* mode (*BC*= *SC*); 2 - boiling (*BC*) and saturation (*SC*) curves in total equilibrium (*INTAS*-99-*EOS*); 3 - total vapour pressure calculated by Fischer [40]; *PCP*– pseudocritical points; *CP*– critical point

In contrast to the non-reacting one-component fluids, uranium dioxide exhibits a coexistence curve and a critical region displaying an essentially different form. As one can see from Figs. 7.9 and 7.11, the heat of vaporisation is a non-monotonic function of temperature, and the U/O ratio in the vapour phase is markedly increasing with temperature (Fig. 7.8).

A characteristic feature of the non-congruent vaporisation in $UO_{2\pm x}$ is the production of a comparatively high maximum vapour pressure ($P_{max} \sim 1\ GPa$), as well as a substantial oxygen enrichment, $(O/U)_{max} \sim 7$, of the vapour phase over the boiling liquid, as shown in Figs. 7.6 and 7.8.

7.4.1 Oxygen potential of liquid UO_2

The partial vapour pressure of oxygen is plotted as a function of inverse temperature in Fig. 7.7 and compared with predictions of the pure ionic model *IM-II* [80], the model of Hyland [74], of Babelot, Ohse and Hoch [108], and of Green and Leibovitz [54].

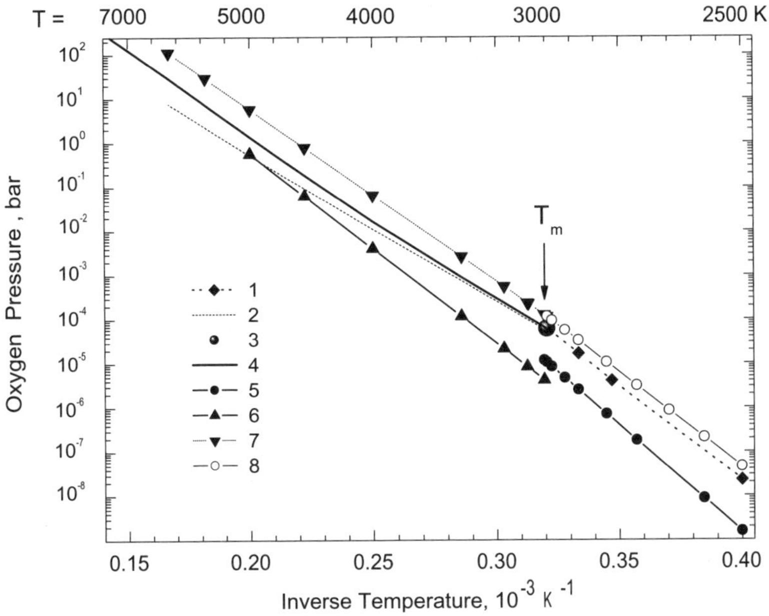

Fig. 7.7. Comparison of the molecular oxygen pressure over solid and liquid $UO_{2.00}$ calculated by *INTAS-99-EOS* with predictions of oxygen-potential models (in P_{O_2} units):

1 - model of Hyland [74] (solid $UO_{2.00}$); 2 - pure ionic model *IM-II* [80] ; 3 - value at T_m used in the calibration of the *INTAS-99-EOS*; 4 - oxygen pressure predicted by the *INTAS-99-EOS* on the *BC*; 5 - model of Babelot, Ohse and Hoch [108] (solid $UO_{2.00}$) ; 6 - *ibid.* (liquid $UO_{2.00}$) ; 7 - model of Green and Leibovitz [54] (liquid $UO_{2.00}$) ; 8 - model of Blackburn [15] (solid $UO_{2.00}$)

The figure also presents results of the non-congruent coexistence from Fischer's [39] work for a neutral composition, and the calculation of the *FCE* coexistence in the *VdW-DH-HSM* for neutral and ionic compositions (*cf.* Fig. 5.1, page 78).

7.4.2 Enthalpy on the vapour-liquid boundary

A remarkable feature of the liquid/vapour phase transition, also in the *FCE* mode, s the non-monotonic dependence on temperature of the heat of vaporisation and vapour compressibility factor at high temperatures.

This feature is not unexpected in chemically reacting liquids [143]. The presence of a maximum in the compressibility factor of the vapour and the vaporisation heat under *FCE* is closely related to the monotonic increase

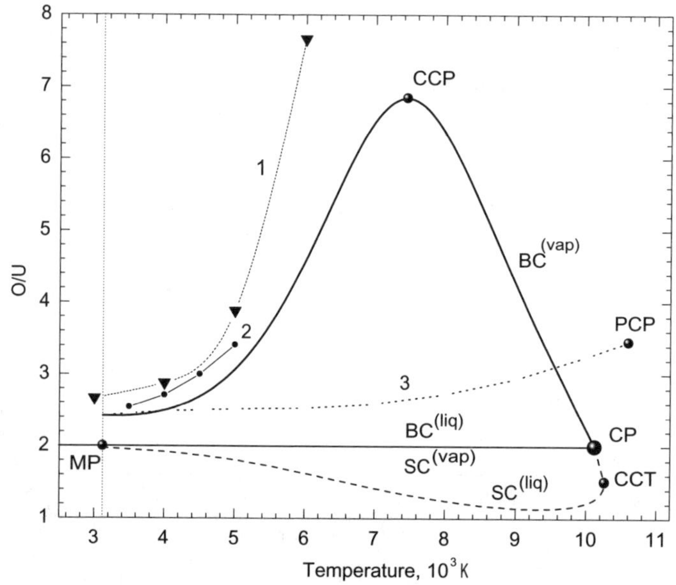

Fig. 7.8. U/O ratio of coexisting phases on the boiling and saturation curves:
1 - full triangles: calculations of Green and Leibovitz [54];
2 - full circles: O/U ratio from the calculated composition of Ohse *et al*[108];
3 – squares: O/U ratio (total) reconstructed from calculations of Fischer [40];
4 – solid and dashed lines – *INTAS*-99-*EOS*:
$BC^{(vap)}$ and $BC^{(liq)}$ – coexisting vapour and liquid at boiling conditions;
$SC^{(vap)}$ and $SC^{(liq)}$ – coexisting vapour and liquid at saturation conditions;
CP – true critical point; *PCP* – pseudo-critical point in Fischer's calculations; *CCT* – point of maximal temperature on *SC* (cricondenterm)

of the UO_2 dissociation in the vapour with temperature. The presence of such a maximum is characteristic for every neutral model within the chemical approach.

The maximal values of the heat of vaporisation and of the vapour compressibility factor predicted by the *INTAS*-99-*EOS* are, respectively:

$$\left\{\frac{L(7000\ K)}{L(T_m)}\right\}_{FCE} \cong 1.37; \quad \left\{\frac{\Delta z(7000\ K)}{\Delta z(T_m)}\right\}_{FCE} \cong 1.26.$$

Even more striking differences between the *FCE* mode and the global equilibrium evaporation regime, due to chemical reactions and non-congruency effects, appear in the behaviour of enthalpy (Fig. 7.9). The heat of non-congruent vaporisation is not a function of temperature or pressure only, for

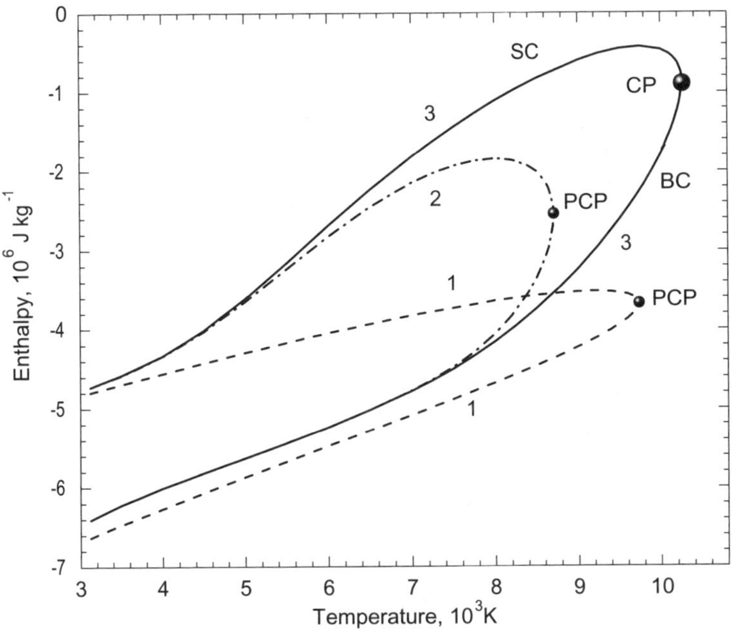

Fig. 7.9. Enthalpy on the gas-liquid boundary. Comparison of three models under the same calibration: 1 - gas-liquid coexistence within the *SSA* mode (UO_2 only); 2 - forced-congruent equilibrium (*FCE*) of $UO_{2.00}$ in multi-molecular model; 3 - total non-congruent equilibrium (*BC*- Boiling curve, *SC*- Saturation curve, *CP* - true critical point of the non-congruent evaporation); *PCP* - pseudo-critical points in the *FCE* and *SSA* mode

the pressure is changing during evaporation at constant temperature (and *viceversa*).

7.4.3 Equilibrium composition and non-ideality effects

The equilibrium constant $K_d = \{n_{\text{UO}} n_{\text{UO}_3}\}^{1/2}/n_{\text{UO}_2}$ in the liquid is essentially determined by the small differences in the intrinsic volumes of the reacting components. In the *FCE* mode, the equality of the dissociation degree in liquid and vapour is enforced, and results in the same increase of the dissociation degree with temperature both in the liquid and in the vapour phase. The non-congruent vaporisation leads to an essential different dissociation in the two phases.

Table 7.4 illustrates the shift of the non-congruent dissociation equilibrium under boiling and saturation conditions, respectively. The ratio of the uranium

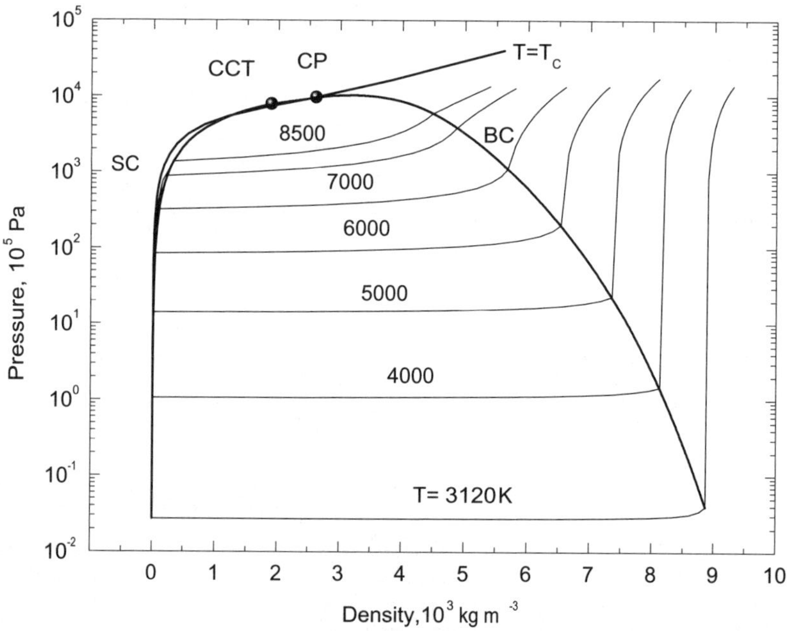

Fig. 7.10. Total vapour pressure on UO_2 as a function of density in the two-phase region: *CP*: critical point; *BC*: boiling curve; *SC*: saturation curve; *CCT*: point of the maximal temperature at the SC. The curves inside the two-phase region are isotherms, the upper curve is the critical isotherm (T_C= 10015 K)

dioxide molecular mass to the calculated mean molecular mass provides the measure of the overall dissociation degree in vapour and liquid. The excess compressibility factor, z-1, calculated from this mean molecular mass, provides an estimate of the non-ideality of the vapour and liquid phase.

7.4.4 Azeotropic compositions

A distinctive feature of the non-congruent evaporation regime in $UO_{2\pm x}$ is the existence of azeotropic curves (*CVC*, congruent vaporizing compositions). One of the azeotropic points, corresponding to *CVC* composition (O/U = 1.94 at $T = 3400\ K$), was used for the calibration of the oxygen potential[5] of liquid $UO_{2\pm x}$. The reconstructed non-congruent gas-liquid *CVC*-coexistence curve in the temperature range $3100\ K \leq T \leq 5000\ K$ is plotted in Fig. 7.12.

[5] The experimental point of Reedy and Chasanov [116] shown in Fig. 7.12 was used for calibration as it is presumed to be an azeotropic point.

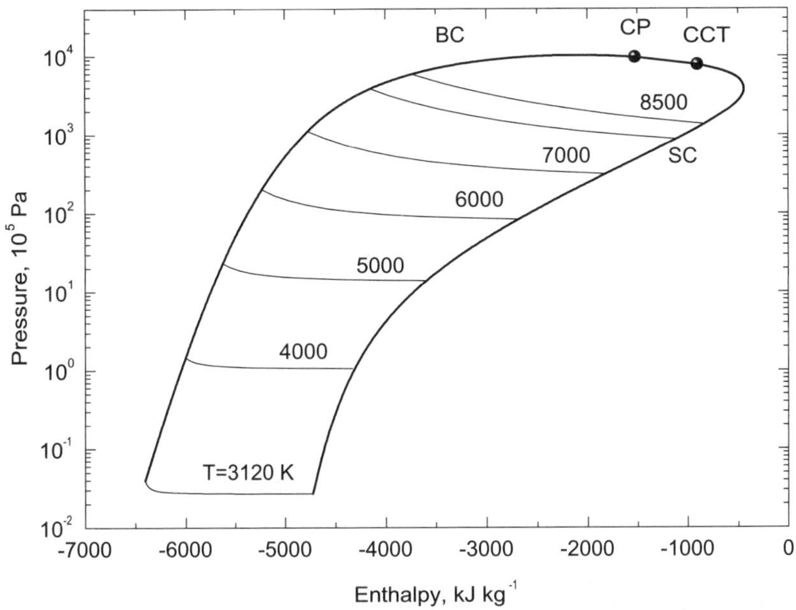

Fig. 7.11. Total vapour pressure of UO_2 as a function of enthalpy in the two-phase region: *CP* - critical point; *BC*- boiling curve; *SC* - saturation curve; *CCT* - point of the maximal temperature at saturation curve. The curves inside the two-phase region are isotherms

One can see that this azeotropic curve has the following properties:

- the azeotrope is located in the hypostoichiometric domain, O/U$<$ 2.00;
- the non-stoichiometry of *CVC*-points monotonously increases with increasing temperature;
- the azeotrope is located in the hypostoichiometric domain, O/U$<$ 2.00;
- the non-stoichiometry of *CVC*-points monotonously increases with increasing temperature;
- the *CVC*-curve consists of two branches - a stable and a metastable one;
- the stable branch of the liquid azeotrope proceeds from the highest calculated temperature, $\sim 5000\ K$, down to the intersection with the liquidus, the hypothetical boundary of equilibrium freezing of $UO_{2\pm x}$;
- the metastable branch extends from the above mentioned intersection point down to lower temperatures.

The model calculations provide the following estimate of the intersection point: $(O/U)^{(CVC)} \simeq 1.967\ ,\ T \simeq 3100\ K$.

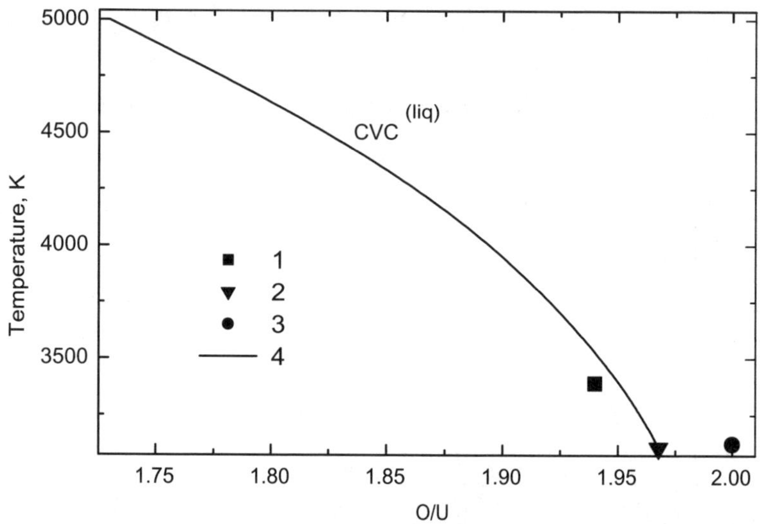

Fig. 7.12. Predicted azeotropes (*CVC*) in liquid urania: 1 – experiment of Reedy and Chasanov [116]; 2 – expected intersection of the liquid azeotrope with the liquidus line; 3 – congruent melting (*CMC*) composition of urania; 4 – prediction of the *INTAS*-99-*EOS*

7.4.5 Critical point in FCE approximation

We recall that the critical point, calculated in the *FCE* mode, is the analogue of the ordinary critical point, where the well-known conditions:

$$\left(\frac{\partial P}{\partial V}\right)_{T,x} = \left(\frac{\partial^2 P}{\partial V^2}\right)_{T,x} = 0; \quad \left(\frac{\partial^3 P}{\partial V^3}\right)_{T,x} < 0$$

are fulfilled. The thermodynamic parameters at the *FCE* critical point, calculated with the *INTAS*-99-*EOS* model, are presented in Table 7.5. Some properties of these parameters are noteworthy:

- According to both simples *TPT* models and *INTAS*-99-*EOS*, the reduced critical temperature in the *FCE* mode is close to that typical for an ordinary Lennard-Jones fluid. The reduced critical temperature $T_C^* = kT/\varepsilon = 1.26$ for simple *TPT-EOS*' is close to the value of kT/ε_0 for the *INTAS*-99-*EOS*.
- On the other hand, the (*FCE* mode) critical pressure and density are considerably higher, not only if compared with the *SSA-VdW-EOS*, but also with the *VdW-DH-HSM*, and with the results of Fischer [39]. Under

FCE mode, uranium dioxide at the critical point is significantly dissociated (actually, the dissociation degree is near to one). Moreover, the dissociation of oxygen plays also an important role, giving rise to an increase in the critical compressibility factor, z_C (Table 7.5).

- The most important dimensionless parameters in the *FCE* mode - the critical compressibility factor and the ratio of the critical density to the melting point density - are considerably higher than in the case of the liquid-gas phase transition in non-reacting fluids.
- For the same reason, a considerable gain (~ 100 %) in the critical entropy is entailed with respect to simple substances (for which the typical value is $S_C/R \simeq 20$ [46]).

The important differences in the critical properties of uranium dioxide with respect to those of ordinary liquids, are not specific, inherent features of the *INTAS*-99-*EOS*. For instance, a good agreement in the predicted degree of dissociation is obtained in all tested versions of the chemical model. In fact, this is the cause of the increase in the heat of vaporisation, and of the subsequent raise of the critical density and pressure, as discussed above.

7.4.6 Critical point under non-congruent evaporation

The critical temperature and pressure of the non-congruent phase transition in UO_2 are also estimated. This critical point essentially differs from that

Table 7.4. Equilibrium Constant $K_d = \{n_{UO} n_{UO_3}\}^{1/2}/n_{UO_2}$, Compressibility Factor, z, and the Ratio of the Molecular Mass of UO_2 to the Mean Molecular Mass $\overline{m} = \sum_i x_i m_i$ in the Liquid and Vapour Phases. Boiling (*BC*) and Saturation (*SC*) Conditions

Temperature, *K*			3120	3500	4000	4500	5000	6000	7000
K_d	Liquid	*BC*	0.0015	0.0041	0.011	0.022	0.038	0.089	0.173
		SC	0.0015	0.004	0.01	0.021	0.035	0.079	0.148
	Vapour	*BC*	0.092	0.129	0.185	0.240	0.301	0.426	0.551
		SC	0.092	0.129	0.183	0.241	0.301	0.427	0.557
$z = \frac{P\overline{m}}{\rho RT}$	Liquid	*BC*	$5\ 10^{-6}$	$3\ 10^{-5}$	0.0001	0.0006	0.002	0.016	0.089
		SC	$3\ 10^{-6}$	$2\ 10^{-5}$	0.0001	0.0004	0.001	0.006	0.022
	Vapour	*BC*	1.000	1.000	0.998	0.996	0.991	0.980	0.968
		SC	1.000	1.000	0.999	0.996	0.991	0.970	0.933
$m_{UO_2}/\overline{m}$	Liquid	*BC*	1.000	1.001	1.001	1.002	1.003	1.005	1.017
		SC	1.002	1.003	1.006	1.009	1.014	1.029	1.049
	Vapour	*BC*	1.012	1.046	1.133	1.302	1.588	2.530	3.475
		SC	1.009	1.021	1.055	1.116	1.205	1.421	1.619

Table 7.5. The *FCE* and True Critical Points: *INTAS*-99-*EOS* in Comparison with Other Calculations

	T_C K	P_C MPa	ρ_C $kg\,m^{-3}$	$z_C^{(a)}$	$z_C^{(b)}$	$L/RT_C^{(c)}$	ρ_m/ρ_C	S_C/R	x_O	$K_d{}^{(d)}$
SSA(UO_2 only):										
Mistura *et al.* [105]	7563	141	1730	0.350		7.00	5.13		0.0	0.0
Simple *VdW*	9367	222	2140	0.359			4.15		0.0	0.0
***FCE* mode**:										
Simple *TPT*	9436	867	3410	0.876		6.01	2.59	54.5	0.16	0.95
VdW+DH	8789	383	2330	0.606	0.447	6.44	3.80	56.3	0.19	0.84
INTAS-99-*EOS*	8710	310	2030	0.570	0.410		4.37	57.4	0.19	0.60
Non-congruent evaporation:										
Fischer [39]	10600	388				$5.0^{(e)}$			0.31	
ibid.$^{(f)}$	10600	158	1560	0.310			5.68			
INTAS-99-*EOS*$^{(g)}$	10015	876	2500	1.135	0.711		3.55	59.9	0.25	0.73

$^{(a)}z = Pm_{UO_2}/\rho RT$;
$^{(b)}z = P/nkT$;
$^{(c)}L_m$ is the UO_2 heat of vaporisation at T_m;
$^{(d)}K_d \equiv \{n_{UO}n_{UO_3}\}^{1/2}/n_{UO_2}$;
$^{(e)}L_m = 440\ kJ\ mol^{-1}$;
$^{(f)}$uranium-bearing species only;
$^{(g)}$true critical point

defined for a gas-liquid phase transition in simple liquids; in particular, one can see that: $(\partial P/\partial V)_C \sim (P/V) \neq 0$. The predicted critical parameters are: $T_c \simeq 10015\ K$, $P_C \simeq 876\ MPa$, $\rho_c \simeq 2500\ kg\ m^{-3}$.

Finally, it is worth remarking that the critical compressibility factor, $z_C^{(a)} = P_C m_{UO_2}/\rho RT_C$, presented in Tables 5.2 and 7.5 is the ratio of the real critical pressure to the ideal-gas pressure of a gas having the same mass density but consisting of UO_2 molecules only. Except for mono-molecular models, such definition has little sense, because it includes both effects of non-ideality and molecular dissociation. Consequently the corresponding difference $z_C^{(a)} - 1$ cannot serve as a measure of non-ideality, in contrast to the compressibility factor $z_C^{(b)} = P_C/n_C kT_C$, whose value is less than one everywhere on the vapour-liquid coexistence curve (see Table 7.4), including the true critical point.

Table 7.6. Selected Properties of the Coexisting Phases along the Boiling Line

Temperature, K		3120	4000	5000	6000	7000	8000
Pressure, MPa		0.00396	0.146	2.31	20.0	111	381
Density, $kg\ m^{-3}$	Liquid	8871	8126	7355	6535	5679	4869
	Vapour	0.041	1.05	9.52	43.7	152.5	468.2
$H,\ kJ\ kg^{-1}$	Liquid	-6406	-6005	-5629	-5239	-4781	-4160
	Vapour	-5390	-4958	-4334	-3452	-2519	-1932
$S,\ kJ\ kg^{-1}K^{-1}$	Liquid	1.162	1.276	1.356	1.430	1.498	1.574
	Vapour	1.777	1.827	2.141	2.769	3.358	3.265
$C_p,\ kJ\ kg^{-1}K^{-1}$	Liquid	0.511	0.399	0.372	0.416	0.523	0.850
	Vapour	0.827	1.535	2.269	2.522	2.535	2.254
$\beta_T,\ 10^{-10}Pa^{-1}$	Liquid	0.40	0.53	0.76	1.17	1.97	3.34
	Vapour	26.05	0.750	0.0507	0.0059	0.0011	0.0003
$\alpha_T,\ 10^{-4}K^{-1}$	Liquid	1.09	0.95	1.08	1.34	1.81	2.85
	Vapour	5.57	7.08	7.02	5.26	4.07	3.54
O/U	Liquid	2.00	2.00	2.00	2.00	2.00	2.00
	Vapour	2.42	2.49	3.05	4.53	6.44	6.28
Ionisation,%	Liquid	9.5	13.0	17.2	21.7	26.3	29.2
	Vapour	3.8	6.7	9.7	5.1	8.6	13.5

7.5 Concluding Remarks

The *INTAS*-99 equation of state of uranium dioxide, the predictions of which are presented in this Chapter, essentially differs from those previously developed. This regards not only the set of adopted models and their fundaments, but also the structure of the equation and the physical and chemical conditions which can be covered. This makes it particularly suitable for a broad gamut of practical applications, especially in reactor safety analysis.

By applying standard computational algorithms all thermodynamic properties of the coexisting phases, including the chemical and ionisation equilibrium in the liquid, can be calculated, so that the complete phase diagram of the condensed and gaseous $UO_{2\pm x}$ up to the critical point can be constructed. Actually, for the first time a calculated phase diagram of uranium oxide was obtained, which correctly reproduces the non-congruent bivariant nature of the evaporation.

One must, however, also mention the intrinsic limits of the adopted approach, three negative aspects of which are evident:

- First, the excitation partition functions of the complex constituent particles are assumed to be equal to their ideal-gas counterparts, taken in this work from Gurvich's handbook [63], whereby the value of the reduced thermodynamic potential, $\Phi^{0}(T) \equiv G(T)/T$ (Appendix A.2) was used for this extraction. No cut-off procedure and no perturbation of the internally bound excited states of complex particles under a dense-liquid surrounding were taken into account.
- Secondly, only single-stage ionisation is included in the present treatment. Therefore, the main ionic species in the model are molecular ions (UO_3^-, UO_2^+ and UO^+). Though the sensitivity studies performed up to now show that the presence of different, additional ionic species has little influence on the results, this aspect might need a more careful analysis.
- Finally, it must be emphasised that the first-order *TPT* neglects fluctuations that become very important in the closest vicinity of the critical point. Consequently, a precise description of this region is impossible with this type of model.

A

Appendix

A.1 Thermophysical Properties of UO_2

This Appendix contains a critical review[1] of experimental data on selected thermodynamic properties of solid, liquid and gaseous UO_2 which are directly or indirectly related to the calibration and/or validation of the equation of state.

When necessary, the presentation of the data is accompanied by some appropriate extrapolations.

A.1.1 Density of UO_2

The preferred method for the determination of density and thermal expansion is by bulk expansion measurement of the T-dependence of the volume (or length) of a macroscopic[2] sample of urania, encapsulated in tungsten. The present *high-temperature* recommendations are based on the work of Christensen [26] (1963) and Drotning [31] (1981).

Conway *et al* [28] measured the *linear* thermal expansion of a rod-shaped specimen positioned in a tungsten support block, located in a resistively heated tungsten tube furnace.

Christensen [26] radiographed encapsulated samples at a series of different temperatures, and determined the associated volumes, from the outline of the UO_2 sample recorded on the film.

Drotning [31] remarked that surface tension and viscosity effects between the *molten* sample and its container can introduce errors into the liquid density values obtained using Christensen's method; to avoid this he used an

[1] The material of this Section is taken from a contribution of G.J.Hyland (University of Warwick, UK) to the INTAS 99-066 Project [77].

[2] This method is preferred in the solid-state since it automatically includes any contributions from Schottky defects which are excluded in *microscopic* approaches based on determination of the T-dependence of the lattice parameters by X-ray or neutron diffraction.

Table A.1. Density Data

Work	Temperature range covered	Number of data points at T > T_m	Sample	Stoichiometry
Conway *et al* [28]	1273–2523 K	12[(a)]	Sintered compact 95.5 TD	Nominally $UO_{2.00}$
Christensen [26]	1553–3373 K	24[(b)]	Single crystal	Nominally $UO_{2.00}$
Drotning [31]	3120–3250 K		Polycrystalline	$UO_{2.02}$ - $UO_{2.04}$

(a) At only three different temperatures
(b) Pooled from five runs

alternative technique known as γ-densitometry, in which the attenuation by the sample of a collimated γ-beam is measured.

- Solid UO_2
  The data of Conway *et al* [28] and Christensen [26] are in good agreement over their common temperature range, and are jointly described by the polinomial:

$$\rho^{(sol)}(t) = 11031.00 - 179.07\,t - 91.36\,t^2 + 2.21\,t^3,\ kg\ m^{-3}, \tag{A.1}$$

  where $t = 10^{-3}T/K$.
  At melting, Eq.(A.1) gives $\rho^{(sol)(T_m)} = 9650\ kg\ m^{-3}$; this recommended value is within the uncertainty associated with Christensen's value ($9670 \pm 130\ kg\ m^{-3}$) at his reported (comparatively low) melting point of 3073 K.
- Molten UO_2
  Christensen's data over $3073 - 3373\ K$, (corrected to $T_m = 3120\ K$) are described by

$$\rho^{(liq)}(t) = 11542.63 - 898.66\,t,\ kg\ m^{-3}\ . \tag{A.2}$$

  At melting, Eq.(A.2) gives $\rho^{(liq)}(t_m) = 8739\ kg\ m^{-3}$, compared with Christensen's measured value of $8740 \pm 160\ kg\ m^{-3}$.
  Drotning's data over $3120\ K - 3250\ K$ are fitted by

$$\rho^{(liq)}(T) = 11717 - 916\ t\ , kg\ m^{-3}, \tag{A.3}$$

  giving $\rho^{(liq)}(\mathrm{T}_m) = 8859\ \ kg\ m^{-3}$, this recommended value is within the uncertainty limits of Christensen's value.

A.1.2 Volume Changes on Melting

Since only Christensen's data extends *across* T_m, it is recommended that this data be used (notwithstanding the possible errors in his liquid value mentioned above).

$$\frac{\Delta\rho}{\rho^{(sol)}} = \left(\frac{\rho^{(liq)} - \rho^{(sol)}}{\rho^{(sol)}}\right) = -9.6\% \ \ (-8.4\%), \tag{A.4}$$

whilst

$$\frac{\Delta V}{V^{(sol)}} = \left(\frac{V^{(liq)} - V^{(sol)}}{V^{(sol)}}\right) = \frac{\rho^{(sol)}}{\rho^{(liq)}} - 1 = +10.6\% \ \ (9.2\%). \tag{A.5}$$

The parenthetical values correspond to using Drotning's value of $\rho^{(liq)}$ and Christensen's value of $\rho^{(sol)}$ (both at T_m).

Corroborations:

- Christensen's solid-state $\rho^{(sol)}(T)$ values are in good agreement with the earlier values obtained by Halden *et al* [66] between 1270 K and T_m, and for large-grain sintered UO_2.
- Schins and Gunnerson [121] have developed a novel technique which predicts the liquid phase density, $\rho^{(liq)}(T_m)$, at the onset of solidification from solid-state properties, via the relation:

$$\rho^{(liq)}(T_m) = \rho_0^{(sol)}(T_o)\left(\frac{\rho^{(sol)}(T_m)}{\rho^{(sol)}(T_o)}\right), \tag{A.6}$$

where $\rho_0^{(sol)}$ is the apparent density at $T_o(T_m)$ as determined by pycnometer (9840 $kg\ m^{-3}$ at $T_o = 300\ K$, where $\rho^{(sol)} = 10970\ kg\ m^{-3}$). Using Christensen's value of $\rho^{(sol)}(T_m) = 9670\ kg\ m^{-3}$, Eq.(A.6) yields $\rho^{(liq)}(T_m) = 8674\ kg\ m^{-3}$ which is within the uncertainty of Christensen's experimental value of $\rho^{(liq)}(T_m) = (8740 \pm 160)\ kg\ m^{-3}$: with this predicted value of expansion on melting:

$$\frac{\Delta\rho}{\rho^{(sol)}} = -10.3\%. \tag{A.7}$$

A.1.3 Thermal Expansion

The isobaric, volumetric coefficient of thermal expansion $\alpha_T(T)$ is defined by

$$\alpha_T = \frac{1}{V}\left(\frac{\partial V}{\partial T}\right)_T = -\frac{10^{-3}}{\rho}\left(\frac{\partial \rho}{\partial t}\right)_T . \tag{A.8}$$

- Solid UO_2
  Eqs.(A.1) and (A.8) yield:

$$\alpha_T^{(sol)}(T \leq T_m) = 10^{-3} \frac{179.07 + 182.72\ t - 6.63\ t^2}{11031.00 - 179.07\ t - 91.36\ t^2 + 2.21\ t^3} \quad \text{(A.9)}$$

giving (as the recommended value):

$$\alpha_T^{(sol)}(T_m) = 7.09\ 10^{-5}\ K^{-1}. \quad \text{(A.10)}$$

- Molten UO_2
  Eqs.(A.3) and (A.8) yield:

$$\alpha_T^{(liq)}(T \geq T_m) = \frac{0.916}{11717 - 916\ t}\ K^{-1} \quad \text{(A.11)}$$

giving (as the recommended value):

$$\alpha_T^{(liq)}(T_m) = 1.03\ 10^{-4} K^{-1}, \quad \text{(A.12)}$$

and an associated coefficient of *linear* expansion $\alpha_T^{(liq)} = 3.45\ 10^{-5}\ K^{-1}$, which essentially agrees with Christensen's cited value ($3.5\ 10^{-5}\ K^{-1}$) between T_m (corrected to 3120 K) and 3420 K (corrected).

A.1.4 Elastic Properties and Adiabatic Compressibility

- Solid UO_2: Single Crystals
  On account of the cubic symmetry of fluorite structured UO_2, there are only 3 independent (second-order) elastic constants c_{11}, c_{12} and c_{44}, in terms of which the Young (E_S), shear (G_S), bulk (B_S) adiabatic moduli and Poisson's ratio (σ_S) are defined as follows:

$$E_S = \frac{(c_{11} - c_{12} + 3c_{44})(c_{11} + 2c_{12})}{2c_{11} + 3c_{12} + c_{44}}, \quad G_S = \frac{1}{5}(c_{11} - c_{12} + 3c_{44}), \quad \text{(A.13)}$$

$$B_S = \frac{1}{3}(c_{11} + 2c_{12}), \qquad \sigma_S = \frac{(c_{11} + 4c_{12} - 2c_{44})}{2(2c_{11} + 3c_{12} + c_{44})}. \quad \text{(A.14)}$$

The c_{kl} are deduced from the speed of sound in different crystallographic direction - obtained either by direct measurement using ultrasonics, or indirectly from the slope of the corresponding long-wave acoustic dispersion curves obtained by neutron diffraction techniques; in both cases, adiabatic (isentropic) [3] values are obtained. The more recent, diffraction based values [27] extend the earlier room temperature ultrasonic data [139, 48]

[3] The adiabatic (subscript $_S$) and isothermal (subscript $_T$) values of Young moduli are related by: $G_T = G_S$, $B_T = B_S C_p (C_p + B_S T \alpha_T^2 V)^{-1}$, where C_p is the heat capacity at constant pressure, α_T is the isobaric coefficient of volume expansion, and V is the specific volume.

almost to the melting point. The ultrasonic room temperature value of B_S is approximately equal to the arithmetic mean of the two diffraction based values. Between room temperature and 2000 K, there are diffraction based values of c_{kl} at only one single temperature near 1200 K, where the deduced value of B_S is 186 GPa. Since, in the particular case of B_S, there is *no difference* between single crystal and polycrystal values, this isolated value of B_S can be meaningful compared with the more abundant values obtained on *polycrystalline* samples (up to 1573 K), whose acquisition we now consider.

- Solid UO_2: Polycrystals

Here two elastic constants - which can be chosen to be Young's modulus, E, and the Poisson's ratio, σ are sufficient to describe the system. Adiabatic (isentropic) values E_S and σ_S can again be obtained from measurement of the (isotropic) longitudinal and transverse ultrasonic velocities, v_ℓ and v_{t}, *via*:

$$E_S = 2\rho(1+\sigma_S)v_t^2, \quad \sigma_S = \frac{\frac{1}{2}v_\ell^2 - v_t^2}{v_\ell^2 - v_t^2}, \tag{A.15}$$

from which B_S and G_S can be obtained *via*:

$$B_S = \frac{E_S}{3(1-2\sigma_S)}, \quad G_S = \frac{E_S}{2(1+\sigma_S)}. \tag{A.16}$$

Values of B_S and G_S at room temperature, obtained in this way, have been published, together with values of E_S up to 1573 K[109].
Alternatively, from data on E_S and G_S obtained from measurement of the flexural and torsional resonant frequencies of prismatic bars of nominally stoichiometric UO_2, values of B_S and σ_S can be determined *via*

$$B_S = \frac{E_S G_S}{3(3G_S - E_S)}, \quad \sigma_S = \left(\frac{E_S}{2G_S}\right) - 1. \tag{A.17}$$

The onset of a strong decrease in E_S and G_S at $T > 1470$ K in the data of Hall [67] between room temperature and 1523 K is confirmed by the subsequent work of Marlowe and Kaznoff [100], which indicates a dramatic decrease - especially in E_S - by 1573 K. Above this temperature it became impossible to detect resonance, owing to the rapid increase in internal friction which accompanies the decrease in E_S and B_S.

Summary of results for B_S

In the range of their common temperatures, the data obtained from polycrystalline samples are scattered in a relatively narrow band about the trend of the more extensive single crystal data. Extrapolating the latter from the highest existing temperature of 2929 K to T_m yields (as recommended value):

$$B_S(T_m) = 92\ GPa, \tag{A.18}$$

the associated adiabatic compressibility , $\beta_S \equiv {B_S}^{-1}$, being

$$\beta_S(T_m) = 11\ 10^{-6}\ MPa^{-1}. \tag{A.19}$$

- Molten UO_2
  Using Christensen's values [26] of the liquid density, ρ, values of the adiabatic bulk modulus, $B_S(T)$, have been obtained [126] from measurement of the (longitudinal) speed of sound, v_ℓ, in molten UO_2 at four temperatures just above T_m, (between $3138 - 3196\ K$), *via*:

$$B_S = \rho v_\ell^2. \tag{A.20}$$

Re-calculation, using the recommended value of the liquid density at T_m, ($8859\ kg\ m^{-3}$), given by Eq.(A.3), yields:

$$B_S(T_m) = 31\ GPa. \tag{A.21}$$

It should be noted that Eq.(A.20) gives an upper limit to B_S, since no account is taken of a possible non-zero shear modulus G_S in the liquid. Including G_S, the longitudinal sound speed, v_ℓ is given by $\rho v_\ell^2 = B_S + \frac{4}{3}G_S$. Correspondingly, ${B_S}^{-1}$ is a lower limit on the adiabatic compressibility, β_S, thus with B_S given by Eq.(A.21), we have:$\beta_S \approx 32\ 10^{-6} MPa^{-1}$.

A.1.5 Heat Capacity of Molten UO_2

The heat capacity, C_P, of liquid UO_2 was measured in a laboratory experiment where sintered microspheres of diameters 0.5 to 1 mm were heated by four tetrahedrally oriented laser beams in an inert-gas-filled autoclave at pressures up to $\sim$ 1000 *bar*. The sample, suspended by a tungsten needle, was heated up to 8000 K during pulses of a few milliseconds duration. Values of $C_P(T)$, obtained from thermal analysis of the cooling curves of the laser heated specimen are given below in Table A.2. Between the melting point T_m and $\sim$ 4000 K, the heat capacity decreases to a value close to that given by the Neumann-Kopp rule for a tri-atomic, harmonic lattice, *i.e.*, $9R$. Near 5000 K, however, the heat capacity again increases, and it appears to saturate at a value $\sim$ 30% higher by 8000 K.

A.1.6 Enthalpy and Entropy of Fusion

There are two independent sets of enthalpy data which extend across T_m: those of Hein and Flagella [68], and of Leibowitz *et al* [94, 93], on slightly hyperstoichometric samples of composition $UO_{2.003}$ and $UO_{2.015}$, respectively. Overall, the agreement is excellent, with only a slight discrepancy in the associated

Table A.2. Heat Capacity Data

T K	C_P $J\ kg^{-1}K^{-1}$
3100	448
3300	390
3500	355
3700	335
3900	323
4100	316
4300	313
4500	313
4700	314
4900	318
5100	322
5300	328
5500	335
5700	343
5900	351
6100	361
6300	370
6500	380
6700	391
6800	396
7000	407
7200	418
7400	429
7600	441
7800	452
8000	463
8200	474

heat of fusion, ΔH_f, namely, 76 $kJ\ mol^{-1}$ and 74 $kJ\ mol^{-1}$, respectively[4]. Accordingly,

$$\Delta H_f = 75\ kJ\ mol^{-1} \tag{A.22}$$

is recommended.

Assuming that the melting of stoichiometric UO_2 occurs at $T_m = 3120\ K$ [114], the associated entropy of fusion, ΔS_f, is given by:

$$\Delta S_f \equiv \Delta H_f / T_m = 24\ J\ mol^{-1}K^{-1}. \tag{A.23}$$

These values of ΔH_f and ΔS_f compare well with those obtained by integrating the directly measured values [119] of the liquid heat capacity tabulated

[4] These corroborated values should be compared with the much larger value of 106 $kJ\ mol^{-1}$ reported by Grossman and Kaznoff [55] obtained by thermal analysis, requiring a *co*-determination of the effective, total hemispherical emittance which is found to be very low ($0.3 < \varepsilon < 0.4$).

in Table A.2, namely $\Delta H_f = 78\ kJ\ mol^{-1}$; $\Delta S_f = 24.98\ J\ mol^{-1}K^{-1}$ (for $T_m = 3123\ K$).

A.1.7 Vapour Pressure over Nominally Stoichiometric UO_2 Near Melting

There is one set [116] of data of the total pressure, P_U, of uranium bearing species (obtained by the transpiration technique) which extends across T_m (from 2615 K to 3390 K, with data at two temperatures below T_m, and two above). The two values at $T < T_m$ are well corroborated by two other [130, 5] independent data sets containing many more points (obtained using the same technique) - especially by that of Tetenbaum and Hunt [130]. Furthermore, a good fit to the P_U values at $T \geq T_m$ is obtained by renormalizing Tetenbaum and Hunt's fit for $T < T_m$ (based on data over $2080 \leq T/K \leq 2705$)[5] by the heat and entropy of fusion: 75 $kJmol^{-1}$ and 24 $J\ mol^{-1}K^{-1}$, Eqs. (A.22), (A.23). Thus

$$\ln\left(\frac{P_U}{MPa}\right)_{T<T_m} = \frac{-72.02}{t} + 17.54 \rightarrow \frac{-63.11}{t} + 14.68 = \ln\left(\frac{P_U}{MPa}\right)_{T>T_m}. \tag{A.24}$$

These two fits yield equal values at T_m, of course, where

$$P_U(T_m) = 3.9\ 10^{-3}\ MPa. \tag{A.25}$$

The highest temperature (3390 K) of the liquid-phase transpiration data is only 110 K below the lowest temperature of the total equilibrium vapour pressure (P_T) measurements of Bober and Singer [16], obtained using a boiling point technique; their data is fitted by

$$\ln\left(\frac{P_T}{MPa}\right) = \frac{-53.07}{t} + 11.63. \tag{A.26}$$

Extrapolating Eqs.(A.24) and (A.25) to the midpoint (3445 K) reveals a remarkable consistency

$$P_U = 2.6\ 10^{-2} MPa, \quad P_T = 2.3\ 10^{-2} MPa. \tag{A.27}$$

Fully meaningful comparison requires, however, the addition to P_U of the contributions from O(g) and $O_2(g)$. Calculating these at $T = 3445\ K$, using the oxygen potential model of Hyland [74], yields: $P(O) = 1.87\,10^{-3}\ MPa$, $P(O_2) = 1.92\ 10^{-4}\ MPa$, whence $P_U + P(O) + P(O_2) = 2.8\ 10^{-2} Mpa$, which is

[5] The higher temperature effusion data [107] of Ohse (extending to 2800 K) is now considered unreliable [137] in consequence of the likely breakdown in the molecular flow conditions - necessary for the validity of the used Knudsen effusion technique - at the relatively high pressures attained.

still within the uncertainty associated with Bober and Singer's value of $P_T = 2.3\ 10^{-2} MPa$.

At T_m, on the other hand, Eq.(A.24) gives $P_U(T_m) = 3.9\ 10^{-3}\ MPa$, whilst extrapolation of Bober and Singer's fit Eq.(A.26) yields:

$$P_T(T_m) = 4.6\ 10^{-3} MPa\ . \tag{A.28}$$

At $T = T_m$, $P(O) = 1.8\ 10^{-4} MPa$, and $P(O_2) = 1.15\ 10^{-5}$MPa, whence

$$P_U + P(O) + P(O_2) = 4.1\ 10^{-3} MPa. \tag{A.29}$$

It is recommended that the mean of Eqs.(A.28) and (A.29) be used to define $P_T(T_m)$: thus

$$P_T(T_m) = 4.35\ 10^{-3} MPa. \tag{A.30}$$

Despite this agreement, it should be noted that the heats of vaporisation predicted by Eq.(A.24) and Eq.(A.26) differ by 84 $kJ\ mol^{-1}$.

It should further be noted that the conditions realised in the experiments of Bober and Singer differ in an important way from those which obtained in the experiments of Ohse *et al* [108] at *ITU*. In the former case, the UO_2 specimen (heated in an autoclave in a quasi-stationary way on a millisecond timescale) evaporates against an external inert gas whose pressure can be varied; the temperature at which molten UO_2 starts to boil is taken to be the temperature at which the vapour pressure equals the pressure of the inert autoclave gas. The pressure *vs.* temperature data so obtained thus define the so-called 'boiling curve'(BC) (see Chapter 4), associated with global equilibrium conditions under which the evaporation is non-congruent $(O/U)^{(vap)} \neq (O/U)_{\text{surface}} = (O/U)_{\text{bulk}}$.

In the *ITU* experiments, on the other hand, the specimen (which is very rapidly laser heated on a *ms* timescale) evaporates into vacuum. Unlike the former, these latter experiments do not yield values of the total equilibrium pressure of neutral vapour species over molten UO_2, in consequence of two effects which work in opposing directions.

(i) Steady-state evaporation from the open UO_2 surface under the highly dynamic conditions under laser heating, can only proceed in the *forced-congruent mode*(FCM), in which $(O/U)^{(vap)} = (O/U)_{bulk} > (O/U)_{surface}$ – the evaporating surface itself becoming progressively hypostoichiometric [20]; in this case, the $P-T$ data so obtained define the saturation curve (SC) , rather than a BC. Correction of the pressure values obtained in FCM to the higher equilibrium pressures corresponding to a sample of UO_2 whose surface has the same stoichiometry as the bulk, can only be done theoretically, within the context of an analysis[6] involving the calculation of all partial pressures - including those of O(g) and $O_2(g)$.

[6] It is to be stressed that the recommended $P-T$ expression given in Ref. [69] is not the result of any such correction, but is simply a fit to purely calculated values of the equilibrium total pressure over stoichiometric, molten UO_2.

(ii) There is an appreciable enhancement of the measured total pressure by positive ions which are thermionically emitted from the laser heated surface - the necessary neutralisation of their space charge being effected by the presence of hot electrons produced by inverse *Bremsstrahlung* absorption of the laser photons [108].

For these reasons, interpretation of the experiment of Bober and Singer [16] is much less problematic, and must be considered to constitute a much more direct way of obtaining values of the total equilibrium pressure over non-congruently vaporising molten UO_2.

Table A.3. Oxygen Vapour Data over Condensed $UO_{2.00}$

T, K	$\frac{\Delta \overline{G}_{O_2}}{kJmol^{-1}}$ [74]	$\frac{P_{O_2}}{MPa}$	$\frac{P_O}{MPa}$	$\frac{\Delta \overline{G}_{O_2}}{kJmol^{-1}}$ [108]
1500	-545	$1.06\ 10^{-20}$	$1.32\ 10^{-16}$	-524
2000	-461	$0.92\ 10^{-13}$	$7.63\ 10^{-11}$	-447
2500	-366	$2.28\ 10^{-9}$	$2.23\ 10^{-7}$	-370
3000	-275	$1.65\ 10^{-6}$	$4.88\ 10^{-5}$	-292
3120(sol)	-252	$6.17\ 10^{-6}$	$1.25\ 10^{-4}$	-273
3120(liq)	-233	$1.28\ 10^{-5}$	$1.80\ 10^{-4}$	-224
3500	-175	$2.48\ 10^{-4}$	$2.62\ 10^{-3}$	-155
4000	-101	$4.86\ 10^{-3}$	$3.25\ 10^{-2}$	- 17
5000	+45	$2.99\ 10^{-1}$	1.19	+121
6000	+193	4.85	13.4	

A.1.8 The Congruently Vaporising Compositions of Urania

- Solid Urania

The total pressure, P_U, of uranium bearing species over solid urania, obtained by Tetenbaum and Hunt [130] using the transpiration technique, exhibit minima, the O/U location of which decreases approximately linearly as T increases, from a value of 1.994 at $T = 2100\ K$ to 1.975 at $T = 2700\ K$. Since the contribution from O(g) and $O_2(g)$ to the total pressure is negligible at these temperatures (see Table A.3), these minima effectively define the congruently vaporising composition (CVC) of solid urania. These CVC values were subsequently corroborated by independent experimental work of Ackermann *et al* [1] based on gravimetric analysis of the residues of sublimation in both tungsten and rhenium effusion cells.

- Molten Urania

  Here the observed constancy of the sample composition at (O/U) = 1.94, during the course of Chasanov and Reedy's transpiration experiments [116] at 3390 *K* is consistent with the *CVC* being located here at this temperature - although the possibility that this low value of the *CVC* might, at least in part, be due to dissolution of tungsten cannot be ruled out, particularly at these high temperatures.

A.2 Individual Components. Tables of Thermodynamic Functions

The thermochemical properties of the elemental and molecular components of the *EOS* model have been critically reviewed and revised by V.Yungman and L.Gorochov (*IVTAN*, Moscow) in the context of the INTAS 99-066 Project [77]. The *IVTAN* database was taken as a reference basis, where, when necessary, corrections and additions have been implemented. The results of this work were first published in the above mentioned reference and, later, in the new (1996) edition of the *IVTAN* Databook [64]. In this Appendix, the updated tables of thermodynamic functions of the pertinent atoms, molecules and ions are reported without quoting sources and inter- or extrapolation methods. For more detailed information the reader is referred to the above mentioned references.

The tables reported here have been organized as follows:

Table	State	Specie	Page
A.4	Condensed phase	UO_2	146
A.5	Ideal gas	U	148
A.6	Ideal gas	U^+	150
A.7	Ideal gas	U^{2+}	152
A.8	Ideal gas	UO	154
A.9	Ideal gas	UO^+	156
A.10	Ideal gas	UO_2	158
A.11	Ideal gas	UO_2^+	160
A.12	Ideal gas	UO_2^{+2}	162
A.13	Ideal gas	UO_2^-	164
A.14	Ideal gas	UO_3	166
A.15	Ideal gas	UO_3^-	168
A.16	Ideal gas	U_2O_2	170
A.17	Ideal gas	U_2O_3	172
A.18	Ideal gas	U_2O_4	174

Table A.4. Thermodynamic Functions of Condensed UO_2 (sol,liq)

$M = 270.0278$ $UO_2(sol, liq) \rightleftharpoons UO_2(g)$ $\Delta_s H^o(0) = 605.036$ $kJ\ mol^{-1}$

T K	$C^o_P(T)$	$\Phi^o(T)$ $J\ K^{-1} mol^{-1}$	$S^o(T)$	$H^o(T) - H^o(0)$ $kJ\ mol^{-1}$	lg $K^o(T)$
298.15	63.600	39.197	77.030	11.280	-96.541
300	63.885	39.431	77.424	11.398	-95.886
400	73.430	51.481	97.333	18.341	-69.493
500	77.295	62.385	114.191	25.903	-53.680
600	79.131	72.241	128.464	33.734	-43.157
700	80.188	81.171	140.748	41.704	-35.655
800	80.988	89.304	151.509	49.764	-30.041
900	81.782	96.758	161.093	57.901	-25.683
1000	82.701	103.631	169.755	66.124	-22.206
1100	83.818	110.008	177.688	74.448	-19.369
1200	85.180	115.958	185.038	82.896	-17.011
1300	86.815	121.539	191.919	91.493	-15.021
1400	88.740	126.800	198.421	100.269	-13.322
1500	90.969	131.783	204.618	109.252	-11.855
1600	93.510	136.522	210.568	118.473	-10.576
1700	96.370	141.048	216.321	127.964	-9.452
1800	99.553	145.386	221.918	137.758	-8.458
1900	103.063	149.559	227.393	147.886	-7.573
2000	106.903	153.585	232.776	158.381	-6.781
2100	111.073	157.483	238.091	169.277	-6.070
2200	115.576	161.267	243.361	180.607	-5.427
2300	120.414	164.950	248.604	192.404	-4.845
2400	125.586	168.545	253.837	204.701	-4.317
2500	131.094	172.061	259.074	217.532	-3.835
2600	136.938	175.509	264.328	230.931	-3.395
2700	143.119	178.896	269.611	244.931	-2.992
2800	149.638	182.231	274.933	259.566	-2.624
2900	156.494	185.520	280.303	274.870	-2.285
3000	163.688	188.770	285.728	290.876	-1.974
3100	171.220	191.986	291.217	307.619	-1.689
3123	*173.000*	*192.721*	*292.490*	*311.577*	*-1.626*
3123	*116.179*	*192.721*	*317.466*	*389.577*	*-1.626*
3200	110.597	195.756	320.226	398.303	-1.457
3300	104.489	199.579	323.533	409.047	-1.250
3400	99.493	203.270	326.575	419.238	-1.056

Thermodynamic Functions of Condensed $UO_2(sol, liq)$ (*end*)

T K	$C_P^o(T)$	$\Phi^o(T)$ J $K^{-1}mol^{-1}$	$S^o(T)$	$H^o(T)-H^o(0)$ kJ mol^{-1}	lg $K^o(T)$
3500	95.445	206.834	329.399	428.977	-0.875
3600	92.204	210.276	332.040	438.353	-0.705
3700	89.651	213.601	334.530	447.441	-0.545
3800	87.684	216.814	336.894	456.303	-0.394
3900	86.214	219.922	339.152	464.994	-0.251
4000	85.165	222.930	341.320	473.560	-0.116
4100	84.473	225.844	343.414	482.039	0.013
4200	84.082	228.667	345.444	490.465	0.134
4300	83.944	231.406	347.421	498.864	0.250
4400	84.017	234.065	349.351	507.260	0.360
4500	84.266	236.648	351.242	515.673	0.465
4600	84.661	239.159	353.098	524.118	0.565
4700	85.177	241.603	354.924	532.609	0.661
4800	85.790	243.982	356.724	541.157	0.752
4900	86.484	246.301	358.499	549.770	0.840
5000	87.242	248.563	360.254	558.456	0.923
5100	88.053	250.770	361.990	567.220	1.003
5200	88.906	252.925	363.708	576.068	1.080
5300	89.792	255.032	365.410	585.003	1.153
5400	90.707	257.091	367.097	594.027	1.223
5500	91.644	259.107	368.769	603.145	1.290
5600	92.602	261.080	370.429	612.357	1.355
5700	93.579	263.013	372.077	621.666	1.417
5800	94.574	264.907	373.713	631.073	1.476
5900	95.587	266.765	375.338	640.581	1.533
6000	96.621	268.588	376.954	650.191	1.588
6200	98.762	272.136	380.156	669.728	1.691
6400	101.023	275.561	383.327	689.704	1.786
6600	103.446	278.874	386.473	710.148	1.874
6800	106.079	282.085	389.599	731.097	1.956
7000	108.977	285.201	392.715	752.597	2.031
7200	112.202	288.231	395.830	774.709	2.100
7400	115.821	291.181	398.952	797.504	2.164
7600	119.905	294.059	402.094	821.069	2.222
7800	124.528	296.869	405.267	845.502	2.276
8000	129.769	299.619	408.485	870.921	2.325

Table A.5. U(g)

$M = 238.029$

T K	$C^o_P(T)$	$\Phi^o(T)$ $J\,K^{-1}mol^{-1}$	$S^o(T)$	$H^o(T)-H^o(0)$ $kJ\ mol^{-1}$
298.15	23.693	177.989	199.787	6.499
300	23.707	178.124	199.934	6.543
400	23.950	184.475	206.807	8.933
500	23.701	189.494	212.128	11.317
600	23.402	193.636	216.422	13.671
700	23.259	197.155	220.016	16.003
800	23.349	200.211	223.125	18.331
900	23.688	202.913	225.892	20.681
1000	24.255	205.339	228.415	23.076
1100	25.022	207.545	230.762	25.539
1200	25.954	209.573	232.978	28.086
1300	27.021	211.455	235.097	30.734
1400	28.192	213.217	237.141	33.494
1500	29.446	214.879	239.129	36.375
1600	30.762	216.455	241.071	39.385
1700	32.123	217.959	242.976	42.529
1800	33.518	219.401	244.852	45.811
1900	34.936	220.790	246.702	49.233
2000	36.368	222.131	248.530	52.798
2100	37.807	223.431	250.340	56.507
2200	39.247	224.695	252.132	60.360
2300	40.680	225.927	253.908	64.356
2400	42.102	227.129	255.669	68.495
2500	43.504	228.306	257.416	72.776
2600	44.880	229.459	259.150	77.195
2700	46.223	230.591	260.869	81.751
2800	47.527	231.702	262.573	86.439
2900	48.785	232.796	264.263	91.255
3000	49.989	233.873	265.938	96.194
3100	51.134	234.934	267.596	101.250
3200	52.215	235.980	269.236	106.418
3300	53.226	237.013	270.859	111.691
3400	54.164	238.032	272.462	117.061
3500	55.026	239.038	274.044	122.521
3600	55.809	240.032	275.606	128.064
3700	56.511	241.015	277.145	133.680
3800	57.133	241.985	278.660	139.363
3900	57.674	242.945	280.151	145.104
4000	58.135	243.894	281.617	150.895

U (g) (end)

T K	$C^o_P(T)$	$\Phi^o(T)$ $J\,K^{-1}mol^{-1}$	$S^o(T)$	$H^o(T)-H^o(0)$ $kJ\ mol^{-1}$
4100	58.517	244.831	283.058	156.729
4200	58.823	245.758	284.472	162.596
4300	59.055	246.675	285.859	168.491
4400	59.215	247.581	287.218	174.405
4500	59.308	248.477	288.550	180.332
4600	59.335	249.362	289.854	186.264
4700	59.302	250.237	291.130	192.196
4800	59.211	251.102	292.378	198.122
4900	59.066	251.957	293.597	204.037
5000	58.872	252.802	294.789	209.934
5100	58.631	253.636	295.952	215.810
5200	58.348	254.461	297.088	221.659
5300	58.024	255.276	298.196	227.476
5400	57.665	256.081	299.277	233.260
5500	57.275	256.876	300.331	239.006
5600	56.857	257.661	301.360	244.713
5700	56.413	258.436	302.362	250.376
5800	55.947	259.202	303.339	255.994
5900	55.462	259.958	304.291	261.565
6000	54.961	260.705	305.219	267.085
6200	53.915	262.170	307.004	277.972
6400	52.828	263.598	308.698	288.642
6600	51.718	264.989	310.306	299.092
6800	50.591	266.344	311.831	309.310
7000	49.475	267.664	313.281	319.313
7200	48.367	268.951	314.658	329.091
7400	47.278	270.204	315.967	338.650
7600	46.213	271.424	317.213	347.993
7800	45.140	272.612	318.389	357.063
8000	44.126	273.770	319.516	365.969
8200	43.149	274.899	320.591	374.679
8400	42.217	275.999	321.620	383.215
8600	41.317	277.071	322.601	391.558
8800	40.444	278.116	323.536	399.698
9000	39.623	279.135	324.436	407.705
9200	38.839	280.129	325.298	415.550
9400	38.090	281.099	326.125	423.242
9600	37.368	282.045	326.916	430.760
9800	36.674	282.967	327.673	438.118
10000	36.010	283.867	328.399	445.326

Table A.6. U^+ (g)

$M = 238.0285 \quad U^+(g) \rightleftharpoons U(g) - e(g) \quad \Delta_r H^o(0) = -597.635\ kJ\ mol^{-1}$

T K	$C_P^o(T)$	$\Phi^o(T)$ $J\ K^{-1} mol^{-1}$	$S^o(T)$	$H^o(T) - H^o(0)$ $kJ\ mol^{-1}$	lg $K^o(T)$
298.15	25.019	177.707	201.481	7.088	104.712
300	25.015	177.854	201.636	7.135	104.059
400	24.959	184.741	208.820	9.632	77.705
500	25.000	190.134	214.394	12.130	61.835
600	25.021	194.569	218.954	14.631	51.216
700	25.031	198.335	222.812	17.133	43.603
800	25.096	201.608	226.157	19.639	37.872
900	25.285	204.504	229.123	22.157	33.399
1000	25.636	207.102	231.803	24.701	29.807
1100	26.165	209.461	234.270	27.290	26.857
1200	26.863	211.625	236.575	29.940	24.391
1300	27.712	213.629	238.758	32.668	22.296
1400	28.685	215.499	240.847	35.486	20.495
1500	29.756	217.257	242.862	38.408	18.929
1600	30.898	218.918	244.818	41.440	17.553
1700	32.085	220.498	246.727	44.589	16.336
1800	33.294	222.007	248.595	47.858	15.250
1900	34.506	223.455	250.428	51.248	14.276
2000	35.703	224.849	252.228	54.758	13.396
2100	36.869	226.195	253.999	58.387	12.597
2200	37.991	227.498	255.740	62.131	11.868
2300	39.060	228.764	257.452	65.984	11.201
2400	40.066	229.994	259.136	69.941	10.588
2500	41.002	231.193	260.791	73.995	10.022
2600	41.865	232.363	262.416	78.139	9.498
2700	42.650	233.506	264.011	82.365	9.012
2800	43.356	234.623	265.575	86.666	8.559
2900	43.982	235.717	267.108	91.034	8.137
3000	44.530	236.788	268.608	95.460	7.741
3100	45.002	237.839	270.076	99.937	7.371
3200	45.400	238.869	271.512	104.458	7.022
3300	45.728	239.879	272.914	109.015	6.694
3400	45.991	240.871	274.283	113.601	6.385
3500	46.191	241.845	275.619	118.211	6.093
3600	46.335	242.801	276.923	122.838	5.817
3700	46.428	243.740	278.193	127.476	5.555
3800	46.473	244.664	279.432	132.122	5.307
3900	46.476	245.571	280.640	136.769	5.070
4000	46.442	246.462	281.816	141.415	4.846

U^+ (g) (end)

T K	$C^o_P(T)$	$\Phi^o(T)$ $J\,K^{-1}mol^{-1}$	$S^o(T)$	$H^o(T)-H^o(0)$ $kJ\ mol^{-1}$	lg $K^o(T)$
4100	46.374	247.338	282.962	146.057	4.632
4200	46.277	248.200	284.078	150.689	4.428
4300	46.156	249.047	285.166	155.311	4.233
4400	46.012	249.880	286.225	159.920	4.047
4500	45.851	250.699	287.258	164.513	3.869
4600	45.674	251.505	288.263	169.089	3.698
4700	45.484	252.298	289.244	173.648	3.535
4800	45.284	253.077	290.199	178.186	3.378
4900	45.075	253.844	291.131	182.704	3.227
5000	44.860	254.599	292.039	187.201	3.083
5100	44.640	255.342	292.926	191.676	2.944
5200	44.416	256.073	293.790	196.129	2.810
5300	44.189	256.793	294.634	200.559	2.681
5400	43.961	257.501	295.458	204.967	2.556
5500	43.733	258.199	296.263	209.351	2.436
5600	43.504	258.885	297.048	213.713	2.321
5700	43.275	259.562	297.816	218.052	2.209
5800	43.047	260.228	298.567	222.368	2.101
5900	42.821	260.884	299.301	226.661	1.996
6000	42.596	261.530	300.019	230.932	1.895
6200	42.151	262.794	301.408	239.407	1.702
6400	41.713	264.022	302.740	247.793	1.521
6600	41.283	265.215	304.017	256.093	1.350
6800	40.861	266.374	305.243	264.307	1.189
7000	40.447	267.502	306.421	272.437	1.036
7200	40.040	268.598	307.555	280.486	0.891
7400	39.639	269.666	308.646	288.453	0.754
7600	39.245	270.706	309.698	296.342	0.623
7800	38.857	271.719	310.713	304.152	0.499
8000	38.475	272.706	311.692	311.885	0.381
8200	38.098	273.668	312.637	319.542	0.267
8400	37.726	274.607	313.551	327.125	0.159
8600	37.360	275.523	314.434	334.633	0.055
8800	36.999	276.417	315.289	342.069	-0.044
9000	36.644	277.290	316.116	349.433	-0.140
9200	36.294	278.143	316.918	356.727	-0.232
9400	35.949	278.976	317.695	363.951	-0.320
9600	35.609	279.791	318.448	371.107	-0.405
9800	35.276	280.587	319.179	378.195	-0.488
10000	34.948	281.366	319.888	385.218	-0.567

Table A.7. U^{2+} (g)

$M = 238.0279 \quad U^{2+}(g) \rightleftharpoons U(g) - 2e(g) \quad \Delta_r H^o(0) = -1597.634\ kJ\ mol^{-1}$

T K	$C_P^o(T)$	$\Phi^o(T)$ $J\ K^{-1} mol^{-1}$	$S^o(T)$	$H^o(T) - H^o(0)$ $kJ\ mol^{-1}$	lg $K^o(T)$
298.15	24.097	178.180	202.226	7.169	279.877
300	24.091	178.329	202.376	7.214	278.136
400	23.869	185.245	209.274	9.611	207.940
500	23.658	190.602	214.578	11.988	165.712
600	23.504	194.967	218.876	14.345	137.488
700	23.487	198.648	222.495	16.693	117.275
800	23.644	201.830	225.640	19.048	102.077
900	23.974	204.634	228.442	21.428	90.226
1000	24.452	207.144	230.992	23.848	80.720
1100	25.043	209.420	233.349	26.322	72.923
1200	25.712	211.507	235.557	28.859	66.409
1300	26.425	213.438	237.643	31.466	60.883
1400	27.155	215.238	239.628	34.145	56.135
1500	27.880	216.928	241.526	36.897	52.010
1600	28.582	218.523	243.348	39.720	48.392
1700	29.247	220.035	245.101	42.612	45.192
1800	29.867	221.475	246.790	45.568	42.340
1900	30.434	222.850	248.421	48.583	39.783
2000	30.945	224.168	249.995	51.653	37.477
2100	31.398	225.435	251.516	54.771	35.386
2200	31.794	226.654	252.986	57.931	33.481
2300	32.134	227.830	254.407	61.127	31.738
2400	32.419	228.966	255.781	64.356	30.137
2500	32.654	230.065	257.109	67.610	28.661
2600	32.842	231.130	258.394	70.885	27.297
2700	32.987	232.163	259.636	74.177	26.031
2800	33.093	233.166	260.838	77.481	24.854
2900	33.163	234.140	262.000	80.794	23.756
3000	33.202	235.088	263.125	84.113	22.730
3100	33.214	236.010	264.214	87.434	21.769
3200	33.202	236.908	265.268	90.755	20.867
3300	33.170	237.783	266.290	94.073	20.018
3400	33.121	238.636	267.279	97.388	19.218
3500	33.057	239.468	268.239	100.697	18.463
3600	32.981	240.280	269.169	103.999	17.749
3700	32.896	241.073	270.071	107.293	17.073
3800	32.804	241.848	270.947	110.578	16.431
3900	32.707	242.605	271.798	113.853	15.823
4000	32.607	243.345	272.625	117.119	15.244

U^{2+} (g) (end)

T K	$C^o_p(T)$	$\Phi^o(T)$ J $K^{-1}mol^{-1}$	$S^o(T)$	$H^o(T)-H^o(0)$ kJ mol^{-1}	lg $K^o(T)$
4100	32.504	244.069	273.429	120.375	14.692
4200	32.401	244.778	274.211	123.620	14.167
4300	32.299	245.471	274.972	126.855	13.665
4400	32.198	246.150	275.714	130.080	13.186
4500	32.099	246.815	276.436	133.295	12.728
4600	32.004	247.467	277.140	136.500	12.289
4700	31.913	248.105	277.828	139.696	11.869
4800	31.826	248.731	278.499	142.883	11.465
4900	31.744	249.346	279.154	146.061	11.078
5000	31.667	249.948	279.795	149.231	10.706
5100	31.595	250.540	280.421	152.395	10.349
5200	31.529	251.120	281.034	155.551	10.005
5300	31.468	251.690	281.634	158.700	9.673
5400	31.413	252.250	282.221	161.845	9.354
5500	31.364	252.800	282.797	164.983	9.046
5600	31.321	253.341	283.362	168.117	8.749
5700	31.283	253.873	283.916	171.248	8.462
5800	31.250	254.395	284.460	174.374	8.184
5900	31.223	254.910	284.994	177.498	7.916
6000	31.201	255.415	285.519	180.619	7.656
6200	31.171	256.403	286.541	186.856	7.161
6400	31.159	257.360	287.530	193.089	6.696
6600	31.163	258.289	288.489	199.321	6.258
6800	31.181	259.191	289.420	205.555	5.845
7000	31.211	260.068	290.324	211.794	5.455
7200	31.250	260.921	291.204	218.040	5.085
7400	31.297	261.751	292.061	224.295	4.735
7600	31.350	262.559	292.896	230.559	4.401
7800	31.406	263.348	293.711	236.835	4.084
8000	31.465	264.117	294.507	243.122	3.782
8200	31.524	264.868	295.285	249.420	3.494
8400	31.582	265.601	296.045	255.732	3.219
8600	31.639	266.317	296.789	262.053	2.955
8800	31.692	267.018	297.517	268.386	2.703
9000	31.740	267.704	298.230	274.730	2.460
9200	31.784	268.375	298.928	281.082	2.228
9400	31.822	269.033	299.612	287.443	2.005
9600	31.854	269.677	300.282	293.811	1.790
9800	31.880	270.308	300.939	300.184	1.583
10000	31.898	270.927	301.583	306.562	1.383

Table A.8. UO (g)

$M = 254.0284 \quad UO(g) \rightleftharpoons U(g) + O(g) \quad \Delta_r H^o(0) = 754.961\ kJ\ mol^{-1}$

T K	$C_P^o(T)$	$\Phi^o(T)$ $J\,K^{-1}mol^{-1}$	$S^o(T)$	$H^o(T) - H^o(0)$ $kJ\ mol^{-1}$	lg $K^o(T)$
298.15	39.276	218.740	252.124	9.954	-127.164
300	39.350	218.947	252.368	10.026	-126.345
400	42.748	228.829	264.184	14.142	-93.330
500	44.770	236.907	273.965	18.529	-73.513
600	45.657	243.790	282.220	23.058	-60.302
700	45.842	249.797	289.278	27.637	-50.866
800	45.693	255.123	295.392	32.215	-43.791
900	45.459	259.902	300.760	36.773	-38.288
1000	45.285	264.231	305.540	41.309	-33.887
1100	45.242	268.185	309.853	45.834	-30.286
1200	45.360	271.824	313.793	50.363	-27.285
1300	45.639	275.194	317.434	54.912	-24.746
1400	46.071	278.334	320.831	59.496	-22.568
1500	46.640	281.275	324.028	64.130	-20.680
1600	47.328	284.042	327.060	68.828	-19.027
1700	48.117	286.658	329.952	73.599	-17.568
1800	48.991	289.141	332.727	78.454	-16.270
1900	49.933	291.506	335.401	83.400	-15.108
2000	50.928	293.766	337.987	88.442	-14.061
2100	51.962	295.932	340.497	93.587	-13.114
2200	53.020	298.013	342.938	98.836	-12.251
2300	54.087	300.018	345.319	104.191	-11.463
2400	55.152	301.954	347.643	109.653	-10.740
2500	56.201	303.828	349.916	115.221	-10.074
2600	57.222	305.643	352.140	120.892	-9.459
2700	58.205	307.406	354.318	126.664	-8.889
2800	59.140	309.119	356.452	132.532	-8.358
2900	60.020	310.788	358.543	138.490	-7.864
3000	60.840	312.414	360.592	144.534	-7.402
3100	61.593	314.000	362.599	150.656	-6.969
3200	62.278	315.550	364.566	156.850	-6.563
3300	62.893	317.065	366.492	163.109	-6.181
3400	63.438	318.546	368.377	169.427	-5.821
3500	63.914	319.996	370.223	175.795	-5.481
3600	64.323	321.417	372.030	182.207	-5.159
3700	64.668	322.809	373.797	188.657	-4.854
3800	64.953	324.173	375.526	195.139	-4.565
3900	65.180	325.512	377.216	201.646	-4.290
4000	65.356	326.825	378.868	208.173	-4.028

UO (g) (end)

T K	$C_P^o(T)$	$\Phi^o(T)$ $J\ K^{-1}mol^{-1}$	$S^o(T)$	$H^o(T)-H^o(0)$ $kJ\ mol^{-1}$	lg $K^o(T)$
4100	65.483	328.114	380.484	214.716	-3.779
4200	65.567	329.380	382.063	221.268	-3.541
4300	65.612	330.623	383.606	227.828	-3.314
4400	65.621	331.845	385.115	234.389	-3.096
4500	65.600	333.045	386.589	240.951	-2.888
4600	65.551	334.225	388.031	247.509	-2.689
4700	65.479	335.384	389.440	254.060	-2.498
4800	65.387	336.525	390.817	260.604	-2.314
4900	65.277	337.647	392.165	267.137	-2.138
5000	65.153	338.750	393.482	273.659	-1.968
5100	65.017	339.836	394.771	280.167	-1.804
5200	64.871	340.905	396.032	286.662	-1.647
5300	64.718	341.957	397.266	293.141	-1.495
5400	64.558	342.992	398.475	299.605	-1.349
5500	64.393	344.012	399.658	306.053	-1.207
5600	64.225	345.016	400.816	312.484	-1.071
5700	64.056	346.005	401.952	318.898	-0.939
5800	63.884	346.979	403.064	325.295	-0.811
5900	63.713	347.939	404.155	331.674	-0.687
6000	63.541	348.885	405.224	338.037	-0.567
6200	63.202	350.736	407.302	350.712	-0.339
6400	62.869	352.535	409.304	363.319	-0.124
6600	62.544	354.285	411.233	375.860	0.078
6800	62.228	355.987	413.095	388.337	0.269
7000	61.923	357.645	414.895	400.752	0.450
7200	61.626	359.259	416.635	413.107	0.621
7400	61.338	360.833	418.320	425.403	0.783
7600	61.058	362.367	419.952	437.643	0.937
7800	60.784	363.864	421.534	449.827	1.083
8000	60.514	365.325	423.070	461.957	1.223
8200	60.249	366.752	424.561	474.033	1.355
8400	59.985	368.146	426.009	486.056	1.482
8600	59.723	369.508	427.418	498.027	1.602
8800	59.460	370.839	428.788	509.946	1.718
9000	59.196	372.142	430.121	521.811	1.828
9200	58.929	373.417	431.419	533.624	1.934
9400	58.659	374.664	432.684	545.383	2.035
9600	58.385	375.886	433.916	557.088	2.132
9800	58.106	377.083	435.117	568.737	2.225
10000	57.821	378.255	436.288	580.330	2.315

Table A.9. UO^+ (g)

$M = 254.0278$ $UO^+(g) \rightleftharpoons U(g) + O(g) - e(g)$ $\Delta_r H^o(0) = 209.961\ kJ\ mol^{-1}$

T K	$C_P^o(T)$	$\Phi^o(T)$ $J\,K^{-1}mol^{-1}$	$S^o(T)$	$H^o(T)-H^o(0)$ $kJ\ mol^{-1}$	lg $K^o(T)$
298.15	39.536	217.314	249.893	9.713	-31.614
300	39.589	217.516	250.138	9.786	-31.390
400	40.742	227.186	261.751	13.826	-22.399
500	40.337	235.040	270.810	17.885	-17.046
600	39.757	241.630	278.112	21.889	-13.507
700	39.399	247.289	284.209	25.844	-10.998
800	39.364	252.240	289.464	29.779	-9.132
900	39.675	256.639	294.115	33.728	-7.691
1000	40.326	260.600	298.326	37.726	-6.548
1100	41.293	264.209	302.213	41.804	-5.621
1200	42.541	267.529	305.857	45.994	-4.854
1300	44.025	270.612	309.320	50.320	-4.211
1400	45.694	273.496	312.642	54.805	-3.665
1500	47.498	276.213	315.856	59.464	-3.196
1600	49.381	278.789	318.981	64.307	-2.789
1700	51.288	281.243	322.032	69.341	-2.434
1800	53.166	283.593	325.017	74.564	-2.121
1900	54.966	285.850	327.940	79.971	-1.845
2000	56.644	288.026	330.803	85.553	-1.599
2100	58.166	290.130	333.604	91.295	-1.380
2200	59.504	292.169	336.341	97.180	-1.182
2300	60.643	294.148	339.012	103.189	-1.004
2400	61.575	296.071	341.614	109.302	-0.842
2500	62.302	297.944	344.143	115.497	-0.696
2600	62.832	299.768	346.597	121.756	-0.561
2700	63.178	301.547	348.976	128.058	-0.438
2800	63.357	303.282	351.277	134.386	-0.325
2900	63.389	304.976	353.501	140.724	-0.221
3000	63.293	306.629	355.649	147.059	-0.125
3100	63.090	308.244	357.721	153.379	-0.035
3200	62.799	309.822	359.720	159.674	0.049
3300	62.437	311.363	361.647	165.937	0.126
3400	62.020	312.870	363.505	172.160	0.199
3500	61.563	314.342	365.296	178.339	0.268
3600	61.078	315.782	367.024	184.472	0.332
3700	60.575	317.189	368.690	190.554	0.392
3800	60.063	318.566	370.299	196.586	0.450
3900	59.549	319.912	371.852	202.567	0.504
4000	59.040	321.230	373.354	208.496	0.555

UO^+ (g) (end)

T K	$C_P^o(T)$	$\Phi^o(T)$ $J\,K^{-1}mol^{-1}$	$S^o(T)$	$H^o(T)-H^o(0)$ $kJ\,mol^{-1}$	lg $K^o(T)$
4100	58.539	322.519	374.805	214.375	0.604
4200	58.050	323.781	376.210	220.205	0.651
4300	57.576	325.016	377.570	225.986	0.695
4400	57.120	326.225	378.889	231.720	0.737
4500	56.682	327.410	380.168	237.410	0.778
4600	56.263	328.570	381.409	243.057	0.817
4700	55.864	329.707	382.614	248.663	0.854
4800	55.486	330.822	383.787	254.231	0.890
4900	55.127	331.914	384.927	259.761	0.925
5000	54.789	332.986	386.037	265.257	0.958
5100	54.470	334.037	387.119	270.720	0.990
5200	54.170	335.068	388.174	276.151	1.021
5300	53.889	336.079	389.203	281.555	1.051
5400	53.625	337.073	390.208	286.930	1.080
5500	53.378	338.048	391.189	292.280	1.108
5600	53.148	339.005	392.149	297.606	1.135
5700	52.933	339.946	393.088	302.910	1.161
5800	52.733	340.870	394.007	308.193	1.186
5900	52.548	341.778	394.907	313.457	1.211
6000	52.375	342.671	395.788	318.703	1.235
6200	52.068	344.412	397.501	329.147	1.280
6400	51.806	346.097	399.149	339.534	1.323
6600	51.585	347.729	400.740	349.872	1.363
6800	51.399	349.311	402.277	360.170	1.401
7000	51.244	350.846	403.765	370.434	1.437
7200	51.117	352.336	405.207	380.669	1.471
7400	51.013	353.784	406.606	390.882	1.503
7600	50.930	355.192	407.965	401.076	1.534
7800	50.863	356.562	409.287	411.255	1.562
8000	50.810	357.896	410.574	421.422	1.589
8200	50.768	359.197	411.828	431.580	1.615
8400	50.735	360.464	413.051	441.731	1.639
8600	50.709	361.701	414.245	451.875	1.662
8800	50.687	362.909	415.410	462.014	1.684
9000	50.667	364.088	416.549	472.149	1.704
9200	50.648	365.241	417.662	482.281	1.724
9400	50.629	366.368	418.751	492.409	1.742
9600	50.607	367.470	419.817	502.532	1.759
9800	50.581	368.549	420.860	512.651	1.776
10000	50.552	369.606	421.882	522.765	1.791

Table A.10. UO_2 (g)

$M = 270.0278 \quad UO_2(g) \rightleftharpoons U(g) + 2O(g) \quad \Delta_r H^o(0) = 1502.656\ kJ\ mol^{-1}$

T K	$C_P^o(T)$	$\Phi^o(T)$ $J\ K^{-1} mol^{-1}$	$S^o(T)$	$H^o(T) - H^o(0)$ $kJ\ mol^{-1}$	lg $K^o(T)$
298.15	58.625	220.266	263.554	12.906	-250.994
300	58.753	220.534	263.917	13.015	-249.362
400	64.108	233.667	281.625	19.183	-183.634
500	66.764	244.766	296.258	25.746	-144.189
600	67.633	254.400	308.527	32.476	-117.892
700	67.645	262.896	318.960	39.244	-99.112
800	67.380	270.481	327.976	45.996	-85.028
900	67.116	277.318	335.896	52.721	-74.076
1000	66.960	283.535	342.959	59.423	-65.315
1100	66.936	289.232	349.338	66.117	-58.147
1200	67.037	294.487	355.166	72.815	-52.174
1300	67.243	299.364	360.540	79.528	-47.120
1400	67.530	303.914	365.533	86.266	-42.788
1500	67.878	308.180	370.203	93.036	-39.032
1600	68.266	312.195	374.596	99.843	-35.745
1700	68.680	315.989	378.747	106.690	-32.844
1800	69.103	319.585	382.685	113.579	-30.265
1900	69.526	323.006	386.433	120.511	-27.956
2000	69.938	326.267	390.009	127.484	-25.877
2100	70.332	329.385	393.431	134.498	-23.995
2200	70.702	332.371	396.712	141.550	-22.283
2300	71.045	335.237	399.862	148.637	-20.719
2400	71.358	337.994	402.892	155.758	-19.284
2500	71.639	340.648	405.811	162.908	-17.963
2600	71.888	343.209	408.626	170.084	-16.742
2700	72.105	345.682	411.343	177.284	-15.611
2800	72.292	348.075	413.969	184.504	-14.559
2900	72.448	350.391	416.509	191.742	-13.579
3000	72.576	352.636	418.967	198.993	-12.663
3100	72.677	354.814	421.348	206.256	-11.805
3200	72.754	356.930	423.657	213.528	-11.000
3300	72.808	358.986	425.897	220.806	-10.243
3400	72.841	360.986	428.071	228.088	-9.529
3500	72.855	362.933	430.183	235.373	-8.854
3600	72.852	364.830	432.235	242.659	-8.216
3700	72.834	366.679	434.231	249.943	-7.612
3800	72.801	368.482	436.173	257.225	-7.038
3900	72.757	370.242	438.063	264.503	-6.494
4000	72.702	371.961	439.905	271.776	-5.975

UO_2 (g) (end)

T K	$C^o_P(T)$	$\Phi^o(T)$ $J\,K^{-1}mol^{-1}$	$S^o(T)$	$H^o(T)-H^o(0)$ $kJ\,mol^{-1}$	lg $K^o(T)$
4100	72.637	373.640	441.699	279.043	-5.480
4200	72.564	375.281	443.449	286.303	-5.009
4300	72.484	376.886	445.155	293.556	-4.558
4400	72.398	378.457	446.820	300.800	-4.127
4500	72.306	379.994	448.446	308.035	-3.715
4600	72.209	381.500	450.035	315.261	-3.320
4700	72.109	382.974	451.586	322.477	-2.940
4800	72.006	384.420	453.104	329.683	-2.576
4900	71.899	385.837	454.587	336.878	-2.226
5000	71.791	387.226	456.039	344.063	-1.890
5100	71.681	388.589	457.459	351.236	-1.566
5200	71.569	389.927	458.850	358.398	-1.253
5300	71.456	391.241	460.212	365.550	-0.952
5400	71.343	392.530	461.547	372.690	-0.662
5500	71.230	393.797	462.855	379.818	-0.382
5600	71.116	395.042	464.137	386.936	-0.111
5700	71.002	396.265	465.395	394.042	0.151
5800	70.889	397.467	466.629	401.136	0.404
5900	70.776	398.650	467.840	408.219	0.649
6000	70.664	399.813	469.028	415.291	0.887
6200	70.441	402.083	471.342	429.402	1.340
6400	70.222	404.283	473.575	443.468	1.766
6600	70.008	406.415	475.732	457.491	2.168
6800	69.798	408.485	477.819	471.472	2.547
7000	69.593	410.495	479.839	485.411	2.906
7200	69.393	412.449	481.797	499.309	3.246
7400	69.199	414.349	483.696	513.168	3.568
7600	69.011	416.198	485.539	526.989	3.875
7800	68.827	417.999	487.329	540.773	4.166
8000	68.650	419.754	489.069	554.520	4.444
8200	68.477	421.465	490.762	568.233	4.709
8400	68.311	423.135	492.410	581.912	4.961
8600	68.149	424.765	494.016	595.558	5.203
8800	67.992	426.357	495.581	609.172	5.434
9000	67.841	427.912	497.107	622.755	5.655
9200	67.695	429.432	498.596	636.308	5.868
9400	67.553	430.920	500.051	649.833	6.071
9600	67.416	432.375	501.471	663.330	6.267
9800	67.283	433.799	502.860	676.800	6.454
10000	67.155	435.194	504.218	690.244	6.635

Table A.11. UO_2^+ (g)

$M = 270.0273$ $UO_2^+(g) \rightleftharpoons U(g) + 2O(g) - e(g)$ $\Delta_r H^o(0) = 977.656\ kJ\ mol^{-1}$

T K	$C_P^o(T)$	$\Phi^o(T)$ $J\,K^{-1}mol^{-1}$	$S^o(T)$	$H^o(T) - H^o(0)$ $kJ\ mol^{-1}$	lg $K^o(T)$
298.15	48.672	223.316	262.338	11.634	-159.184
300	48.760	223.558	262.640	11.725	-158.125
400	52.940	235.219	277.266	16.819	-115.485
500	56.141	244.879	289.439	22.280	-89.919
600	58.616	253.198	299.903	28.023	-72.893
700	60.532	260.540	309.089	33.985	-60.749
800	62.006	267.129	317.272	40.115	-51.655
900	63.136	273.117	324.644	46.374	-44.595
1000	64.009	278.610	331.343	52.733	-38.958
1100	64.695	283.686	337.477	59.170	-34.355
1200	65.246	288.408	343.131	65.668	-30.526
1300	65.703	292.821	348.372	72.216	-27.292
1400	66.094	296.965	353.256	78.806	-24.526
1500	66.435	300.872	357.828	85.433	-22.132
1600	66.740	304.567	362.125	92.092	-20.041
1700	67.017	308.074	366.180	98.780	-18.199
1800	67.269	311.409	370.017	105.495	-16.564
1900	67.501	314.591	373.661	112.233	-15.104
2000	67.714	317.631	377.129	118.994	-13.790
2100	67.909	320.544	380.437	125.776	-12.604
2200	68.086	323.339	383.600	132.575	-11.526
2300	68.247	326.025	386.631	139.392	-10.543
2400	68.393	328.612	389.538	146.224	-9.642
2500	68.522	331.105	392.333	153.070	-8.814
2600	68.637	333.512	395.023	159.928	-8.050
2700	68.737	335.838	397.615	166.797	-7.342
2800	68.823	338.090	400.116	173.675	-6.686
2900	68.896	340.270	402.533	180.561	-6.074
3000	68.957	342.385	404.870	187.454	-5.503
3100	69.005	344.437	407.132	194.352	-4.969
3200	69.042	346.431	409.323	201.255	-4.469
3300	69.068	348.369	411.448	208.160	-3.998
3400	69.085	350.255	413.510	215.068	-3.555
3500	69.092	352.091	415.513	221.977	-3.137
3600	69.090	353.880	417.459	228.886	-2.741
3700	69.080	355.624	419.352	235.795	-2.367
3800	69.063	357.325	421.194	242.702	-2.012
3900	69.039	358.986	422.988	249.607	-1.675
4000	69.009	360.608	424.735	256.510	-1.355

UO_2^+ (g) (end)

T K	$C_P^o(T)$	$\Phi^o(T)$ $J\,K^{-1}mol^{-1}$	$S^o(T)$	$H^o(T)-H^o(0)$ $kJ\ mol^{-1}$	lg $K^o(T)$
4100	68.973	362.193	426.439	263.409	-1.050
4200	68.932	363.742	428.100	270.304	-0.759
4300	68.886	365.258	429.722	277.195	-0.481
4400	68.836	366.741	431.305	284.081	-0.215
4500	68.782	368.193	432.851	290.962	0.039
4600	68.724	369.615	434.362	297.837	0.282
4700	68.664	371.009	435.840	304.707	0.515
4800	68.600	372.374	437.285	311.570	0.739
4900	68.535	373.714	438.699	318.427	0.954
5000	68.467	375.027	440.082	325.277	1.161
5100	68.398	376.316	441.438	332.120	1.360
5200	68.327	377.581	442.765	338.956	1.551
5300	68.255	378.823	444.066	345.785	1.736
5400	68.182	380.043	445.341	352.607	1.914
5500	68.108	381.242	446.591	359.422	2.085
5600	68.034	382.420	447.818	366.229	2.251
5700	67.959	383.578	449.022	373.028	2.411
5800	67.884	384.716	450.203	379.821	2.566
5900	67.808	385.836	451.363	386.605	2.716
6000	67.733	386.938	452.502	393.382	2.861
6200	67.583	389.089	454.720	406.914	3.137
6400	67.434	391.173	456.863	420.416	3.396
6600	67.287	393.196	458.936	433.888	3.641
6800	67.143	395.159	460.943	447.331	3.871
7000	67.002	397.066	462.887	460.745	4.088
7200	66.864	398.921	464.773	474.132	4.294
7400	66.729	400.726	466.603	487.491	4.489
7600	66.598	402.483	468.381	500.823	4.674
7800	66.471	404.195	470.109	514.130	4.849
8000	66.348	405.864	471.790	527.412	5.016
8200	66.228	407.492	473.427	540.670	5.174
8400	66.113	409.081	475.022	553.904	5.326
8600	66.001	410.632	476.576	567.115	5.470
8800	65.893	412.148	478.092	580.304	5.607
9000	65.789	413.630	479.572	593.473	5.739
9200	65.688	415.080	481.017	606.620	5.865
9400	65.591	416.498	482.428	619.748	5.985
9600	65.498	417.886	483.808	632.857	6.101
9800	65.408	419.245	485.158	645.948	6.211
10000	65.321	420.576	486.478	659.021	6.317

Table A.12. UO_2^{+2} (g)

$$M = 270.0267$$
$$UO_2^{+2}(g) \rightleftharpoons U(g) + 2O(g) - 2e(g) \quad \Delta_r H^o(0) = -173.543\ kJ\ mol^{-1}$$

T K	$C_P^o(T)$	$\Phi^o(T)$ $J\,K^{-1}mol^{-1}$	$S^o(T)$	$H^o(T)-H^o(0)$ $kJ\ mol^{-1}$	lg $K^o(T)$
298.15	47.850	216.120	254.486	11.439	42.864
300	47.936	216.357	254.783	11.528	42.673
400	51.920	227.822	269.150	16.531	34.901
500	54.693	237.312	281.054	21.871	30.168
600	56.583	245.469	291.203	27.441	26.961
700	57.890	252.647	300.030	33.168	24.629
800	58.817	259.066	307.824	39.006	22.849
900	59.491	264.878	314.792	44.923	21.438
1000	59.995	270.189	321.088	50.899	20.289
1100	60.379	275.081	326.825	56.918	19.333
1200	60.678	279.615	332.092	62.972	18.521
1300	60.915	283.841	336.958	69.052	17.823
1400	61.106	287.799	341.480	75.153	17.215
1500	61.262	291.520	345.701	81.272	16.680
1600	61.391	295.031	349.659	87.405	16.205
1700	61.498	298.355	353.384	93.549	15.780
1800	61.590	301.511	356.902	99.704	15.397
1900	61.668	304.515	360.234	105.867	15.050
2000	61.736	307.381	363.399	112.037	14.734
2100	61.796	310.120	366.413	118.214	14.445
2200	61.852	312.745	369.289	124.396	14.180
2300	61.904	315.264	372.039	130.584	13.935
2400	61.955	317.685	374.675	136.777	13.709
2500	62.006	320.015	377.205	142.975	13.499
2600	62.061	322.262	379.638	149.178	13.304
2700	62.121	324.431	381.981	155.387	13.122
2800	62.188	326.527	384.242	161.603	12.952
2900	62.265	328.555	386.425	167.825	12.793
3000	62.353	330.519	388.538	174.056	12.644
3100	62.454	332.424	390.584	180.296	12.503
3200	62.570	334.272	392.569	186.547	12.371
3300	62.704	336.068	394.496	192.811	12.246
3400	62.856	337.814	396.370	199.089	12.128
3500	63.028	339.514	398.195	205.383	12.017
3600	63.222	341.168	399.973	211.695	11.911
3700	63.438	342.781	401.708	218.028	11.811
3800	63.677	344.354	403.403	224.384	11.716
3900	63.940	345.890	405.060	230.764	11.625
4000	64.226	347.389	406.682	237.172	11.539

UO_2^{+2}(g) (end)

T K	$C_P^o(T)$	$\Phi^o(T)$ $J\ K^{-1} mol^{-1}$	$S^o(T)$	$H^o(T)-H^o(0)$ $kJ\ mol^{-1}$	lg $K^o(T)$
4100	64.537	348.855	408.272	243.610	11.457
4200	64.871	350.288	409.831	250.081	11.379
4300	65.229	351.691	411.362	256.585	11.304
4400	65.610	353.064	412.866	263.127	11.233
4500	66.012	354.410	414.345	269.708	11.164
4600	66.436	355.728	415.800	276.330	11.099
4700	66.879	357.022	417.234	282.996	11.036
4800	67.340	358.291	418.647	289.707	10.975
4900	67.818	359.537	420.040	296.464	10.917
5000	68.310	360.761	421.415	303.271	10.861
5100	68.816	361.963	422.773	310.127	10.807
5200	69.332	363.146	424.114	317.034	10.755
5300	69.858	364.309	425.440	323.994	10.705
5400	70.391	365.453	426.750	331.006	10.656
5500	70.929	366.579	428.047	338.072	10.609
5600	71.470	367.688	429.330	345.192	10.564
5700	72.012	368.781	430.600	352.366	10.520
5800	72.553	369.858	431.857	359.594	10.477
5900	73.090	370.919	433.101	366.877	10.436
6000	73.622	371.966	434.334	374.212	10.395
6200	74.664	374.017	436.766	389.042	10.318
6400	75.664	376.015	439.152	404.075	10.244
6600	76.609	377.964	441.495	419.303	10.175
6800	77.489	379.866	443.795	434.714	10.108
7000	78.295	381.725	446.053	450.294	10.043
7200	79.022	383.543	448.269	466.027	9.982
7400	79.664	385.322	450.443	481.897	9.922
7600	80.220	387.064	452.575	497.887	9.865
7800	80.688	388.770	454.665	513.979	9.810
8000	81.070	390.444	456.713	530.157	9.756
8200	81.368	392.084	458.719	546.402	9.704
8400	81.585	393.694	460.682	562.698	9.653
8600	81.725	395.275	462.604	579.030	9.604
8800	81.793	396.826	464.484	595.383	9.556
9000	81.795	398.350	466.322	611.743	9.509
9200	81.736	399.848	468.119	628.097	9.463
9400	81.622	401.319	469.876	644.434	9.418
9600	81.459	402.765	471.592	660.743	9.374
9800	81.252	404.187	473.270	677.014	9.332
10000	81.007	405.585	474.909	693.241	9.289

Table A.13. UO_2^- (g)

$M = 270.0284$ $UO_2^-(g) \rightleftharpoons U(g) + 2O(g) + e(g)$ $\Delta_r H^o(0) = 1592.656\ kJ\ mol^{-1}$

T K	$C_P^o(T)$	$\Phi^o(T)$ $J\ K^{-1} mol^{-1}$	$S^o(T)$	$H^o(T) - H^o(0)$ $kJ\ mol^{-1}$	lg $K^o(T)$
298.15	56.169	232.127	278.489	13.823	-267.369
300	56.215	232.414	278.836	13.927	-265.634
400	58.171	246.164	295.298	19.654	-195.709
500	59.398	257.348	308.421	25.537	-153.676
600	60.207	266.794	319.327	31.519	-125.605
700	60.806	274.981	328.654	37.571	-105.521
800	61.319	282.210	336.808	43.678	-90.435
900	61.812	288.687	344.059	49.834	-78.683
1000	62.314	294.556	350.597	56.041	-69.267
1100	62.830	299.926	356.560	62.298	-61.552
1200	63.354	304.877	362.050	68.607	-55.112
1300	63.873	309.473	367.141	74.968	-49.655
1400	64.374	313.764	371.893	81.381	-44.970
1500	64.848	317.789	376.351	87.842	-40.903
1600	65.284	321.582	380.550	94.349	-37.339
1700	65.677	325.168	384.520	100.898	-34.189
1800	66.025	328.571	388.284	107.483	-31.383
1900	66.325	331.809	391.862	114.101	-28.869
2000	66.579	334.898	395.271	120.747	-26.601
2100	66.789	337.850	398.525	127.415	-24.546
2200	66.957	340.680	401.636	134.103	-22.674
2300	67.086	343.395	404.615	140.805	-20.960
2400	67.181	346.006	407.472	147.519	-19.387
2500	67.245	348.520	410.216	154.241	-17.935
2600	67.281	350.944	412.854	160.967	-16.593
2700	67.293	353.284	415.394	167.696	-15.347
2800	67.284	355.546	417.841	174.425	-14.187
2900	67.258	357.735	420.202	181.152	-13.104
3000	67.217	359.856	422.481	187.876	-12.091
3100	67.163	361.912	424.684	194.595	-11.140
3200	67.098	363.907	426.816	201.308	-10.247
3300	67.025	365.844	428.879	208.015	-9.405
3400	66.946	367.728	430.879	214.713	-8.611
3500	66.861	369.560	432.818	221.404	-7.859
3600	66.772	371.344	434.701	228.085	-7.148
3700	66.680	373.081	436.529	234.758	-6.472
3800	66.585	374.774	438.306	241.421	-5.831
3900	66.490	376.425	440.034	248.075	-5.220
4000	66.394	378.037	441.716	254.719	-4.638

UO_2^- (g) (end)

T K	$C_p^o(T)$	$\Phi^o(T)$ $J\,K^{-1}mol^{-1}$	$S^o(T)$	$H^o(T)-H^o(0)$ $kJ\,mol^{-1}$	lg $K^o(T)$
4100	66.298	379.610	443.355	261.354	-4.083
4200	66.203	381.147	444.951	267.979	-3.552
4300	66.108	382.649	446.508	274.594	-3.045
4400	66.014	384.117	448.026	281.200	-2.559
4500	65.922	385.554	449.509	287.797	-2.093
4600	65.832	386.960	450.957	294.385	-1.646
4700	65.743	388.337	452.372	300.963	-1.216
4800	65.656	389.685	453.755	307.533	-0.804
4900	65.571	391.007	455.108	314.095	-0.406
5000	65.488	392.302	456.432	320.648	-0.024
5100	65.408	393.572	457.728	327.192	0.345
5200	65.329	394.818	458.997	333.729	0.701
5300	65.253	396.041	460.241	340.258	1.045
5400	65.178	397.241	461.460	346.780	1.376
5500	65.106	398.420	462.655	353.294	1.697
5600	65.036	399.577	463.828	359.801	2.007
5700	64.968	400.715	464.978	366.301	2.307
5800	64.902	401.832	466.107	372.795	2.598
5900	64.838	402.931	467.216	379.282	2.880
6000	64.776	404.012	468.305	385.763	3.153
6200	64.658	406.120	470.428	398.706	3.676
6400	64.547	408.162	472.479	411.626	4.168
6600	64.443	410.141	474.463	424.525	4.634
6800	64.345	412.061	476.385	437.404	5.074
7000	64.253	413.926	478.249	450.264	5.492
7200	64.167	415.738	480.058	463.105	5.888
7400	64.086	417.500	481.815	475.931	6.265
7600	64.009	419.215	483.523	488.740	6.625
7800	63.938	420.886	485.185	501.535	6.967
8000	63.870	422.513	486.803	514.315	7.294
8200	63.807	424.101	488.379	527.083	7.606
8400	63.747	425.650	489.916	539.839	7.905
8600	63.691	427.162	491.415	552.582	8.191
8800	63.637	428.639	492.879	565.315	8.466
9000	63.587	430.082	494.309	578.037	8.729
9200	63.539	431.494	495.706	590.750	8.982
9400	63.494	432.874	497.072	603.453	9.226
9600	63.452	434.226	498.408	616.148	9.460
9800	63.412	435.549	499.716	628.834	9.685
10000	63.373	436.845	500.997	641.512	9.903

Table A.14. UO_3(g)

$M = 286.0272 \quad UO_3(g) \rightleftharpoons U(g) + 3O(g) \quad \Delta_r H^o(0) = 2068.251\ kJ\ mol^{-1}$

T K	$C_P^o(T)$	$\Phi^o(T)$ $J\,K^{-1}mol^{-1}$	$S^o(T)$	$H^o(T) - H^o(0)$ $kJ\ mol^{-1}$	lg $K^o(T)$
298.15	64.529	258.899	309.694	15.145	-344.863
300	64.649	259.213	310.093	15.264	-342.614
400	70.080	274.441	329.485	22.017	-252.040
500	73.712	287.101	345.541	29.220	-197.646
600	76.117	298.009	359.207	36.719	-161.359
700	77.748	307.617	371.071	44.417	-135.426
800	78.890	316.216	381.532	52.253	-115.968
900	79.713	324.001	390.874	60.185	-100.829
1000	80.323	331.117	399.305	68.188	-88.713
1100	80.787	337.670	406.984	76.245	-78.797
1200	81.146	343.744	414.029	84.342	-70.531
1300	81.431	349.404	420.536	92.471	-63.535
1400	81.659	354.703	426.579	100.626	-57.535
1500	81.845	359.685	432.220	108.802	-52.333
1600	81.998	364.385	437.507	116.994	-47.779
1700	82.126	368.834	442.482	125.200	-43.759
1800	82.235	373.058	447.179	133.419	-40.183
1900	82.327	377.077	451.628	141.647	-36.982
2000	82.408	380.911	455.853	149.884	-34.099
2100	82.479	384.576	459.875	158.128	-31.488
2200	82.543	388.087	463.714	166.379	-29.112
2300	82.603	391.455	467.384	174.637	-26.941
2400	82.661	394.693	470.901	182.900	-24.950
2500	82.719	397.809	474.277	191.169	-23.115
2600	82.780	400.813	477.522	199.444	-21.420
2700	82.845	403.712	480.647	207.725	-19.848
2800	82.916	406.514	483.662	216.013	-18.387
2900	82.997	409.225	486.573	224.309	-17.025
3000	83.088	411.850	489.388	232.613	-15.752
3100	83.192	414.396	492.114	240.927	-14.560
3200	83.312	416.866	494.757	249.252	-13.440
3300	83.448	419.265	497.323	257.589	-12.387
3400	83.603	421.598	499.816	265.942	-11.394
3500	83.777	423.867	502.242	274.311	-10.456
3600	83.973	426.078	504.605	282.698	-9.569
3700	84.191	428.231	506.908	291.106	-8.729
3800	84.431	430.331	509.157	299.537	-7.931
3900	84.696	432.381	511.353	307.993	-7.174
4000	84.984	434.382	513.501	316.477	-6.452

UO_3 (g) (end)

T	$C^o_p(T)$	$\Phi^o(T)$	$S^o(T)$	$H^o(T) - H^o(0)$	$-$ lg $K^o(T)$
K		$J\,K^{-1}mol^{-1}$		$kJ\,mol^{-1}$	
4100	85.296	436.337	515.603	324.991	-5.765
4200	85.632	438.249	517.663	333.537	-5.109
4300	85.991	440.120	519.682	342.118	-4.483
4400	86.372	441.951	521.663	350.736	-3.885
4500	86.776	443.744	523.609	359.393	-3.311
4600	87.200	445.501	525.521	368.092	-2.762
4700	87.644	447.223	527.401	376.834	-2.236
4800	88.106	448.913	529.251	385.621	-1.730
4900	88.585	450.571	531.072	394.455	-1.244
5000	89.078	452.199	532.867	403.338	-0.777
5100	89.584	453.798	534.636	412.271	-0.328
5200	90.102	455.370	536.380	421.256	0.105
5300	90.628	456.914	538.102	430.292	0.523
5400	91.162	458.434	539.801	439.382	0.925
5500	91.700	459.928	541.478	448.525	1.313
5600	92.242	461.399	543.135	457.722	1.688
5700	92.784	462.848	544.773	466.973	2.050
5800	93.325	464.274	546.391	476.278	2.401
5900	93.863	465.680	547.991	485.638	2.740
6000	94.396	467.065	549.573	495.051	3.067
6200	95.438	469.777	552.685	514.035	3.693
6400	96.439	472.415	555.731	533.223	4.281
6600	97.384	474.985	558.714	552.607	4.834
6800	98.265	477.491	561.634	572.173	5.355
7000	99.072	479.936	564.494	591.908	5.848
7200	99.799	482.324	567.296	611.796	6.314
7400	100.442	484.658	570.039	631.822	6.756
7600	100.998	486.940	572.725	651.967	7.175
7800	101.467	489.174	575.355	672.215	7.572
8000	101.849	491.360	577.929	692.548	7.951
8200	102.147	493.503	580.448	712.949	8.311
8400	102.364	495.602	582.912	733.401	8.655
8600	102.504	497.661	585.322	753.890	8.982
8800	102.573	499.680	587.680	774.399	9.295
9000	102.575	501.661	589.985	794.914	9.595
9200	102.516	503.606	592.239	815.425	9.881
9400	102.403	505.515	594.443	835.917	10.155
9600	102.240	507.390	596.597	856.382	10.418
9800	102.033	509.233	598.703	876.810	10.671
10000	101.789	511.043	600.762	897.193	10.913

Table A.15. UO_3^- (g)

$M = 286.0278$ $UO_3^-(g) \rightleftharpoons U(g) + 3O(g) + e(g)$ $\Delta_r H^o(0) = 2268.251\ kJ\ mol^{-1}$

T K	$C_P^o(T)$	$\Phi^o(T)$ $J\ K^{-1} mol^{-1}$	$S^o(T)$	$H^o(T) - H^o(0)$ $kJ\ mol^{-1}$	lg $K^o(T)$
298.15	67.754	267.148	319.736	15.679	-380.322
300	67.876	267.474	320.156	15.805	-377.851
400	73.214	283.273	340.467	22.878	-278.289
500	76.840	296.435	357.217	30.391	-218.456
600	79.466	307.783	371.471	38.213	-178.511
700	81.431	317.786	383.875	46.262	-149.944
800	82.915	326.747	394.850	54.483	-128.495
900	84.041	334.869	404.684	62.833	-111.794
1000	84.904	342.303	413.586	71.282	-98.420
1100	85.579	349.158	421.711	79.808	-87.467
1200	86.120	355.520	429.181	88.394	-78.330
1300	86.568	361.455	436.092	97.029	-70.591
1400	86.950	367.018	442.522	105.705	-63.951
1500	87.284	372.255	448.533	114.417	-58.190
1600	87.583	377.200	454.176	123.161	-53.144
1700	87.854	381.886	459.494	131.933	-48.686
1800	88.102	386.338	464.522	140.731	-44.719
1900	88.329	390.580	469.292	149.553	-41.165
2000	88.538	394.630	473.828	158.396	-37.962
2100	88.730	398.505	478.153	167.260	-35.061
2200	88.905	402.220	482.284	176.142	-32.419
2300	89.063	405.788	486.240	185.040	-30.004
2400	89.206	409.219	490.033	193.954	-27.787
2500	89.334	412.525	493.678	202.881	-25.744
2600	89.447	415.714	497.184	211.820	-23.855
2700	89.545	418.795	500.561	220.770	-22.103
2800	89.630	421.773	503.819	229.729	-20.474
2900	89.702	424.657	506.966	238.695	-18.954
3000	89.761	427.452	510.008	247.669	-17.533
3100	89.808	430.163	512.952	256.647	-16.201
3200	89.844	432.795	515.804	265.630	-14.950
3300	89.870	435.352	518.569	274.616	-13.773
3400	89.885	437.839	521.252	283.603	-12.662
3500	89.891	440.260	523.858	292.592	-11.613
3600	89.889	442.617	526.390	301.581	-10.621
3700	89.879	444.915	528.853	310.570	-9.679
3800	89.861	447.156	531.250	319.557	-8.786
3900	89.836	449.342	533.583	328.542	-7.937
4000	89.805	451.477	535.857	337.524	-7.128

UO_3^- (g) (end)

T K	$C_P^o(T)$	$\Phi^o(T)$ $J\ K^{-1} mol^{-1}$	$S^o(T)$	$H^o(T)-H^o(0)$ $kJ\ mol^{-1}$	lg $K^o(T)$
4100	89.769	453.562	538.075	346.503	-6.357
4200	89.727	455.600	540.237	355.477	-5.621
4300	89.681	457.593	542.348	364.448	-4.918
4400	89.630	459.543	544.409	373.413	-4.245
4500	89.576	461.451	546.423	382.374	-3.601
4600	89.518	463.320	548.391	391.328	-2.983
4700	89.457	465.150	550.315	400.277	-2.390
4800	89.394	466.944	552.198	409.220	-1.821
4900	89.328	468.703	554.041	418.156	-1.274
5000	89.260	470.428	555.845	427.085	-0.747
5100	89.190	472.120	557.612	436.008	-0.240
5200	89.119	473.781	559.343	444.923	0.249
5300	89.047	475.411	561.040	453.832	0.720
5400	88.974	477.012	562.704	462.733	1.175
5500	88.900	478.585	564.336	471.626	1.614
5600	88.825	480.131	565.937	480.513	2.039
5700	88.750	481.650	567.508	489.391	2.449
5800	88.675	483.144	569.051	498.263	2.846
5900	88.599	484.613	570.566	507.126	3.231
6000	88.524	486.058	572.055	515.983	3.604
6200	88.373	488.879	574.955	533.672	4.315
6400	88.224	491.613	577.758	551.332	4.985
6600	88.077	494.265	580.471	568.962	5.617
6800	87.933	496.839	583.098	586.563	6.214
7000	87.791	499.340	585.645	604.135	6.779
7200	87.653	501.772	588.116	621.680	7.315
7400	87.518	504.138	590.516	639.197	7.823
7600	87.387	506.442	592.848	656.687	8.307
7800	87.260	508.687	595.116	674.152	8.768
8000	87.136	510.875	597.324	691.591	9.207
8200	87.017	513.010	599.474	709.007	9.627
8400	86.901	515.094	601.570	726.399	10.027
8600	86.789	517.129	603.613	743.768	10.411
8800	86.681	519.117	605.607	761.115	10.778
9000	86.577	521.061	607.554	778.440	11.130
9200	86.477	522.962	609.456	795.746	11.468
9400	86.379	524.822	611.315	813.031	11.792
9600	86.286	526.643	613.132	830.298	12.104
9800	86.195	528.426	614.910	847.546	12.404
10000	86.108	530.173	616.651	864.776	12.693

Table A.16. U_2O_2 (g)

$M = 508.0568 \quad U_2O_2(g) \rightleftharpoons 2U(g) + 2O(g) \quad \Delta_r H^o(0) = 2033.322\ kJ\ mol^{-1}$

T K	$C_P^o(T)$	$\Phi^o(T)$ $J\ K^{-1} mol^{-1}$	$S^o(T)$	$H^o(T) - H^o(0)$ $kJ\ mol^{-1}$	lg $K^o(T)$
298.15	75.080	280.357	334.927	16.270	-337.803
300	75.252	280.695	335.392	16.409	-335.594
400	81.690	297.297	358.041	24.298	-246.617
500	84.682	311.363	376.633	32.635	-193.206
600	86.121	323.576	392.215	41.183	-157.588
700	86.818	334.357	405.549	49.834	-132.144
800	87.168	343.998	417.167	58.536	-113.058
900	87.377	352.710	427.447	67.264	-98.213
1000	87.545	360.652	436.662	76.010	-86.336
1100	87.713	367.948	445.014	84.772	-76.616
1200	87.896	374.693	452.653	93.553	-68.515
1300	88.096	380.964	459.697	102.352	-61.659
1400	88.315	386.824	466.233	111.173	-55.780
1500	88.554	392.324	472.334	120.016	-50.682
1600	88.816	397.505	478.058	128.884	-46.219
1700	89.107	402.404	483.451	137.780	-42.278
1800	89.432	407.049	488.553	146.707	-38.772
1900	89.793	411.467	493.398	155.668	-35.632
2000	90.193	415.680	498.013	164.667	-32.804
2100	90.633	419.707	502.424	173.708	-30.241
2200	91.113	423.563	506.652	182.795	-27.909
2300	91.632	427.265	510.713	191.931	-25.776
2400	92.188	430.824	514.625	201.122	-23.818
2500	92.776	434.252	518.400	210.370	-22.014
2600	93.393	437.559	522.050	219.678	-20.346
2700	94.037	440.754	525.587	229.050	-18.798
2800	94.701	443.845	529.019	238.486	-17.359
2900	95.383	446.840	532.354	247.990	-16.015
3000	96.077	449.745	535.599	257.563	-14.759
3100	96.780	452.565	538.761	267.206	-13.581
3200	97.487	455.307	541.845	276.919	-12.475
3300	98.195	457.976	544.855	286.704	-11.433
3400	98.899	460.574	547.797	296.558	-10.449
3500	99.596	463.108	550.674	306.483	-9.520
3600	100.282	465.579	553.490	316.477	-8.641
3700	100.953	467.993	556.246	326.539	-7.806
3800	101.607	470.351	558.947	336.667	-7.014
3900	102.239	472.657	561.595	346.860	-6.260
4000	102.848	474.913	564.191	357.114	-5.542

U_2O_2 (g) (end)

T K	$C_P^o(T)$	$\Phi^o(T)$ $J\ K^{-1} mol^{-1}$	$S^o(T)$	$H^o(T)-H^o(0)$ $kJ\ mol^{-1}$	lg $K^o(T)$
4100	103.431	477.121	566.738	367.428	-4.858
4200	103.984	479.285	569.237	377.799	-4.204
4300	104.507	481.405	571.690	388.224	-3.579
4400	104.996	483.485	574.098	398.700	-2.981
4500	105.451	485.525	576.463	409.222	-2.408
4600	105.871	487.527	578.785	419.789	-1.858
4700	106.253	489.493	581.066	430.395	-1.331
4800	106.599	491.424	583.307	441.038	-0.824
4900	106.906	493.322	585.508	451.714	-0.337
5000	107.176	495.187	587.671	462.418	0.132
5100	107.407	497.022	589.796	473.148	0.584
5200	107.601	498.826	591.883	483.898	1.019
5300	107.759	500.601	593.934	494.667	1.440
5400	107.880	502.348	595.950	505.449	1.845
5500	107.966	504.068	597.930	516.241	2.236
5600	108.018	505.762	599.876	527.041	2.615
5700	108.038	507.429	601.788	537.844	2.981
5800	108.026	509.073	603.667	548.648	3.335
5900	107.984	510.692	605.513	559.448	3.678
6000	107.915	512.287	607.328	570.243	4.010
6200	107.697	515.410	610.863	591.806	4.644
6400	107.386	518.447	614.277	613.316	5.241
6600	106.995	521.401	617.576	634.755	5.803
6800	106.538	524.277	620.764	656.110	6.335
7000	106.025	527.078	623.845	677.367	6.838
7200	105.469	529.807	626.824	698.517	7.314
7400	104.880	532.469	629.705	719.552	7.767
7600	104.266	535.064	632.494	740.467	8.196
7800	103.637	537.597	635.195	761.258	8.605
8000	102.999	540.070	637.810	781.921	8.994
8200	102.357	542.485	640.346	802.457	9.366
8400	101.717	544.845	642.805	822.864	9.721
8600	101.084	547.151	645.191	843.144	10.060
8800	100.460	549.405	647.507	863.299	10.384
9000	99.848	551.611	649.758	883.329	10.695
9200	99.251	553.768	651.946	903.239	10.993
9400	98.670	555.880	654.074	923.031	11.278
9600	98.105	557.947	656.146	942.708	11.553
9800	97.559	559.972	658.163	962.274	11.816
10000	97.032	561.955	660.129	981.733	12.069

Table A.17. U_2O_3 (g)

$M = 524.0562 \quad U_2O_3(g) \rightleftharpoons 2U(g) + 3O(g) \quad \Delta_r H^o(0) = 2661.117 \ kJ \ mol^{-1}$

T K	$C_P^o(T)$	$\Phi^o(T)$ $J \ K^{-1} mol^{-1}$	$S^o(T)$	$H^o(T) - H^o(0)$ $kJ \ mol^{-1}$	lg $K^o(T)$
298.15	92.808	299.831	372.996	21.814	-441.570
300	92.917	300.284	373.571	21.986	-438.681
400	98.052	322.173	401.041	31.547	-322.317
500	101.730	340.246	423.341	41.547	-252.460
600	104.306	355.702	442.129	51.857	-205.871
700	106.166	369.234	458.355	62.385	-172.586
800	107.592	381.284	472.628	73.075	-147.619
900	108.761	392.154	485.370	83.895	-128.198
1000	109.778	402.060	496.883	94.822	-112.661
1100	110.702	411.165	507.390	105.847	-99.948
1200	111.567	419.592	517.060	116.961	-89.353
1300	112.391	427.439	526.022	128.159	-80.387
1400	113.185	434.782	534.381	139.438	-72.701
1500	113.960	441.686	542.216	150.796	-66.038
1600	114.719	448.202	549.595	162.230	-60.206
1700	115.469	454.373	556.573	173.739	-55.059
1800	116.213	460.236	563.194	185.323	-50.481
1900	116.953	465.822	569.497	196.981	-46.384
2000	117.691	471.158	575.515	208.714	-42.694
2100	118.426	476.265	581.275	220.519	-39.353
2200	119.158	481.165	586.801	232.399	-36.314
2300	119.885	485.874	592.113	244.351	-33.536
2400	120.604	490.408	597.231	256.375	-30.988
2500	121.313	494.780	602.169	268.471	-28.642
2600	122.008	499.003	606.940	280.638	-26.474
2700	122.685	503.086	611.558	292.872	-24.464
2800	123.340	507.041	616.031	305.174	-22.595
2900	123.970	510.874	620.370	317.540	-20.853
3000	124.572	514.595	624.583	329.967	-19.225
3100	125.142	518.209	628.678	342.453	17.700
3200	125.678	521.724	632.659	354.994	-16.269
3300	126.177	525.144	636.534	367.587	-14.921
3400	126.639	528.476	640.308	380.228	-13.651
3500	127.061	531.724	643.985	392.914	-12.452
3600	127.442	534.892	647.570	405.639	-11.317
3700	127.783	537.985	651.066	418.401	-10.242
3800	128.083	541.006	654.478	431.194	-9.222
3900	128.342	543.959	657.809	444.016	-8.252
4000	128.562	546.846	661.061	456.861	-7.329

U_2O_3 (g) (end)

T K	$C_P^o(T)$	$\Phi^o(T)$ $J\,K^{-1}mol^{-1}$	$S^o(T)$	$H^o(T)-H^o(0)$ $kJ\,mol^{-1}$	lg $K^o(T)$
4100	128.743	549.670	664.238	469.727	-6.449
4200	128.887	552.435	667.342	482.609	-5.610
4300	128.996	555.143	670.376	495.503	-4.808
4400	129.070	557.796	673.343	508.407	-4.042
4500	129.112	560.396	676.244	521.316	-3.307
4600	129.123	562.945	679.081	534.228	-2.604
4700	129.107	565.446	681.858	547.140	-1.929
4800	129.064	567.899	684.576	560.049	-1.281
4900	128.996	570.308	687.237	572.952	-0.658
5000	128.906	572.672	689.842	585.847	-0.058
5100	128.796	574.995	692.394	598.732	0.518
5200	128.668	577.277	694.893	611.606	1.074
5300	128.523	579.519	697.343	624.465	1.610
5400	128.363	581.723	699.744	637.310	2.126
5500	128.189	583.891	702.098	650.137	2.625
5600	128.004	586.022	704.406	662.947	3.107
5700	127.809	588.119	706.670	675.738	3.573
5800	127.605	590.182	708.891	688.509	4.024
5900	127.393	592.213	711.070	701.259	4.460
6000	127.175	594.212	713.209	713.987	4.882
6200	126.725	598.118	717.372	739.378	5.689
6400	126.262	601.908	721.388	764.676	6.447
6600	125.792	605.587	725.266	789.882	7.161
6800	125.321	609.163	729.015	814.993	7.836
7000	124.854	612.639	732.641	840.011	8.474
7200	124.392	616.022	736.151	864.935	9.078
7400	123.941	619.315	739.553	889.768	9.651
7600	123.500	622.522	742.853	914.512	10.196
7800	123.072	625.649	746.055	939.169	10.713
8000	122.658	628.698	749.166	963.742	11.206
8200	122.258	631.674	752.190	988.233	11.676
8400	121.873	634.578	755.131	1012.646	12.125
8600	121.504	637.415	757.995	1036.983	12.554
8800	121.149	640.187	760.784	1061.249	12.964
9000	120.809	642.898	763.503	1085.444	13.357
9200	120.484	645.548	766.154	1109.573	13.733
9400	120.173	648.142	768.742	1133.639	14.094
9600	119.876	650.681	771.269	1157.644	14.440
9800	119.592	653.167	773.738	1181.591	14.773
10000	119.321	655.603	776.151	1205.481	15.093

Table A.18. U_2O_4 (g)

$M = 540.0556 \quad U_2O_4(g) \rightleftharpoons 2U(g) + 4O(g) \quad \Delta_r H^o(0) = 3422.912\ kJ\ mol^{-1}$

T K	$C_p^o(T)$	$\Phi^o(T)$ $J\ K^{-1}mol^{-1}$	$S^o(T)$	$H^o(T)-H^o(0)$ $kJ\ mol^{-1}$	lg $K^o(T)$
298.15	104.142	287.010	352.304	19.468	-567.128
300	104.485	287.414	352.949	19.660	-563.406
400	118.645	307.929	385.138	30.883	-413.473
500	126.392	326.184	412.537	43.176	-323.470
600	130.371	342.579	435.974	56.037	-263.456
700	132.391	357.402	456.239	69.186	-220.588
800	133.485	370.889	473.996	82.485	-188.438
900	134.170	383.238	489.760	95.870	-163.435
1000	134.690	394.610	503.924	109.314	-143.434
1100	135.157	405.141	516.783	122.806	-127.071
1200	135.619	414.943	528.563	136.345	-113.435
1300	136.096	424.106	539.437	149.931	-101.898
1400	136.593	432.709	549.541	163.565	-92.008
1500	137.108	440.816	558.983	177.250	-83.436
1600	137.634	448.481	567.848	190.987	-75.935
1700	138.164	455.751	576.208	204.777	-69.314
1800	138.689	462.665	584.120	218.619	-63.428
1900	139.203	469.257	591.632	232.514	-58.159
2000	139.699	475.556	598.785	246.459	-53.416
2100	140.170	481.588	605.613	260.453	-49.122
2200	140.612	487.375	612.144	274.492	-45.216
2300	141.022	492.936	618.403	288.574	-41.648
2400	141.397	498.290	624.413	302.696	-38.375
2500	141.735	503.451	630.192	316.853	-35.361
2600	142.037	508.434	635.757	331.042	-32.577
2700	142.303	513.249	641.123	345.259	-29.997
2800	142.533	517.909	646.303	359.501	-27.599
2900	142.730	522.423	651.308	373.764	-25.364
3000	142.894	526.801	656.149	388.046	-23.275
3100	143.028	531.049	660.837	402.342	-21.319
3200	143.134	535.177	665.380	416.651	-19.483
3300	143.214	539.189	669.786	430.968	-17.757
3400	143.271	543.093	674.062	445.293	-16.129
3500	143.305	546.895	678.216	459.622	-14.592
3600	143.321	550.599	682.253	473.953	-13.139
3700	143.318	554.211	686.180	488.285	-11.762
3800	143.300	557.734	690.002	502.616	-10.455
3900	143.268	561.173	693.723	516.945	-9.214
4000	143.223	564.533	697.350	531.269	-8.033

U_2O_4 (g) (end)

T K	$C_P^o(T)$	$\Phi^o(T)$ $J\ K^{-1} mol^{-1}$	$S^o(T)$	$H^o(T)-H^o(0)$ $kJ\ mol^{-1}$	lg $K^o(T)$
4100	143.168	567.815	700.886	545.589	-6.907
4200	143.103	571.025	704.335	559.903	-5.833
4300	143.029	574.164	707.702	574.209	-4.808
4400	142.949	577.237	710.989	588.508	-3.827
4500	142.862	580.245	714.200	602.799	-2.888
4600	142.770	583.191	717.339	617.081	-1.989
4700	142.674	586.078	720.409	631.353	-1.126
4800	142.573	588.908	723.411	645.615	-0.298
4900	142.470	591.683	726.350	659.867	0.497
5000	142.364	594.406	729.227	674.109	1.263
5100	142.256	597.077	732.046	688.340	1.999
5200	142.146	599.699	734.807	702.560	2.709
5300	142.035	602.274	737.513	716.769	3.393
5400	141.923	604.803	740.167	730.967	4.052
5500	141.811	607.288	742.770	745.154	4.689
5600	141.698	609.730	745.325	759.329	5.304
5700	141.586	612.131	747.832	773.493	5.898
5800	141.473	614.492	750.293	787.646	6.473
5900	141.361	616.814	752.711	801.788	7.029
6000	141.250	619.099	755.085	815.919	7.568
6200	141.029	623.561	759.713	844.146	8.595
6400	140.811	627.886	764.187	872.330	9.562
6600	140.598	632.082	768.517	900.471	10.472
6800	140.389	636.157	772.711	928.570	11.331
7000	140.186	640.117	776.778	956.627	12.144
7200	139.988	643.968	780.724	984.645	12.913
7400	139.795	647.716	784.557	1012.623	13.642
7600	139.608	651.367	788.283	1040.563	14.335
7800	139.426	654.924	791.907	1068.466	14.994
8000	139.250	658.393	795.434	1096.334	15.621
8200	139.079	661.777	798.871	1124.166	16.219
8400	138.913	665.082	802.220	1151.966	16.790
8600	138.753	668.309	805.487	1179.732	17.335
8800	138.598	671.463	808.675	1207.467	17.857
9000	138.448	674.547	811.788	1235.172	18.356
9200	138.303	677.564	814.830	1262.847	18.834
9400	138.163	680.516	817.802	1290.493	19.293
9600	138.027	683.406	820.710	1318.112	19.733
9800	137.896	686.238	823.554	1345.705	20.156
10000	137.769	689.012	826.339	1373.271	20.563

A.3 Estimated Molecular and Ionic Interaction Constants

This section contains some theoretical considerations used in the initial estimations[7] of the particle interaction constants used as a *first guess* in the procedure of the *EOS* constant fitting. Parameterization of the *EOS* is made in terms of a number of molecular constants pertaining to the considered atomic, molecular and ionic species, which are assumed to compose fluid UO_2. Yet, only for certain species are these constants available as empirical data. For others, they must be deduced through an acceptable extrapolation.

In order to establish a physical perspective on this problem let us assume the case of the van der Waals model. At low densities the pressure P can be presented in the virial expansion form:

$$P = nRT(1 + B_2(T)n + B_3(T)n^2 + \ldots). \tag{A.31}$$

Here $B_i(T)$ are virial coefficients, which are functions of temperature only. Their values can be determined if the potential energy of the intermolecular interaction is known. In particular, the second virial coefficient $B_2(T)$ is related to the pair interaction potential Φ, which depends on the relative position of two particles. Let $\Phi_{ij}(r, \omega)$ be the interaction potentials of the i-th and j-th particles in a M-component mixture (r is the intermolecular distance and ω is the angular variable), then:

$$B_2 = \sum_{i=1}^{M} \sum_{j=1}^{M} x_i x_j B_2^{(i,j)}(T), \tag{A.32}$$

where x_i and x_j are the mole fractions of the component i and j, respectively, and

$$B_2^{(i,j)} = -2\pi N_\omega \int d\omega \int d\mathbf{r} \left[\exp\left(-\frac{\Phi_{ij}(\mathbf{r},\omega)}{kT} \right) - 1 \right], \tag{A.33}$$

where k is the Boltzmann constant, $N_\omega = N_A / \int d\omega$, and N_A the Avogadro number. Eqs.(A.32) and (A.33) are well-known formulae of statistical mechanics of imperfect gas [71] valid only for neutral species. Since at high temperatures we have:

$$\Phi_{ij}/kT \ll 1 \tag{A.34}$$

Eq.(A.33) can be simplified:

$$B_2^{(i,j)} \approx b_{ij} - a_{ij}/RT\ , \tag{A.35}$$

where

[7] The material of this Section was originally presented by A.M.Semenov (Power Engineering Institute (Technical University), Moscow, Russia) within the *INTAS* 99-066 Project [77].

$$b_{ij} = (2/3)\pi N_A \sigma_{ij}^3; \quad a_{ij} = 2\pi N_\omega \int d\omega \int_{\sigma_{ij}}^{\infty} r^2 dr \left[-\Phi_{ij} + \frac{1}{2kT} (\Phi_{ij})^2 \right]. \quad \text{(A.36)}$$

By taking into account Eq.(A.35), we can write Eq.(A.33) in the form: $B_2 \approx b - a/RT$ Eq.(A.35) with:

$$b = \sum_{i=1}^{M} \sum_{j=1}^{M} x_i x_j b_{ij}; \quad a = \sum_{i=1}^{M} \sum_{j=1}^{M} x_i x_j a_{ij}. \quad \text{(A.37)}$$

On the other hand, at the zero density limit the *EOS* has the form of the virial expansion of Eq.(A.31), and:

$$B_2 = b - \frac{a}{RT} \quad \text{(A.38)}$$

as in Eq.(A.36). Consequently, Eqs.(A.36) and (A.37) represent a rough approximation for the parameters a and b of the *EOS* Eq.(2.39). In the general case of a long-range molecular interaction, three additional contributions, electrostatic Φ^{es}, induced Φ^{ind} (polarisation) and dispersion Φ^{disp} [71], must be included:

$$\Phi_{ij} = \Phi_{ij}^{es} + \Phi_{ij}^{ind} + \Phi_{ij}^{disp}. \quad \text{(A.39)}$$

For electrically neutral particles i and j the following expansions hold:

$$\Phi_{ij}^{es} = -\frac{D_i D_j}{r^3} \varphi_{ij}^{(DD)}(\omega_i, \omega_j) - \frac{D_i Q_j}{r^5} \varphi_{ij}^{(DQ)}(\omega_i, \omega_j) - \frac{Q_i Q_j}{r^5} \varphi_{ij}^{(QQ)}(\omega_i, \omega_j) + ...$$

$$\Phi_{ij}{}^{disp} = -\frac{C_6{}^{(i,j)}}{r^6} - \frac{C_8{}^{(i,j)}}{r^8} + \ldots$$

If one of the pair particles has a charge $e\, Z_i$, then the additional contribution to the induced interaction is:

$$\Phi_{ij}^{ind} = -\frac{e^2 Z_i^2 \alpha_j}{2r^4}. \quad \text{(A.40)}$$

Substituting Eq.(A.39) into Eq.(A.37) and leaving in the integrand Eq.(A.37) only the dominant terms of the expansion, leads to the approximation:

$$a_{ij} = \frac{2}{3}\pi N_A \frac{1}{\sigma_{ij}^3} (C_6^{(i,j)} + D_i^2 \alpha_j + D_j^2 \alpha_i + \frac{D_i^2 D_j^2}{3kT}). \quad \text{(A.41)}$$

Eq.(A.40) gives the following additional contribution to Eq.(A.41):

$$a_{ij} = 2\pi N_A^2 \frac{e^2 Z_i^2 \alpha_j}{\sigma_{ij}}. \quad \text{(A.42)}$$

In these expansions, D_i and Q_i are the dipole and quadrupole moments, α_i is the polarisability of the i-th particle; $C_6^{(i,j)}$ and $C_8^{(i,j)}$ are the constants of the dispersion interaction; $\omega = \{\vartheta_i, \varphi_i\}$, ϑ_i and φ_i are the polar and azimuthal angles between the i-th molecule axis and the direct line joining the i-th and j-th molecules centres; $\varphi^{(DQ)}$ and $\varphi^{(QQ)}$ are functions of the angular variables corresponding to $\varphi^{(DD)}$. Yet, this approach is quite contraversial because it is based on Eqs.(A.39) and (A.40), which are valid for long-ranged intermolecular forces only, and are extrapolated here to short intermolecular distances. If, however, some information is available on the molecular interaction at intermediate intermolecular distances, we may still attempt to use them.

A.3.1 Polarisability of Atomic O and U

The polarisability of *atomic oxygen* is known both from experiment and from quantum-mechanical calculations. In many reference books (see, for example, [113]), the value $\alpha_o = 5.41(5)$ is given. (hereafter all values are given in *a.u.*, unless other units are indicated). For atomic uranium the exact value of polarisability is unknown, however, it is possible to make some estimations by using:

(i) Slater's rules for atomic polarisability [71]: $\alpha = \alpha_S$,

(ii) the equation $\alpha_r = (r_S)^3$, where r_S is the Slater radius of the atom [71].

(iii) the relationship, which follows, respectively, from the London and Slater- Kirkwood equations for the dispersion constant C_6 [71, 84]:

$$C_6^{(i,j)} = \frac{3}{2}\alpha_i\alpha_j\frac{\mathcal{E}_i\mathcal{E}_j}{\mathcal{E}_i + \mathcal{E}_j}; \tag{A.43}$$

$$C_6^{(i,j)} = \frac{3}{2}\alpha_i\alpha_j\left(\sqrt{\frac{\alpha_i}{N_i^e}} + \sqrt{\frac{\alpha_j}{N_j^e}}\right)^{-1}. \tag{A.44}$$

The parameter $\mathcal{E}_i$ in Eq.(A.43) is either the ionisation potential I_i, or a resonance excitation energy level ΔE_i of the i-th atom, and N^e in Eq.(A.44) is the corresponding number of valence electrons. The equations can be extended to non-identical atoms. Comparing the right hand sides of Eqs.(A.43) and (A.44), gives:

$$\alpha = N_e/\mathcal{E}^2. \tag{A.45}$$

According to the definition of $\mathcal{E}$, Eq.(A.45) yields: $\alpha_{\mathcal{E}} = N_e/\Delta E^2$; $\alpha_I = N_e/I^2$. Values of $\alpha^{(sol)}$, α_r , α_I and $\alpha_{\mathcal{E}}$ are presented in Table A.19 for 20 atoms, for which experimental data, α_{exp}, are available [113], as well as for several atoms, including uranium, for which no experimental data exist. In this Table are also shown the electronic configurations of the atoms and the variation in the number of valence electrons, N_e. The contributions to this value from the outer electronic shells s^2, p^6, and so on, are taken to be 1, 5... *etc.*. For some atoms (in particular, for uranium) the excitation energies, ΔE, are uncertain.

One can see from Table A.19 that there is a considerable disagreement between different estimations of α for the same atom. In most cases it is

Table A.19. Polarisability of Atoms

Atom	Electronic configuration	Number of valence electrons	α_S	α_r	α_I	$\alpha_\mathcal{E}$	α_{Exp}
H	$1s$	1	4.0	1.0	4.0	7.1	4.5
Li	$2s$	1	140	29	26	216	162
Na	$3s$	1	303	69	28	168	162
K	$4s$	1	1401	241	39	286	287
Rb	$5s$	1	2496	385	42	304	310
Cs	$6s$	1	3595	515	256	386	385
B	$2s^2 2p$	1	26	3.6	11	58	20.5
Al	$3s^2 3p$	1	142	17	21	75	56
C	$2s^2 2p^2$	2	14	2	12	927	11.08
Si	$3s^2 3p^2$	2	96	10	22	2428	37
O	$2s^2 2p^4$	4	5.6	0.7	16	35	5.41
Sc	$3d^1 4s^2$	1+(2-1) = 2	895	95	34	726	140
Ti	$3d^2 4s^2$	2+(2-1) = 3	743	82	48	3362	148
Fe	$3d^6 4s^2$	6+(2-1) = 7	357	49	84	7025	88.0
Co	$3d^7 4s^2$	7+(2-1) = 8	303	43	96	6957	74.0
He	$1s^2$	2-1 = 1	0.96	0.2	1.2	1.9	1.38
Ne	$2s^2 2p^6$	6-1 = 5	2.7	0.3	8.0	13	2.68
Ar	$3s^2 3p^6$	6-1 = 5	27	2.4	15	28	11.08
Kr	$4s^2 4p^6$	6-1 = 5	58	5	19	38	16.74
Xe	$5s^2 5p^6$	6-1 = 5	107	7	25	54	27.06
La	$5d^1 6s^2$	1+(2-1) = 2	2783	203	48	13532	
Th	$3d^2 7s^2$	2+(2-1) = 3	2370	152	68		
Pu	$5f^6 7s^2$	6+(2-1) = 7	2149	205	141		
U	$5f^3 \ldots 6d^1 7s^2$	3+1+(2-1) = 5	2771	176	96		

not possible to use the best estimation method. For uranium (and for most of other atoms) α_S is strongly over-estimated, whilst α_r and α_I differ by a factor of two. This difference gives an idea of the error of the estimation of $\alpha_U \approx \alpha_r$:

$$\alpha_U = 175 \pm 100.$$

A.3.2 Polarisability of U_mO_n Molecules

Intuitively, we can expect that the polarisability of polyatomic molecules is a combination of the atomic polarisabilies α_i. Two hypotheses can be examined:

1. The polarisabilities are additive,

$$\alpha \approx \alpha_{ad} = \sum_{i=1}^{N_e} \alpha_i. \quad \text{(A.46)}$$

or

2. the squared polarisabilities are additive,

$$\alpha \approx \alpha_{sq} = \left(\sum_{i=1}^{N_M} \alpha_i^2 \right)^{1/2} , \tag{A.47}$$

where N_M is the number of atoms in the molecule.

In Table A.20 are presented the polarisabilities of 24 polyatomic molecules calculated from Eqs.(A.46) and (A.47), and compared with experimental data [113]. One can see that the estimated α_{ad} are often greater than the experimental values, whilst α_{sq} are smaller. The mean value is thus empirically recommendable:

$$\alpha = \frac{1}{2} \left(\alpha_{ad} + \alpha_{sq} \right) . \tag{A.48}$$

The average deviation is about 15–20%.

Evaluations for U_mO_n molecules based on Eqs. (A.46)-(A.48), with $\alpha_{_O}$ and $\alpha_{_U}$ as in Section A.3, give:

$$\alpha_{_{UO}} \approx \alpha_{UO_2} \approx \alpha_{UO_3} = 180(100).$$

The polarisability of diatomic oxygen is [113]:

$$\alpha_{O_2} = 10.79\,(1)\,.$$

The recommended polarisabilities of O and U atoms and of U_mO_n-type molecules are presented in Table A.21.

A.3.3 Dispersion Constants

For two identical atoms (O or U) both Eqs.(A.43) and (A.44) given Eq.(A.45), give the same result: $C_6 = (3/4)(N_e)^2/\mathcal{E}^3$. However, for other particles, whose polarisabilities are evaluated on the basis of $\alpha_{_O}$ and $\alpha_{_U}$, the results for $C_6^{(i,j)}$ from Eqs.(A.43) and (A.44) are different. The values given by Eq.(A.44) are, in our opinion, more reliable. These values are presented in Table A.22. For the evaluation we used the variant $\mathcal{E} = I$, since the excitation energies for U_mO_n are uncertain. The values of the ionisation potentials for all particles are adopted as in [63, 113]. The valence electron numbers are: 6 for U, 2 for O, 1 for O_2, 4 for UO, 2 for UO_2, and 1 for UO_3. The accuracy of data from Table A.22 is unfortunately very low (probably within a factor of two).

A.3.4 Dipole Moments of U_mO_n Molecules

Atomic U and O, as well as homonuclear diatomic molecules have no dipole moment. The UO_2 molecule has a centre of symmetry (see, for example, [63]),

Table A.20. Polarisabilities of Molecules

Molecule	α_{ad}	α_{sq}	α	α_{exp}
H_2	9.0	6.4	7.7	5.33
N_2	15.0	10.6	12.8	11.88
O_2	10.8	7.7	9.2	10.79
Cl_2	30.0	21.2	25.6	31.11
HF	8.3	5.9	7.1	16.60
HCl	19.5	15.7	17.6	17.75
HBr	30.5	26.4	28.4	24.36
HJ	28.5	24.4	26.5	36.7
N_2O	20.4	11.9	16.2	20.25
CO	17.2	13.1	15.1	13.16
CO_2	22.6	14.0	18.3	17.88
SO_2	30.8	21.4	26.1	25.10
H_2S	29.0	20.1	25.0	25.50
CS_2	51.8	30.6	41.2	58.98
NH_3	21.0	10.8	15.9	15.25
$(CN)_2$	38.6	19.8	29.2	33.8
HCN	23.8	14.7	19.2	17.48
CH_4	29.8	14.8	22.3	17.55
C_2H_6	50.6	20.0	35.3	30.17
CH_3Cl	40.3	20.6	30.5	30.77
CH_2Cl_2	50.8	27.7	39.3	43.73
$CHCl_3$	61.3	28.9	45.1	55.54
CCL_4	71.8	32.2	52.0	70.86
CH_3OH	35.2	15.8	25.5	21.8

Table A.21. Polarisability of U_mO_n Particles

O	O_2	U	UO	UO_2	UO_3
5.41	10.79	175	180	180	180

Table A.22. Dispersion Coefficients $C_6^{(i,j)}$ for Interaction of U_mO_n Particles

	O	O_2	U	UO	UO_2	UO_3
O	13	16	200	200	100	100
O_2		21	300	300	200	200
U			4000	4000	3000	3000
UO				4000	3000	2000
UO_2					2000	2000
UO_3						2000

and, hence, its dipole moment is also zero. Both UO and UO_3 molecules have non-zero dipole moments, but their values are unknown. A.M.Semenov [77] suggested the following empirical formula for a rough estimation of the dipole moment for diatomic molecules AB:

$$D = er_e \frac{|\alpha_A - \alpha_B|}{\alpha_A + \alpha_B}, \tag{A.49}$$

where e is the electron charge, r_e is the equilibrium intermolecular distance in the molecule, and α_A and α_B are the polarisabilities of the A and B atoms. For some polar diatomic molecules with ionic bonds and strongly different polarisabilities, this formula gives quite good results (see Table A.23). For the UO molecule ($r_e = 1.82$ Å [63]) the estimation Eq.(A.49) gives: $D_{UO} \approx 7\ Db$. This value is probably underestimated, since the charge separation in UO may be larger than those in the molecules presented in Table A.23. The error of D_{UO} may be within a factor of two.

Table A.23. Dipole Moments of Some Diatomic Molecules

Molecule	Estimation Eq.(A.49)	Experimental data
NaF	7.5	8.16
LiF	6.1	6.33
CsF	9.4	7.88
CsCl	11.0	10.39
RbF	9.1	8.55
RbI	11.1	11.48

The molecule UO_3 has a T-form structure, and the U-O distances are the same as in UO (1.79 and 1.76 Å). Therefore, we have adopted:

$$D_{UO_3} \approx D_{UO} \approx 3^{+3}_{-1}.$$

A.3.5 Sizes of U_mO_n Molecules

As sources for the evaluation of the effective diameter of a molecule one can use:

1. spectroscopic and structural data on interatomic distances in this molecule,
2. experimental data on the density, $\rho^{(sol)}$, of a pure substance, build from these molecules, in the condensed (usually crystalline) state:

$$\rho^{(sol)} = \eta m / \left(\frac{1}{6} \pi \sigma^3 \right),$$

   where m is the mass of the molecule, η the packing parameter (for the close packing of hard spheres $\eta = 0.74$). Hence it is possible to obtain an estimate of molecule-in-crystal diameters:

$$\sigma^{(sol)} = \left(\frac{6\eta m}{\pi \rho^{(sol)}} \right)^{1/3}. \qquad \text{(A.50)}$$

3. Finally, the molecular size can be deduced from the short-range potential of molecular interaction in the case of simple substances.

<u>Molecule O_2</u>

The equilibrium distance between the oxygen atoms in this molecule is [113], [63]:

$$r_e(O_2) = 1.21 \text{ Å}.$$

If the O atoms are taken as spheres of diameter $\sigma_O = r_e(O_2)$, then the 'spectroscopic diameter' of the O_2 molecule is given by:

$$\sigma_{O_2} = 2r_e(O_2) = 2.4 \text{ Å}. \qquad \text{(A.51)}$$

The density of solid oxygen at $T = 21\ K$ is $\rho^{(sol)} = 1425\ kg\ m^{-3}$ [127]. From Eq.(A.50), with $\eta = 0.74$, we obtain:

$$\sigma_{O_2}^{(sol)} = 3.7 \text{ Å}. \qquad \text{(A.52)}$$

If the interaction of oxygen molecules may be described by means of a Lennard-Jones potential Eq.(2.47), the value of the parameter σ can be obtained from experimental data on viscosity and thermal conductivity of gaseous oxygen over a wide interval of temperature [71],[85]:

$$\sigma_{O_2} = 3.6 \text{ Å}. \qquad \text{(A.53)}$$

The estimations Eqs.(A.52) and (A.53) are in good agreement and are correctly correlated: in fact, the effective diameter of molecule-in-crystal $\sigma^{(sol)}$ (A.52) must satisfy the inequality:

$$\sigma < \sigma^{(sol)} < r_m,$$

where σ is potential parameter of Eq.(A.52), and r_m is the equilibrium intermolecular distance, realised at the minimum of the $\Phi_2(r)$ function; $\Phi_2(r_m) = -\varepsilon$. From Eq.(A.52), $r_m = 2^{1/6}\sigma = 1.122\sigma = 4.04$ Å. However, the value from Eq.(A.52) is strongly under-estimated and not suitable as an estimate of the parameter σ. We have thus adopted $\sigma_{O_2} = 3.6$ Å.

Atom O

We obtained in the preceeding paragraph:

$$\sigma_O^{(sol)} = r_e(O_2) = 1.21 \text{ Å}. \tag{A.54}$$

Another estimate is based on the Slater radius of this atom, $\overline{r_O} = 0.89$ Å (see Table A.19):

$$\sigma_{O\,S} = 2\overline{r_O} = 1.8 \text{ Å}. \tag{A.55}$$

The value of Eq.(A.54) is the size of an O atom-in-molecule, while Eq.(A.55) gives the value of a free O atom (which is greater than in a bound state). Another estimate can be obtained from Eq.(A.53), with simple geometrical considerations:

$$\sigma_O^{(sol)} = \frac{1}{2}\sigma_{O_2}^{(sol)} = 1.8 \text{ Å}. \tag{A.56}$$

The coincidence of the above estimates is probably accidental. All attempts to directly establish the size of an oxygen atom were unsuccessful, since atomic oxygen does not exist as individual substance. Finally, we adopted:

$$\sigma_O = 1.8 \text{ Å}.$$

Atom U

The Slater radius of U-atom is (see Table A.19), $\overline{r_U} = 5.6$ Å. The diameter of a free uranium atom is thus:

$$\sigma_{U\,S} = 2\overline{r_U} = 11.2 \text{ Å}. \tag{A.57}$$

This large value is, of course, most unlikely a reasonable estimate of the potential parameter σ_U. The atom-in-crystal diameter which follows from Eq.(A.50) and the density $\rho^{(sol)} = 19040$ kg m^{-3} of crystalline uranium at 20 $^\circ C$ [45] is:

$$\sigma_U^{(sol)} = 3.0 \text{ Å}. \tag{A.58}$$

The estimation of Eq.(A.58) is based on the assumption of close-packing (η = 0.74). The Ashcroft's rule: $\eta \approx 0.5$ [7] is considered as more adequate for liquid and solid metals. With this value of η, σ_U should decrease by 12% below the value of Eq.(A.58). It is similar to the case of O_2 considered above: the interaction potential parameter σ Eq.(A.53) is smaller than the 'crystalline' diameter of the molecule Eq.(A.52), evaluated for close-packing structure. Like to the O_2 case, we reduced the value of the molecule-in-crystal diameter $\sigma^{(sol)}$ Eq.(A.50) by 2-3% from the estimate based on the potential. Extending this approach to uranium, we finally obtained from Eq.(A.58):

$$\sigma_U = 2.9 \ \text{Å}.$$

Molecule UO_2

This molecule has a linear symmetrical shape, with distances between U and O atoms, r(U-O) = 1.80(5) Å [63]. If we adopt $r(\text{U-O}) = r_U + r_O$, where r_U and r_O are the atomic radii in molecule, and assume, on the basis of Eq.(A.55) and (A.57), that $r_U/r_O = 11.2/1.8 = 6.2$, then $r_U = 1.55$ Å and $r_O = 0.25$ Å. Consequently, taking into account the form of the UO_2 molecule, we can write:

$$2r_U < \sigma_{UO_2} < 2[r(U-O) + r_O],$$

or

$$3.1 \ \text{Å} < \sigma_{\mathrm{UO_2}} < 4.1 \ \text{Å}. \qquad \text{(A.59)}$$

The density of crystal UO_2 is $\rho^{(sol)} = 10950 \ kg \ m^{-3}$ [45]. Then from eq.(A.50) for $\eta = 0.74$ the diameter-in-crystal is:

$$\sigma^{(sol)}_{UO_2} = 3.9 \ \text{Å}. \qquad \text{(A.60)}$$

After reducing this value by 2-3%, we find the estimate for the UO_2 interaction potential parameter σ:

$$\sigma_{UO_2} = 3.8 \ \text{Å}.$$

This is in good agreement with the inequality Eq.(A.59).

Molecule UO_3

In the T-shaped molecule UO_3, the distances between U and O are [63]: $r(\text{U-O}^{(1)}) = r(\text{U-O}^{(2)}) = 1.76$ Å; $r(\text{U-O}^{(3)}) = 1.79$ Å. Proceeding as we did for UO_2, we find:

$$r\left(U - O^{(3)}\right) + r_U + r_O < \sigma^{(sol)}_{UO_3} < r\left(U - O^{(1)}\right) + r\left(U - O^{(2)}\right) + 2r_O.$$

Thus

$$3.6 < \sigma^{(sol)}_{UO_3} < 4.1 \ \text{Å}. \qquad \text{(A.61)}$$

The crystal density of UO_3 is $\rho^{(sol)} = 8020 \ kg \ m^{-3}$ for the α-modification, and $\rho^{(sol)} = 8340 \ kg \ m^{-3}$ for β-modification [45]. Then the diameter of the molecule-in-crystal, again for $\eta = 0.74$, is:

$$\sigma^{(sol)}_{UO_3} = 4.3 \ \text{Å}. \qquad \text{(A.62)}$$

The decrease of this value by 5%: $\sigma_{UO_3} = 4.1$ Å, brings the potential parameter in accordance with the inequality of Eq.(A.61).

Molecule UO

The equilibrium distance r(U-O) in this molecule is $r_e = 1.82(5)$ Å [63]. Thus:

$$r(U-O) < \sigma_{UO} < r(U-O) + r_U + r_O,$$

or

$$1.8 \text{ Å} < \sigma_{UO}^{(sol)} < 3.6 \text{ Å}. \tag{A.63}$$

UO does not exist as a crystal. If we plot the virtual 'crystalline' diameters $\sigma^{(sol)}$ of the clusters UO_n from Eqs.(A.58), (A.60), and Eq.(A.62) ($n = 0$, 2, 3) versus n, then this dependence may be approximated by a straight line within an error of less than 0.05 Å. Interpolating this function to $n = 1$, gives to the 'crystalline' diameter of the UO molecule:

$$\sigma_{UO}^{(sol)} = 3.5 \text{ Å}. \tag{A.64}$$

This value satisfies the inequality Eq.(A.63). Reducing it, as above, by 3%, one gets finally: $\sigma_{UO} = 3.4$ Å.

The results are presented in Table A.24. The estimated error of the molecule diameter σ_i is about 0.1- 0.2 Å.

Table A.24. Sizes of U_mO_n-type Species (in Å)

O	O_2	U	UO	UO_2	UO_3
1.8	3.6	2.9	3.5	3.8	4.3

A.3.6 Conclusions

The results presented in Tables 8.3, 8.4 and 8.6 as well as the estimates of the UO and UO_3 dipole moments (Table A.23) are used as initial estimates of parameters of the *EOS*. The effect of the uncertainty of these parameters was investigated in an extended sensitivity study, which showed that these data provide an acceptable basis for a validation of the *EOS* model of UO_2 in the temperature and pressure range where experimental data are available. More crucial is the effect of this uncertainty at high temperatures approaching the critical region. However, any further improvement in this sense requires more accurate information not only on the molecular parameters, but also a corresponding formal accuracy of the effective interatomic potentials between the U_mO_n molecules.

A.4 Thermodynamic Tables

Index of the Tables

The tables contain the following thermodynamic functions:

ρ	$kg\ m^{-3}$	–	Density
H	$kJ\ kg^{-1}$	–	Enthalpy
G	$kJ\ kg^{-1}$	–	Gibbs potential
S	$kJ\ kg^{-1}\ K^{-1}$	–	Entropy
C_P	$kJ\ kg^{-1}\ K^{-1}$	–	Isobaric heat capacity
C_P/C_V		–	Heat capacity ratio
β_T	bar^{-1}	–	Isothermal compressibility
α_T	K^{-1}	–	Volume thermal expansion coefficient
V_S	$m\ s^{-1}$	–	Velocity of sound
μ_O/RT		–	Reduced chemical potential of atomic oxygen calculated relative to the energy of an atomic ideal gas at $T = 0\ K$

and are collected according to the following criteria:

Table	Title	Interval	Pages
	Saturation Conditions		
A.25	Liquid Coexisting with the Saturated Vapour	$T = 3000 \ldots 10000\ K$, O/U= 1.75 ... 2.25	188– 212
A.26	Saturated Vapour	$T = 3000 \ldots 10000\ K$, O/U= 1.75 ... 2.25	213–237
	Boiling Conditions		
A.27	Boiling Liquid	$T = 3000 \ldots 10000\ K$, O/U= 1.75 ... 2.25	238–262
A.28	Vapour Coexisting with the Boiling Liquid	$T = 3000 \ldots 10000\ K$; O/U= 1.75 ... 2.25	263–287
	Single-Phase Regions		
A.29	Liquid Phase	$O/U = 1.75 \ldots 2.25$, $P = 1 \ldots 10000\ bar$	288 – 310
A.30	Vapour Phase	$O/U = 1.75 \ldots 2.25$, $P = 1 \ldots 10000\ bar$	311 – 347

Table A.25. Liquid Coexisting with the Saturated Vapour ($(O/U)^{(vap)} = 1.75$)

T K	P bar	ρ $kg\,m^{-3}$	H $10^{3} kJ\,kg^{-1}$	G $kJ\,kg^{-1}$	S $kJ\,kg^{-1}\,K^{-1}$	C_P	C_P/C_V	β_T $10^{-4} bar^{-1}$	α_T $10^{-3} K^{-1}$	V_S $m\,s^{-1}$	O/U
3000	0.01491	9123	-6.262	-9672	1.137	0.494	1.281	0.037	0.111	1941	1.910
3120	0.02949	9014	-6.186	-9792	1.156	0.493	1.257	0.039	0.106	1897	1.903
3500	0.1788	8711	-5.954	-10180	1.208	0.435	1.233	0.044	0.094	1804	1.880
4000	1.078	8375	-5.673	-10701	1.257	0.364	1.255	0.050	0.088	1729	1.844
4500	4.347	8074	-5.397	-11213	1.292	0.333	1.305	0.058	0.090	1672	1.799
5000	13.59	7790	-5.100	-11694	1.319	0.327	1.366	0.067	0.096	1615	1.740
5500	35.52	7513	-4.770	-12134	1.339	0.334	1.426	0.079	0.104	1551	1.667
6000	80.78	7235	-4.414	-12538	1.354	0.347	1.474	0.094	0.112	1474	1.584
6500	164.0	6949	-4.045	-12921	1.366	0.356	1.534	0.113	0.122	1398	1.498
7000	303.8	6648	-3.679	-13304	1.375	0.369	1.584	0.138	0.134	1314	1.415
7500	522.4	6326	-3.325	-13705	1.384	0.387	1.639	0.172	0.148	1227	1.340
8000	848.5	5979	-2.989	-14139	1.394	0.416	1.712	0.221	0.169	1139	1.277
8500	1322	5598	-2.670	-14620	1.406	0.469	1.826	0.296	0.203	1050	1.227
9000	2004	5169	-2.363	-15170	1.423	0.574	2.029	0.426	0.267	960	1.195
9500	3012	4654	-2.058	-15842	1.451	0.806	2.442	0.690	0.401	872	1.189
9800	3886	4263	-1.868	-16370	1.480	1.097	2.906	1.009	0.562	822	1.212
9900	4255	4106	-1.801	-16589	1.494	1.244	3.117	1.168	0.640	806	1.229
10000	4690	3925	-1.730	-16848	1.512	1.429	3.352	1.363	0.733	792	1.254

Table A.25. Liquid Coexisting with the Saturated Vapour ($(O/U)^{(vap)} = 1.80$)

T K	P bar	ρ $kg\,m^{-3}$	H $10^{3} kJ\,kg^{-1}$	G $kJ\,kg^{-1}$	S $kJ\,kg^{-1}\,K^{-1}$	C_P	C_P/C_V	β_T $10^{-4} bar^{-1}$	α_T $10^{-3} K^{-1}$	V_S $m\,s^{-1}$	O/U
3000	0.01441	9098	-6.303	-9718	1.138	0.495	1.281	0.037	0.111	1942	1.928
3120	0.02857	8989	-6.229	-9839	1.157	0.493	1.257	0.039	0.106	1898	1.922
3500	0.1748	8684	-5.999	-10232	1.209	0.436	1.233	0.044	0.094	1805	1.899
4000	1.064	8347	-5.720	-10756	1.259	0.367	1.255	0.050	0.088	1729	1.864
4500	4.329	8046	-5.443	-11270	1.295	0.336	1.306	0.058	0.090	1672	1.819
5000	13.62	7762	-5.144	-11753	1.322	0.330	1.367	0.068	0.096	1614	1.760
5500	35.76	7486	-4.813	-12194	1.342	0.337	1.427	0.079	0.104	1550	1.686
6000	81.57	7209	-4.454	-12598	1.357	0.350	1.476	0.094	0.113	1473	1.602
6500	166.1	6924	-4.083	-12981	1.369	0.359	1.536	0.114	0.123	1397	1.515
7000	308.2	6625	-3.714	-13364	1.378	0.372	1.586	0.139	0.134	1313	1.431
7500	531.2	6304	-3.359	-13765	1.387	0.390	1.642	0.173	0.149	1226	1.356
8000	864.9	5957	-3.021	-14200	1.397	0.420	1.715	0.223	0.170	1137	1.292
8500	1351	5577	-2.700	-14684	1.410	0.476	1.831	0.299	0.206	1048	1.242
9000	2055	5146	-2.392	-15240	1.428	0.586	2.039	0.431	0.271	959	1.210
9500	3104	4627	-2.086	-15926	1.457	0.830	2.462	0.699	0.410	872	1.208
9800	4029	4227	-1.895	-16475	1.488	1.139	2.933	1.023	0.576	823	1.235
9900	4426	4064	-1.827	-16708	1.503	1.295	3.142	1.183	0.655	808	1.255
10000	4902	3873	-1.754	-16989	1.523	1.489	3.367	1.375	0.747	795	1.284

Table A.25. Liquid Coexisting with the Saturated Vapour ($(O/U)^{(vap)} = 1.85$)

T K	P bar	ρ $kg\,m^{-3}$	H $10^{3} kJ\,kg^{-1}$	G $kJ\,kg^{-1}$	S $kJ\,kg^{-1}\,K^{-1}$	C_P $kJ\,kg^{-1}\,K^{-1}$	C_P/C_V	β_T $10^{-2} bar^{-1}$	α_T $10^{-3} K^{-1}$	V_S $m\,s^{-1}$	O/U
3000	0.01399	9075	-6.340	-9757	1.139	0.495	1.282	0.037	0.111	1943	1.944
3120	0.02784	8966	-6.267	-9880	1.158	0.494	1.258	0.039	0.106	1899	1.938
3500	0.1718	8660	-6.040	-10277	1.211	0.438	1.233	0.044	0.095	1805	1.917
4000	1.057	8322	-5.761	-10805	1.261	0.369	1.256	0.051	0.089	1729	1.882
4500	4.328	8020	-5.484	-11322	1.297	0.339	1.307	0.058	0.091	1671	1.838
5000	13.69	7736	-5.185	-11807	1.324	0.333	1.368	0.068	0.097	1613	1.778
5500	36.05	7460	-4.852	-12249	1.345	0.339	1.429	0.080	0.105	1549	1.704
6000	82.45	7185	-4.492	-12654	1.360	0.352	1.477	0.095	0.114	1472	1.619
6500	168.2	6901	-4.119	-13038	1.372	0.361	1.538	0.114	0.124	1396	1.531
7000	312.9	6602	-3.748	-13420	1.382	0.374	1.588	0.140	0.135	1312	1.446
7500	540.3	6282	-3.390	-13822	1.391	0.393	1.644	0.175	0.150	1225	1.370
8000	881.5	5936	-3.051	-14258	1.401	0.424	1.719	0.224	0.172	1136	1.306
8500	1380	5556	-2.729	-14745	1.414	0.482	1.837	0.302	0.208	1047	1.257
9000	2107	5124	-2.420	-15308	1.432	0.598	2.049	0.435	0.275	959	1.226
9500	3200	4599	-2.114	-16009	1.463	0.856	2.482	0.709	0.419	873	1.226
9800	4179	4190	-1.921	-16581	1.496	1.183	2.958	1.037	0.589	825	1.258
9900	4608	4020	-1.852	-16829	1.513	1.347	3.163	1.200	0.669	811	1.281
10000	5132	3817	-1.777	-17137	1.536	1.548	3.372	1.384	0.759	799	1.317

Table A.25. Liquid Coexisting with the Saturated Vapour ($(O/U)^{(vap)} = 1.875$)

T K	P bar	ρ $kg\,m^{-3}$	H $10^{3} kJ\,kg^{-1}$	G $kJ\,kg^{-1}$	S $kJ\,kg^{-1}\,K^{-1}$	C_P	C_P/C_V	β_T $10^{-4} bar^{-1}$	α_T $10^{-3} K^{-1}$	V_S $m\,s^{-1}$	O/U
3000	0.01382	9065	-6.357	-9775	1.139	0.495	1.282	0.038	0.111	1943	1.951
3120	0.02755	8956	-6.284	-9899	1.159	0.495	1.258	0.039	0.107	1899	1.945
3500	0.1708	8649	-6.058	-10298	1.211	0.439	1.233	0.044	0.095	1805	1.925
4000	1.055	8310	-5.780	-10828	1.262	0.371	1.256	0.051	0.089	1729	1.891
4500	4.333	8008	-5.504	-11346	1.298	0.340	1.307	0.059	0.091	1671	1.846
5000	13.73	7724	-5.204	-11832	1.326	0.334	1.369	0.068	0.097	1613	1.787
5500	36.21	7448	-4.871	-12275	1.346	0.340	1.430	0.080	0.105	1548	1.712
6000	82.92	7173	-4.510	-12681	1.362	0.354	1.478	0.095	0.114	1471	1.628
6500	169.4	6890	-4.136	-13065	1.374	0.362	1.538	0.115	0.124	1396	1.539
7000	315.2	6591	-3.764	-13448	1.383	0.376	1.589	0.140	0.136	1311	1.454
7500	544.9	6272	-3.406	-13849	1.392	0.395	1.646	0.175	0.151	1224	1.377
8000	890.0	5926	-3.066	-14286	1.403	0.426	1.721	0.225	0.173	1135	1.313
8500	1395	5545	-2.743	-14775	1.415	0.486	1.839	0.303	0.209	1047	1.264
9000	2133	5112	-2.434	-15341	1.434	0.604	2.054	0.438	0.278	958	1.233
9500	3249	4585	-2.127	-16051	1.466	0.869	2.492	0.713	0.423	873	1.235
9800	4257	4171	-1.933	-16635	1.500	1.205	2.970	1.043	0.596	826	1.270
9900	4704	3997	-1.864	-16891	1.518	1.373	3.172	1.202	0.676	813	1.295
10000	5256	3787	-1.787	-17215	1.543	1.578	3.371	1.386	0.763	801	1.334

Table A.25. Liquid Coexisting with the Saturated Vapour ($(O/U)^{(vap)} = 1.90$)

T K	P bar	ρ $kg\,m^{-3}$	H $10^{3} kJ\,kg^{-1}$	G $kJ\,kg^{-1}$	S $kJ\,kg^{-1}\,K^{-1}$	C_P $kJ\,kg^{-1}\,K^{-1}$	C_P/C_V	β_T $10^{-4} bar^{-1}$	α_T $10^{-3} K^{-1}$	V_S $m\,s^{-1}$	O/U
3000	0.01369	9056	-6.372	-9791	1.140	0.495	1.282	0.038	0.111	1944	1.957
3120	0.02731	8946	-6.300	-9916	1.159	0.495	1.258	0.039	0.107	1899	1.952
3500	0.1700	8639	-6.075	-10316	1.212	0.440	1.234	0.044	0.095	1805	1.932
4000	1.054	8299	-5.798	-10849	1.263	0.372	1.256	0.051	0.089	1729	1.899
4500	4.341	7996	-5.522	-11370	1.299	0.342	1.308	0.059	0.092	1671	1.855
5000	13.78	7712	-5.222	-11857	1.327	0.336	1.369	0.068	0.098	1613	1.795
5500	36.39	7436	-4.889	-12301	1.348	0.342	1.430	0.080	0.106	1548	1.721
6000	83.41	7161	-4.528	-12707	1.363	0.355	1.479	0.096	0.115	1471	1.636
6500	170.5	6878	-4.153	-13091	1.375	0.364	1.539	0.115	0.125	1395	1.547
7000	317.6	6580	-3.780	-13474	1.385	0.377	1.590	0.141	0.136	1311	1.461
7500	549.6	6262	-3.421	-13876	1.394	0.396	1.647	0.176	0.151	1223	1.384
8000	898.5	5916	-3.080	-14314	1.404	0.429	1.722	0.226	0.173	1135	1.320
8500	1410	5535	-2.757	-14804	1.417	0.489	1.842	0.304	0.210	1046	1.271
9000	2160	5101	-2.447	-15374	1.436	0.610	2.059	0.440	0.280	958	1.241
9500	3299	4571	-2.140	-16092	1.469	0.882	2.502	0.718	0.428	873	1.244
9800	4338	4151	-1.946	-16689	1.504	1.227	2.982	1.050	0.602	827	1.282
9900	4804	3973	-1.875	-16955	1.523	1.400	3.180	1.207	0.682	814	1.309
10000	5388	3755	-1.798	-17296	1.550	1.608	3.367	1.388	0.768	804	1.353

Table A.25. Liquid Coexisting with the Saturated Vapour ($(O/U)^{(vap)} = 1.92$)

T K	P bar	ρ $kg\,m^{-3}$	H $10^3 kJ\,kg^{-1}$	G $kJ\,kg^{-1}$	S $kJ\,kg^{-1}$	C_P K^{-1}	C_P/C_V	β_T $10^{-4} bar^{-1}$	α_T $10^{-3} K^{-1}$	V_S $m\,s^{-1}$	O/U
3000	0.01360	9049	-6.383	-9803	1.140	0.495	1.282	0.038	0.111	1944	1.962
3120	0.02718	8939	-6.312	-9928	1.159	0.496	1.258	0.039	0.107	1900	1.957
3500	0.1697	8631	-6.088	-10331	1.212	0.441	1.234	0.044	0.095	1805	1.937
4000	1.055	8291	-5.812	-10865	1.263	0.373	1.257	0.051	0.090	1729	1.905
4500	4.351	7987	-5.536	-11387	1.300	0.343	1.309	0.059	0.092	1670	1.861
5000	13.83	7702	-5.237	-11876	1.328	0.337	1.370	0.068	0.098	1612	1.802
5500	36.55	7427	-4.903	-12321	1.349	0.343	1.431	0.081	0.106	1547	1.727
6000	83.82	7152	-4.541	-12727	1.364	0.356	1.480	0.096	0.115	1470	1.642
6500	171.4	6870	-4.166	-13112	1.376	0.365	1.540	0.115	0.125	1394	1.553
7000	319.6	6572	-3.793	-13495	1.386	0.378	1.591	0.141	0.136	1310	1.467
7500	553.3	6253	-3.433	-13898	1.395	0.398	1.648	0.176	0.152	1223	1.390
8000	905.4	5908	-3.091	-14336	1.406	0.430	1.724	0.227	0.174	1134	1.325
8500	1423	5527	-2.768	-14827	1.419	0.492	1.844	0.305	0.211	1046	1.276
9000	2182	5092	-2.458	-15400	1.438	0.615	2.063	0.442	0.281	958	1.247
9500	3340	4560	-2.150	-16124	1.471	0.893	2.509	0.722	0.431	873	1.252
9800	4404	4135	-1.955	-16733	1.508	1.245	2.990	1.054	0.607	828	1.292
9900	4886	3954	-1.884	-17007	1.528	1.421	3.185	1.211	0.687	816	1.321
10000	5497	3729	-1.806	-17363	1.556	1.632	3.360	1.387	0.770	806	1.368

Table A.25. Liquid Coexisting with the Saturated Vapour ($(O/U)^{(vap)} = 1.94$)

T K	P bar	ρ $kg\,m^{-3}$	H $10^{3} kJ\,kg^{-1}$	G $kJ\,kg^{-1}$	S	C_P $kJ\,kg^{-1}\,K^{-1}$	C_P/C_V	β_T $10^{-4} bar^{-1}$	α_T $10^{-3} K^{-1}$	V_S $m\,s^{-1}$	O/U
3000	0.01354	9043	-6.394	-9814	1.140	0.496	1.283	0.038	0.111	1944	1.967
3120	0.02708	8933	-6.323	-9940	1.159	0.496	1.258	0.039	0.107	1900	1.962
3500	0.1695	8624	-6.099	-10344	1.213	0.442	1.234	0.044	0.095	1805	1.943
4000	1.056	8282	-5.825	-10881	1.264	0.375	1.257	0.051	0.090	1729	1.911
4500	4.363	7978	-5.550	-11404	1.301	0.344	1.309	0.059	0.092	1670	1.867
5000	13.38	7693	-5.250	-11894	1.329	0.338	1.371	0.069	0.098	1612	1.808
5500	36.71	7418	-4.917	-12340	1.350	0.344	1.432	0.081	0.106	1547	1.733
6000	84.24	7143	-4.555	-12747	1.365	0.357	1.480	0.096	0.115	1470	1.648
6500	172.4	6861	-4.179	-13132	1.377	0.366	1.541	0.116	0.125	1394	1.558
7000	321.6	6564	-3.805	-13516	1.387	0.379	1.592	0.141	0.137	1310	1.472
7500	557.1	6245	-3.444	-13918	1.397	0.399	1.649	0.177	0.152	1222	1.395
8000	912.3	5900	-3.102	-14357	1.407	0.432	1.725	0.228	0.175	1134	1.331
8500	1435	5519	-2.779	-14850	1.420	0.494	1.846	0.306	0.212	1045	1.281
9000	2203	5084	-2.468	-15425	1.440	0.620	2.067	0.444	0.283	957	1.253
9500	3381	4548	-2.160	-16157	1.473	0.903	2.517	0.725	0.435	873	1.259
9800	4472	4119	-1.965	-16777	1.511	1.263	2.998	1.059	0.612	829	1.302
9900	4970	3935	-1.893	-17059	1.532	1.442	3.189	1.214	0.691	817	1.332
10000	5613	3702	-1.813	-17433	1.562	1.655	3.351	1.386	0.772	808	1.384

Table A.25. Liquid Coexisting with the Saturated Vapour ($(O/U)^{(vap)} = 1.95$)

T K	P bar	ρ $kg\,m^{-3}$	H $10^{3} kJ\,kg^{-1}$	G $kJ\,kg^{-1}$	S $kJ\,kg^{-1}\,K^{-1}$	C_P $kJ\,kg^{-1}\,K^{-1}$	C_P/C_V	β_T $10^{-4} bar^{-1}$	α_T $10^{-3} K^{-1}$	V_S $m\,s^{-1}$	O/U
3000	0.01352	9040	-6.399	-9820	1.140	0.496	1.283	0.038	0.111	1944	1.969
3120	0.02706	8930	-6.328	-9945	1.160	0.496	1.258	0.039	0.107	1900	1.964
3500	0.1695	8621	-6.105	-10350	1.213	0.443	1.234	0.044	0.095	1805	1.945
4000	1.057	8278	-5.831	-10888	1.264	0.375	1.257	0.051	0.090	1729	1.913
4500	4.370	7974	-5.556	-11413	1.301	0.345	1.309	0.059	0.092	1670	1.870
5000	13.91	7689	-5.257	-11903	1.329	0.339	1.371	0.069	0.098	1612	1.811
5500	36.79	7413	-4.924	-12349	1.350	0.344	1.432	0.081	0.106	1547	1.736
6000	84.45	7139	-4.561	-12757	1.366	0.357	1.481	0.096	0.115	1470	1.651
6500	172.9	6857	-4.185	-13142	1.378	0.366	1.541	0.116	0.125	1394	1.561
7000	322.6	6560	-3.811	-13526	1.388	0.380	1.593	0.142	0.137	1309	1.475
7500	559.0	6241	-3.450	-13929	1.397	0.399	1.650	0.177	0.152	1222	1.398
8000	915.7	5896	-3.108	-14368	1.408	0.433	1.726	0.228	0.175	1134	1.333
8500	1441	5515	-2.784	-14862	1.421	0.496	1.848	0.307	0.213	1045	1.284
9000	2214	5079	-2.473	-15438	1.441	0.622	2.069	0.445	0.284	957	1.256
9500	3402	4543	-2.165	-16173	1.475	0.909	2.521	0.727	0.437	874	1.262
9800	4506	4111	-1.969	-16799	1.513	1.273	3.002	1.061	0.615	830	1.307
9900	5014	3925	-1.898	-17086	1.534	1.453	3.191	1.216	0.693	818	1.338
10000	5674	3687	-1.817	-17470	1.565	1.667	3.345	1.385	0.772	809	1.392

Table A.25. Liquid Coexisting with the Saturated Vapour ($(O/U)^{(vap)} = 1.96$)

T K	P bar	ρ $kg\,m^{-3}$	H $10^{3} kJ\,kg^{-1}$	G $kJ\,kg^{-1}$	S $kJ\,kg^{-1}\,K^{-1}$	C_P	C_P/C_V	β_T $10^{-4} bar^{-1}$	α_T $10^{-3} K^{-1}$	V_S $m\,s^{-1}$	O/U
3000	0.01351	9037	-6.403	-9824	1.140	0.496	1.283	0.038	0.111	1944	1.971
3120	0.02704	8927	-6.332	-9951	1.160	0.497	1.259	0.039	0.107	1900	1.966
3500	0.1696	8617	-6.111	-10356	1.213	0.443	1.234	0.044	0.095	1805	1.948
4000	1.058	8274	-5.837	-10895	1.265	0.376	1.258	0.051	0.090	1729	1.916
4500	4.377	7969	-5.563	-11421	1.302	0.345	1.310	0.059	0.092	1670	1.873
5000	13.94	7684	-5.264	-11912	1.330	0.339	1.371	0.069	0.099	1611	1.814
5500	36.88	7409	-4.930	-12359	1.351	0.345	1.432	0.081	0.106	1546	1.739
6000	84.67	7135	-4.568	-12767	1.366	0.358	1.481	0.096	0.115	1469	1.654
6500	173.4	6853	-4.192	-13152	1.379	0.367	1.541	0.116	0.125	1393	1.564
7000	323.6	6556	-3.817	-13536	1.388	0.380	1.593	0.142	0.137	1309	1.478
7500	560.9	6237	-3.456	-13939	1.398	0.400	1.650	0.177	0.152	1222	1.401
8000	919.2	5892	-3.113	-14379	1.408	0.434	1.727	0.228	0.175	1133	1.336
8500	1447	5511	-2.789	-14873	1.422	0.497	1.849	0.307	0.213	1045	1.287
9000	2225	5075	-2.478	-15451	1.441	0.625	2.071	0.446	0.285	957	1.258
9500	3423	4537	-2.170	-16189	1.476	0.914	2.525	0.729	0.439	874	1.266
9800	4541	4103	-1.974	-16821	1.515	1.282	3.006	1.063	0.617	830	1.312
9900	5058	3914	-1.902	-17113	1.536	1.463	3.192	1.217	0.696	818	1.345
10000	5736	3673	-1.820	-17507	1.569	1.678	3.339	1.383	0.773	811	1.400

Table A.25. Liquid Coexisting with the Saturated Vapour ($(O/U)^{(vap)} = 1.97$)

T K	P bar	ρ $kg\,m^{-3}$	H $10^{3} kJ\,kg^{-1}$	G $kJ\,kg^{-1}$	S $kJ\,kg^{-1}\,K^{-1}$	C_P	C_P/C_V	β_T $10^{-4} bar^{-1}$	α_T $10^{-3} K^{-1}$	V_S $m\,s^{-1}$	O/U
3000	0.01350	9035	-6.408	-9829	1.140	0.496	1.283	0.038	0.111	1944	1.973
3120	0.02704	8924	-6.337	-9955	1.160	0.497	1.259	0.039	0.107	1900	1.968
3500	0.1697	8614	-6.116	-10362	1.213	0.444	1.234	0.044	0.096	1805	1.950
4000	1.060	8271	-5.843	-10902	1.265	0.377	1.258	0.051	0.090	1728	1.919
4500	4.386	7965	-5.569	-11429	1.302	0.346	1.310	0.059	0.092	1670	1.876
5000	13.97	7680	-5.270	-11920	1.330	0.340	1.371	0.069	0.099	1611	1.817
5500	36.97	7405	-4.937	-12368	1.351	0.345	1.433	0.081	0.107	1546	1.742
6000	84.89	7130	-4.574	-12776	1.367	0.358	1.481	0.096	0.115	1469	1.657
6500	173.9	6849	-4.198	-13162	1.379	0.367	1.542	0.116	0.126	1393	1.567
7000	324.6	6552	-3.823	-13546	1.389	0.381	1.593	0.142	0.137	1309	1.481
7500	562.9	6233	-3.462	-13949	1.398	0.401	1.651	0.178	0.153	1222	1.403
8000	922.7	5888	-3.119	-14389	1.409	0.435	1.727	0.229	0.175	1133	1.339
8500	1453	5507	-2.794	-14884	1.422	0.498	1.850	0.308	0.214	1044	1.289
9000	2236	5070	-2.483	-15463	1.442	0.627	2.073	0.447	0.286	957	1.261
9500	3444	4531	-2.175	-16205	1.477	0.920	2.529	0.731	0.440	874	1.270
9800	4577	4095	-1.979	-16844	1.517	1.291	3.009	1.065	0.620	831	1.317
9900	5103	3904	-1.906	-17140	1.539	1.474	3.193	1.219	0.698	819	1.351
10000	5800	3658	-1.824	-17545	1.572	1.690	3.332	1.382	0.773	812	1.409

Table A.25. Liquid Coexisting with the Saturated Vapour ($(O/U)^{(vap)} = 1.98$)

T K	P bar	ρ $kg\,m^{-3}$	H $10^{3} kJ\,kg^{-1}$	G $kJ\,kg^{-1}$	S $kJ\,kg^{-1}$	C_P K^{-1}	C_P/C_V	β_T $10^{-4} bar^{-1}$	α_T $10^{-3} K^{-1}$	V_S $m\,s^{-1}$	O/U
3000	0.01350	9032	-6.412	-9833	1.140	0.496	1.283	0.038	0.111	1944	1.974
3120	0.02705	8921	-6.341	-9960	1.160	0.497	1.259	0.039	0.107	1900	1.970
3500	0.1699	8611	-6.121	-10367	1.213	0.444	1.235	0.044	0.096	1805	1.952
4000	1.061	8267	-5.848	-10909	1.265	0.377	1.258	0.051	0.090	1728	1.921
4500	4.394	7961	-5.575	-11436	1.302	0.346	1.310	0.059	0.093	1669	1.879
5000	14.00	7675	-5.277	-11929	1.330	0.340	1.372	0.069	0.099	1611	1.820
5500	37.06	7400	-4.943	-12377	1.352	0.346	1.433	0.081	0.107	1546	1.745
6000	85.11	7126	-4.581	-12786	1.367	0.359	1.482	0.096	0.116	1469	1.660
6500	174.4	6844	-4.204	-13171	1.380	0.368	1.542	0.116	0.126	1393	1.570
7000	325.6	6547	-3.829	-13556	1.390	0.381	1.594	0.142	0.137	1309	1.483
7500	564.8	6229	-3.467	-13959	1.399	0.401	1.651	0.178	0.153	1221	1.406
8000	926.2	5884	-3.124	-14400	1.409	0.436	1.728	0.229	0.176	1133	1.341
8500	1460	5503	-2.800	-14895	1.423	0.500	1.851	0.309	0.214	1044	1.292
9000	2247	5066	-2.488	-15476	1.443	0.630	2.076	0.448	0.287	957	1.264
9500	3465	4526	-2.180	-16222	1.478	0.925	2.533	0.733	0.442	874	1.273
9800	4612	4087	-1.983	-16866	1.519	1.300	3.013	1.067	0.622	831	1.322
9900	5149	3894	-1.910	-17168	1.541	1.485	3.194	1.220	0.700	820	1.357
10000	5866	3642	-1.827	-17585	1.576	1.701	3.324	1.380	0.773	813	1.418

Table A.25. Liquid Coexisting with the Saturated Vapour ($(O/U)^{(vap)} = 1.99$)

T K	P bar	ρ $kg\,m^{-3}$	H $10^{3} kJ\,kg^{-1}$	G $kJ\,kg^{-1}$	S $kJ\,kg^{-1}\,K^{-1}$	C_P	C_P/C_V	β_T $10^{-4} bar^{-1}$	α_T $10^{-3} K^{-1}$	V_S $m\,s^{-1}$	O/U
3000	0.01351	9030	-6.416	-9837	1.141	0.497	1.283	0.038	0.111	1945	1.976
3120	0.02707	8919	-6.346	-9964	1.160	0.497	1.259	0.039	0.107	1900	1.972
3500	0.1701	8608	-6.126	-10373	1.213	0.445	1.235	0.044	0.096	1805	1.954
4000	1.063	8263	-5.854	-10916	1.265	0.378	1.258	0.051	0.090	1728	1.924
4500	4.404	7957	-5.581	-11444	1.303	0.347	1.311	0.059	0.093	1669	1.882
5000	14.03	7671	-5.283	-11937	1.331	0.341	1.372	0.069	0.099	1611	1.823
5500	37.15	7396	-4.950	-12386	1.352	0.346	1.433	0.081	0.107	1546	1.748
6000	85.33	7122	-4.587	-12795	1.368	0.359	1.482	0.097	0.116	1469	1.662
6500	174.9	6840	-4.210	-13181	1.380	0.368	1.542	0.116	0.126	1393	1.573
7000	326.6	6543	-3.835	-13566	1.390	0.382	1.594	0.142	0.138	1308	1.486
7500	566.7	6226	-3.473	-13969	1.400	0.402	1.651	0.178	0.153	1221	1.409
8000	929.8	5880	-3.130	-14410	1.410	0.436	1.729	0.229	0.176	1132	1.344
8500	1466	5499	-2.805	-14906	1.424	0.501	1.852	0.309	0.215	1044	1.295
9000	2259	5062	-2.493	-15489	1.444	0.632	2.078	0.448	0.288	957	1.267
9500	3486	4520	-2.185	-16238	1.479	0.931	2.537	0.735	0.444	874	1.277
9800	4649	4078	-1.988	-16889	1.521	1.310	3.016	1.069	0.624	832	1.327
9900	5196	3883	-1.915	-17196	1.544	1.496	3.194	1.221	0.701	821	1.363
10000	5934	3626	-1.831	-17626	1.579	1.712	3.315	1.377	0.773	815	1.427

Table A.25. Liquid Coexisting with the Saturated Vapour ($(O/U)^{(vap)} = 2.00$)

T K	P bar	ρ $kg\,m^{-3}$	H $10^{3} kJ\,kg^{-1}$	G $kJ\,kg^{-1}$	S $kJ\,kg^{-1}\,K^{-1}$	C_P $kJ\,kg^{-1}\,K^{-1}$	C_P/C_V	β_T $10^{-4} bar^{-1}$	α_T $10^{-3} K^{-1}$	V_S $m\,s^{-1}$	O/U
3000	0.01353	9027	-6.420	-9841	1.141	0.497	1.283	0.038	0.111	1945	1.978
3120	0.02711	8916	-6.349	-9969	1.160	0.498	1.259	0.039	0.107	1900	1.973
3500	0.1704	8605	-6.130	-10378	1.214	0.446	1.235	0.044	0.096	1805	1.956
4000	1.066	8259	-5.859	-10922	1.266	0.379	1.259	0.051	0.091	1728	1.926
4500	4.414	7953	-5.587	-11452	1.303	0.348	1.311	0.059	0.093	1669	1.884
5000	14.06	7667	-5.289	-11946	1.331	0.341	1.372	0.069	0.099	1611	1.826
5500	37.24	7392	-4.956	-12395	1.353	0.347	1.433	0.081	0.107	1546	1.751
6000	85.56	7118	-4.593	-12804	1.368	0.360	1.482	0.097	0.116	1468	1.665
6500	175.4	6836	-4.216	-13190	1.381	0.369	1.543	0.116	0.126	1392	1.575
7000	327.6	6540	-3.840	-13575	1.391	0.382	1.594	0.143	0.138	1308	1.489
7500	568.7	6222	-3.478	-13979	1.400	0.403	1.652	0.178	0.153	1221	1.411
8000	933.3	5877	-3.135	-14420	1.411	0.437	1.729	0.230	0.176	1132	1.346
8500	1472	5495	-2.810	-14917	1.424	0.503	1.853	0.310	0.215	1044	1.297
9000	2270	5057	-2.498	-15501	1.445	0.635	2.080	0.449	0.289	957	1.270
9500	3508	4514	-2.189	-16254	1.480	0.936	2.541	0.737	0.446	874	1.281
9800	4685	4070	-1.992	-16912	1.522	1.319	3.019	1.071	0.627	832	1.332
9900	5243	3872	-1.919	-17224	1.546	1.506	3.194	1.222	0.703	822	1.370
10000	6005	3609	-1.834	-17668	1.583	1.724	3.304	1.374	0.772	816	1.437

Table A.25. Liquid Coexisting with the Saturated Vapour ($(O/U)^{(vap)} = 2.01$)

T K	P bar	ρ $kg\,m^{-3}$	H $10^{3} kJ\,kg^{-1}$	G $kJ\,kg^{-1}$	S $kJ\,kg^{-1}\,K^{-1}$	C_P	C_P/C_V	β_T $10^{-4} bar^{-1}$	α_T $10^{-3} K^{-1}$	V_S $m\,s^{-1}$	O/U
3000	0.01355	9025	-6.423	-9845	1.141	0.497	1.283	0.038	0.111	1945	1.979
3120	0.02716	8914	-6.353	-9972	1.160	0.498	1.259	0.039	0.107	1900	1.975
3500	0.1708	8602	-6.135	-10383	1.214	0.446	1.235	0.044	0.096	1805	1.958
4000	1.068	8256	-5.865	-10928	1.266	0.379	1.259	0.051	0.091	1728	1.929
4500	4.424	7949	-5.593	-11459	1.304	0.348	1.311	0.059	0.093	1669	1.887
5000	14.10	7663	-5.295	-11954	1.332	0.342	1.373	0.069	0.099	1610	1.829
5500	37.34	7387	-4.962	-12403	1.353	0.347	1.434	0.081	0.107	1545	1.754
6000	85.79	7114	-4.599	-12813	1.369	0.360	1.483	0.097	0.116	1468	1.668
6500	175.9	6832	-4.222	-13200	1.381	0.369	1.543	0.117	0.126	1392	1.578
7000	328.6	6536	-3.846	-13585	1.391	0.383	1.595	0.143	0.138	1308	1.491
7500	570.6	6218	-3.484	-13989	1.401	0.403	1.652	0.178	0.153	1220	1.414
8000	936.8	5873	-3.140	-14431	1.411	0.438	1.730	0.230	0.177	1132	1.349
8500	1478	5491	-2.815	-14928	1.425	0.504	1.854	0.310	0.216	1043	1.300
9000	2281	5053	-2.503	-15513	1.446	0.637	2.082	0.450	0.289	956	1.273
9500	3530	4509	-2.194	-16270	1.482	0.942	2.545	0.738	0.448	874	1.284
9800	4723	4061	-1.997	-16935	1.524	1.329	3.023	1.073	0.629	833	1.337
9900	5291	3861	-1.923	-17253	1.549	1.517	3.194	1.222	0.705	823	1.376
10000	6078	3592	-1.837	-17711	1.587	1.735	3.293	1.371	0.771	818	1.447

Table A.25. Liquid Coexisting with the Saturated Vapour ($(O/U)^{(vap)} = 2.02$)

T K	P bar	ρ $kg\,m^{-3}$	H $10^{3} kJ\,kg^{-1}$	G $kJ\,kg^{-1}$	S $kJ\,kg^{-1}\,K^{-1}$	C_P $kJ\,kg^{-1}\,K^{-1}$	C_P/C_V	β_T $10^{-4} bar^{-1}$	α_T $10^{-3} K^{-1}$	V_S $m\,s^{-1}$	O/U
3000	0.01359	9023	-6.426	-9848	1.141	0.497	1.283	0.038	0.111	1945	1.981
3120	0.02722	8912	-6.357	-9976	1.160	0.498	1.259	0.039	0.107	1900	1.976
3500	0.1712	8599	-6.139	-10388	1.214	0.447	1.235	0.044	0.096	1805	1.960
4000	1.071	8252	-5.870	-10934	1.266	0.380	1.259	0.051	0.091	1728	1.931
4500	4.435	7945	-5.599	-11466	1.304	0.349	1.311	0.059	0.093	1669	1.890
5000	14.13	7658	-5.301	-11962	1.332	0.342	1.373	0.069	0.099	1610	1.831
5500	37.43	7383	-4.968	-12412	1.353	0.348	1.434	0.081	0.107	1545	1.757
6000	86.03	7109	-4.605	-12822	1.369	0.361	1.483	0.097	0.116	1468	1.671
6500	176.4	6828	-4.228	-13209	1.382	0.370	1.543	0.117	0.126	1392	1.581
7000	329.7	6532	-3.852	-13594	1.392	0.383	1.595	0.143	0.138	1308	1.494
7500	572.6	6214	-3.489	-13999	1.401	0.404	1.653	0.179	0.154	1220	1.416
8000	940.4	5869	-3.145	-14441	1.412	0.439	1.731	0.230	0.177	1132	1.351
8500	1485	5487	-2.820	-14939	1.426	0.505	1.855	0.311	0.216	1043	1.303
9000	2292	5049	-2.508	-15526	1.446	0.640	2.084	0.451	0.290	956	1.276
9500	3551	4503	-2.199	-16286	1.483	0.948	2.549	0.740	0.449	874	1.288
9800	4760	4053	-2.001	-16958	1.526	1.338	3.025	1.075	0.631	833	1.343
9900	5341	3850	-1.927	-17282	1.551	1.528	3.193	1.223	0.706	824	1.383
10000	6154	3573	-1.840	-17757	1.592	1.746	3.281	1.367	0.770	819	1.457

Table A.25. Liquid Coexisting with the Saturated Vapour ($(O/U)^{(vap)} = 2.03$)

T K	P bar	ρ $kg\,m^{-3}$	H $10^{3} kJ\,kg^{-1}$	G $kJ\,kg^{-1}$	S	C_P $kJ\,kg^{-1}\,K^{-1}$	C_P/C_V	β_T $10^{-4} bar^{-1}$	α_T $10^{-3} K^{-1}$	V_S $m\,s^{-1}$	O/U
3000	0.01363	9021	-6.429	-9851	1.141	0.498	1.283	0.038	0.112	1945	1.982
3120	0.02730	8910	-6.360	-9980	1.160	0.499	1.259	0.039	0.107	1900	1.978
3500	0.1717	8597	-6.143	-10392	1.214	0.447	1.235	0.044	0.096	1805	1.962
4000	1.074	8249	-5.875	-10940	1.266	0.381	1.259	0.051	0.091	1728	1.933
4500	4.447	7941	-5.605	-11473	1.304	0.350	1.312	0.059	0.093	1669	1.892
5000	14.17	7654	-5.307	-11970	1.333	0.343	1.373	0.069	0.099	1610	1.834
5500	37.53	7379	-4.974	-12421	1.354	0.348	1.434	0.082	0.107	1545	1.760
6000	86.26	7105	-4.611	-12831	1.370	0.361	1.483	0.097	0.116	1468	1.674
6500	176.9	6824	-4.234	-13218	1.382	0.370	1.544	0.117	0.126	1392	1.584
7000	330.7	6528	-3.858	-13604	1.392	0.384	1.596	0.143	0.138	1307	1.497
7500	574.5	6210	-3.495	-14009	1.402	0.404	1.653	0.179	0.154	1220	1.419
8000	943.9	5865	-3.151	-14451	1.413	0.440	1.731	0.231	0.177	1131	1.354
8500	1491	5483	-2.825	-14950	1.426	0.507	1.856	0.311	0.217	1043	1.305
9000	2304	5044	-2.513	-15538	1.447	0.642	2.086	0.452	0.291	956	1.279
9500	3573	4497	-2.204	-16302	1.484	0.953	2.552	0.742	0.451	875	1.292
9800	4799	4044	-2.005	-16981	1.528	1.347	3.028	1.077	0.633	834	1.348
9900	5391	3839	-1.931	-17312	1.554	1.539	3.192	1.223	0.708	824	1.390
10000	6234	3554	-1.842	-17804	1.596	1.757	3.267	1.363	0.768	821	1.468

Table A.25. Liquid Coexisting with the Saturated Vapour ($(O/U)^{(vap)} = 2.04$)

T K	P bar	ρ $kg\,m^{-3}$	H $10^{3} kJ\,kg^{-1}$	G $kJ\,kg^{-1}$	S $kJ\,kg^{-1}\,K^{-1}$	C_P	C_P/C_V	β_T $10^{-4} bar^{-1}$	α_T $10^{-3} K^{-1}$	V_S $m\,s^{-1}$	O/U
3000	0.01368	9020	-6.432	-9854	1.141	0.498	1.284	0.038	0.112	1945	1.983
3120	0.02739	8908	-6.363	-9983	1.160	0.499	1.259	0.039	0.107	1900	1.979
3500	0.1722	8594	-6.147	-10397	1.214	0.448	1.236	0.044	0.096	1805	1.964
4000	1.077	8245	-5.880	-10946	1.267	0.381	1.260	0.051	0.091	1728	1.936
4500	4.459	7937	-5.610	-11480	1.305	0.350	1.312	0.059	0.093	1668	1.895
5000	14.21	7650	-5.313	-11978	1.333	0.343	1.374	0.069	0.100	1610	1.837
5500	37.63	7375	-4.980	-12429	1.354	0.349	1.435	0.082	0.108	1545	1.762
6000	86.50	7101	-4.617	-12840	1.370	0.362	1.484	0.097	0.116	1468	1.676
6500	177.4	6820	-4.240	-13228	1.383	0.370	1.544	0.117	0.127	1391	1.586
7000	331.7	6524	-3.863	-13613	1.393	0.384	1.596	0.143	0.138	1307	1.499
7500	576.5	6206	-3.500	-14018	1.402	0.405	1.654	0.179	0.154	1220	1.421
8000	947.5	5861	-3.156	-14461	1.413	0.441	1.732	0.231	0.178	1131	1.356
8500	1497	5479	-2.830	-14961	1.427	0.508	1.857	0.312	0.217	1043	1.308
9000	2315	5040	-2.518	-15550	1.448	0.645	2.088	0.453	0.292	956	1.281
9500	3596	4491	-2.208	-16319	1.485	0.959	2.556	0.744	0.453	875	1.295
9800	4837	4035	-2.010	-17005	1.530	1.357	3.031	1.078	0.635	834	1.353
9900	5443	3827	-1.935	-17342	1.556	1.549	3.190	1.224	0.709	825	1.397
10000	6318	3534	-1.845	-17853	1.601	1.767	3.251	1.358	0.766	823	1.480

Table A.25. Liquid Coexisting with the Saturated Vapour ($(O/U)^{(vap)} = 2.05$)

T K	P bar	ρ $kg\,m^{-3}$	H $10^{3} kJ\,kg^{-1}$	G $kJ\,kg^{-1}$	S $kJ\,kg^{-1}\,K^{-1}$	C_P	C_P/C_V	β_T $10^{-4} bar^{-1}$	α_T $10^{-3} K^{-1}$	V_S $m\,s^{-1}$	O/U
3000	0.01373	9018	-6.435	-9857	1.141	0.498	1.284	0.038	0.112	1945	1.984
3120	0.02750	8906	-6.366	-9986	1.160	0.500	1.260	0.039	0.107	1900	1.980
3500	0.1728	8591	-6.151	-10401	1.214	0.449	1.236	0.044	0.096	1805	1.965
4000	1.080	8242	-5.884	-10952	1.267	0.382	1.260	0.051	0.091	1728	1.938
4500	4.472	7933	-5.615	-11487	1.305	0.351	1.312	0.059	0.094	1668	1.897
5000	14.25	7646	-5.319	-11985	1.333	0.344	1.374	0.069	0.100	1610	1.840
5500	37.74	7371	-4.986	-12437	1.355	0.349	1.435	0.082	0.108	1544	1.765
6000	86.74	7097	-4.623	-12849	1.371	0.362	1.484	0.097	0.117	1467	1.679
6500	177.9	6816	-4.245	-13237	1.383	0.371	1.544	0.117	0.127	1391	1.589
7000	332.8	6520	-3.869	-13623	1.393	0.385	1.596	0.143	0.139	1307	1.502
7500	578.5	6203	-3.505	-14028	1.403	0.406	1.654	0.179	0.154	1219	1.424
8000	951.1	5858	-3.161	-14471	1.414	0.442	1.733	0.231	0.178	1131	1.359
8500	1504	5475	-2.835	-14971	1.428	0.509	1.859	0.312	0.218	1043	1.310
9000	2326	5036	-2.523	-15563	1.449	0.648	2.090	0.454	0.293	956	1.284
9500	3618	4486	-2.213	-16335	1.486	0.965	2.560	0.746	0.455	875	1.299
9800	4877	4027	-2.014	-17028	1.532	1.366	3.033	1.080	0.638	835	1.359
9900	5495	3815	-1.939	-17373	1.559	1.560	3.188	1.224	0.711	826	1.404
10000	6406	3513	-1.847	-17906	1.606	1.778	3.234	1.352	0.764	825	1.492

Table A.25. Liquid Coexisting with the Saturated Vapour ($(O/U)^{(vap)} = 2.06$)

T K	P bar	ρ $kg\,m^{-3}$	H $10^{3} kJ\,kg^{-1}$	G $kJ\,kg^{-1}$	S $kJ\,kg^{-1}\,K^{-1}$	C_P	C_P/C_V	β_T $10^{-4} bar^{-1}$	α_T $10^{-3} K^{-1}$	V_S $m\,s^{-1}$	O/U
3000	0.01379	9017	-6.437	-9859	1.141	0.498	1.284	0.038	0.112	1945	1.985
3120	0.02762	8904	-6.369	-9989	1.160	0.500	1.260	0.039	0.107	1900	1.982
3500	0.1734	8589	-6.154	-10405	1.214	0.449	1.236	0.044	0.096	1805	1.967
4000	1.084	8239	-5.889	-10958	1.267	0.383	1.260	0.051	0.091	1727	1.940
4500	4.485	7929	-5.621	-11494	1.305	0.351	1.313	0.060	0.094	1668	1.900
5000	14.29	7642	-5.325	-11993	1.334	0.345	1.374	0.069	0.100	1609	1.842
5500	37.84	7367	-4.992	-12446	1.355	0.350	1.435	0.082	0.108	1544	1.768
6000	86.99	7093	-4.629	-12857	1.371	0.363	1.484	0.097	0.117	1467	1.682
6500	178.4	6812	-4.251	-13246	1.384	0.371	1.545	0.117	0.127	1391	1.592
7000	333.8	6516	-3.874	-13632	1.394	0.385	1.597	0.144	0.139	1306	1.504
7500	580.4	6199	-3.511	-14037	1.404	0.406	1.655	0.180	0.155	1219	1.426
8000	954.7	5854	-3.166	-14481	1.414	0.442	1.733	0.232	0.178	1131	1.361
8500	1510	5471	-2.840	-14982	1.428	0.511	1.860	0.313	0.218	1042	1.313
9000	2333	5031	-2.528	-15575	1.450	0.650	2.092	0.455	0.294	956	1.287
9500	3641	4480	-2.218	-16351	1.488	0.970	2.564	0.747	0.457	875	1.303
9800	4917	4018	-2.018	-17052	1.534	1.376	3.035	1.082	0.640	836	1.364
9900	5549	3803	-1.943	-17404	1.562	1.571	3.186	1.223	0.712	827	1.411
10000	6499	3491	-1.850	-17961	1.611	1.788	3.215	1.346	0.761	827	1.505

Table A.25. Liquid Coexisting with the Saturated Vapour ($(O/U)^{(vap)} = 2.08$)

T K	P bar	ρ $kg\,m^{-3}$	H $10^{3} kJ\,kg^{-1}$	G $kJ\,kg^{-1}$	S $kJ\,kg^{-1}$	C_P K^{-1}	C_P/C_V	β_T $10^{-4} bar^{-1}$	α_T $10^{-3} K^{-1}$	V_S $m\,s^{-1}$	O/U
3000	0.01394	9014	-6.441	-9864	1.141	0.499	1.284	0.038	0.112	1945	1.987
3120	0.02790	8901	-6.373	-9994	1.160	0.501	1.260	0.039	0.108	1900	1.984
3500	0.1749	8584	-6.161	-10413	1.215	0.451	1.236	0.044	0.097	1805	1.970
4000	1.091	8232	-5.898	-10968	1.268	0.384	1.261	0.051	0.092	1727	1.944
4500	4.514	7922	-5.631	-11507	1.306	0.353	1.313	0.060	0.094	1668	1.905
5000	14.37	7634	-5.336	-12008	1.334	0.346	1.375	0.070	0.100	1609	1.847
5500	38.05	7359	-5.003	-12462	1.356	0.351	1.436	0.082	0.108	1544	1.773
6000	87.48	7085	-4.640	-12875	1.372	0.364	1.485	0.097	0.117	1467	1.687
6500	179.5	6805	-4.262	-13263	1.385	0.372	1.545	0.118	0.127	1390	1.597
7000	335.9	6509	-3.885	-13650	1.395	0.386	1.597	0.144	0.139	1306	1.510
7500	584.4	6191	-3.521	-14056	1.405	0.407	1.656	0.180	0.155	1219	1.431
8000	961.9	5846	-3.176	-14500	1.416	0.444	1.735	0.232	0.179	1130	1.366
8500	1523	5464	-2.850	-15003	1.430	0.514	1.862	0.314	0.219	1042	1.318
9000	2361	5023	-2.537	-15599	1.451	0.655	2.096	0.457	0.296	956	1.293
9500	3686	4468	-2.227	-16383	1.490	0.982	2.571	0.751	0.460	875	1.310
9800	4998	3999	-2.027	-17101	1.538	1.395	3.039	1.084	0.644	837	1.375
9900	5660	3779	-1.950	-17468	1.567	1.592	3.179	1.222	0.714	830	1.426
10000	6705	3440	-1.853	-18085	1.623	1.808	3.170	1.331	0.753	832	1.534

Table A.25. Liquid Coexisting with the Saturated Vapour ($(O/U)^{(vap)} = 2.10$)

T K	P bar	ρ $kg\,m^{-3}$	H $10^{3}kJ\,kg^{-1}$	G $kJ\,kg^{-1}$	S $kJ\,kg^{-1}$	C_P K^{-1}	C_P/C_V	β_T $10^{-4}bar^{-1}$	α_T $10^{-3}K^{-1}$	V_S $m\,s^{-1}$	O/U
3000	0.01412	9011	-6.445	-9868	1.141	0.500	1.284	0.038	0.112	1945	1.989
3120	0.02823	8898	-6.378	-9998	1.160	0.502	1.260	0.039	0.108	1900	1.986
3500	0.1767	8580	-6.167	-10420	1.215	0.452	1.237	0.044	0.097	1805	1.973
4000	1.100	8226	-5.906	-10978	1.268	0.385	1.261	0.051	0.092	1727	1.948
4500	4.545	7914	-5.641	-11520	1.306	0.354	1.314	0.060	0.094	1667	1.909
5000	14.46	7626	-5.346	-12022	1.335	0.347	1.375	0.070	0.100	1608	1.853
5500	38.28	7351	-5.014	-12477	1.357	0.352	1.436	0.082	0.108	1543	1.778
6000	87.99	7078	-4.652	-12891	1.373	0.365	1.486	0.098	0.117	1466	1.692
6500	180.6	6797	-4.273	-13281	1.386	0.373	1.546	0.118	0.127	1390	1.602
7000	338.1	6501	-3.896	-13668	1.396	0.387	1.598	0.144	0.139	1305	1.515
7500	588.4	6184	-3.531	-14075	1.406	0.409	1.657	0.181	0.155	1218	1.436
8000	969.2	5839	-3.186	-14520	1.417	0.446	1.736	0.233	0.179	1130	1.371
8500	1536	5456	-2.859	-15024	1.431	0.516	1.864	0.315	0.220	1041	1.323
9000	2384	5014	-2.547	-15623	1.453	0.661	2.100	0.459	0.298	955	1.298
9500	3733	4457	-2.236	-16415	1.493	0.994	2.579	0.755	0.464	876	1.317
9800	5083	3981	-2.035	-17150	1.542	1.414	3.042	1.087	0.647	838	1.386
9900	5777	3753	-1.957	-17535	1.573	1.613	3.171	1.221	0.715	832	1.441
10000	6951	3379	-1.854	-18234	1.638	1.828	3.111	1.311	0.741	838	1.569

Table A.25. Liquid Coexisting with the Saturated Vapour ($(O/U)^{(vap)} = 2.125$)

T K	P bar	ρ $kg\,m^{-3}$	H $10^{3} kJ\,kg^{-1}$	G $kJ\,kg^{-1}$	S	C_P $kJ\,kg^{-1}\,K^{-1}$	C_P/C_V	β_T $10^{-4} bar^{-1}$	α_T $10^{-3} K^{-1}$	V_S $m\,s^{-1}$	O/U
3000	0.01438	9009	-6.449	-9872	1.141	0.500	1.284	0.038	0.112	1945	1.991
3120	0.02871	8895	-6.382	-10003	1.161	0.503	1.261	0.039	0.108	1900	1.988
3500	0.1792	8575	-6.174	-10427	1.215	0.453	1.237	0.044	0.097	1805	1.976
4000	1.113	8219	-5.916	-10990	1.269	0.387	1.262	0.052	0.092	1726	1.953
4500	4.587	7905	-5.653	-11535	1.307	0.355	1.315	0.060	0.095	1667	1.915
5000	14.58	7617	-5.359	-12040	1.336	0.348	1.376	0.070	0.101	1608	1.859
5500	38.56	7341	-5.028	-12497	1.358	0.353	1.437	0.082	0.109	1543	1.785
6000	88.64	7068	-4.665	-12912	1.374	0.366	1.486	0.098	0.118	1466	1.699
6500	181.9	6787	-4.287	-13302	1.387	0.374	1.547	0.118	0.128	1389	1.608
7000	340.8	6492	-3.909	-13690	1.397	0.389	1.599	0.145	0.140	1305	1.521
7500	593.5	6175	-3.544	-14097	1.407	0.410	1.658	0.181	0.156	1217	1.442
8000	978.3	5830	-3.198	-14544	1.418	0.448	1.738	0.234	0.180	1129	1.377
8500	1552	5447	-2.871	-15050	1.433	0.520	1.867	0.316	0.221	1041	1.329
9000	2414	5003	-2.558	-15653	1.455	0.667	2.105	0.461	0.300	955	1.305
9500	3792	4442	-2.247	-16456	1.496	1.008	2.588	0.759	0.469	876	1.327
9800	5192	3957	-2.045	-17213	1.548	1.438	3.044	1.089	0.652	840	1.401
9900	5933	3719	-1.966	-17623	1.581	1.639	3.157	1.217	0.716	835	1.462
10000	7386	3266	-1.849	-18506	1.666	1.851	2.995	1.272	0.716	849	1.634

Table A.25. Liquid Coexisting with the Saturated Vapour ($(O/U)^{(vap)} = 2.15$)

T	P	ρ	H	G	S	C_P	C_P/C_V	β_T	α_T	V_S	O/U
K	bar	$kg\,m^{-3}$	$10^{3} kJ\,kg^{-1}$	$kJ\,kg^{-1}$	$kJ\,kg^{-1}\,K^{-1}$			$10^{-4} bar^{-1}$	$10^{-3} K^{-1}$	$m\,s^{-1}$	
3000	0.01467	9006	-6.452	-9875	1.141	0.501	1.285	0.038	0.112	1945	1.992
3120	0.02927	8892	-6.386	-10008	1.161	0.504	1.261	0.039	0.108	1900	1.990
3500	0.1820	8570	-6.181	-10435	1.215	0.455	1.238	0.044	0.098	1805	1.979
4000	1.126	8212	-5.925	-11001	1.269	0.389	1.263	0.052	0.093	1726	1.957
4500	4.632	7897	-5.664	-11549	1.308	0.357	1.315	0.060	0.095	1666	1.920
5000	14.70	7608	-5.372	-12057	1.337	0.349	1.377	0.070	0.101	1607	1.865
5500	38.86	7332	-5.041	-12515	1.359	0.354	1.438	0.083	0.109	1542	1.791
6000	89.30	7059	-4.678	-12932	1.375	0.367	1.487	0.098	0.118	1465	1.705
6500	183.3	6778	-4.300	-13323	1.388	0.375	1.548	0.118	0.128	1389	1.614
7000	343.5	6483	-3.922	-13712	1.399	0.390	1.600	0.145	0.140	1304	1.527
7500	598.6	6166	-3.556	-14120	1.408	0.412	1.659	0.182	0.156	1217	1.448
8000	987.5	5821	-3.210	-14567	1.420	0.450	1.739	0.235	0.181	1128	1.383
8500	1569	5437	-2.883	-15076	1.434	0.523	1.870	0.318	0.222	1040	1.335
9000	2444	4993	-2.570	-15683	1.457	0.674	2.111	0.464	0.302	955	1.312
9500	3852	4427	-2.259	-16497	1.499	1.023	2.597	0.763	0.473	877	1.336
9800	5306	3932	-2.055	-17278	1.553	1.463	3.045	1.091	0.656	842	1.416
9900	6101	3682	-1.975	-17717	1.590	1.665	3.138	1.212	0.715	839	1.484
10000	8434	2759	-1.711	-19507	1.780	1.885	2.605	1.253	0.634	868	1.876

Table A.25. Liquid Coexisting with the Saturated Vapour ($(O/U)^{(vap)} = 2.20$)

T K	P bar	ρ $kg\,m^{-3}$	H $10^{3} kJ\,kg^{-1}$	G $kJ\,kg^{-1}$	S $kJ\,kg^{-1}\,K^{-1}$	C_P $kJ\,kg^{-1}\,K^{-1}$	C_P/C_V	β_T $10^{-4} bar^{-1}$	α_T $10^{-3} K^{-1}$	V_S $m\,s^{-1}$	O/U
3000	0.01537	9003	-6.457	-9880	1.141	0.503	1.285	0.038	0.113	1944	1.994
3120	0.03059	8887	-6.392	-10014	1.161	0.506	1.262	0.039	0.108	1900	1.992
3500	0.1888	8561	-6.191	-10447	1.216	0.458	1.239	0.045	0.098	1804	1.984
4000	1.158	8198	-5.942	-11022	1.270	0.392	1.264	0.052	0.093	1725	1.965
4500	4.734	7880	-5.685	-11577	1.309	0.359	1.317	0.060	0.096	1665	1.930
5000	14.96	7590	-5.396	-12089	1.339	0.352	1.378	0.070	0.102	1606	1.876
5500	39.48	7314	-5.066	-12551	1.361	0.356	1.439	0.083	0.110	1541	1.803
6000	90.67	7040	-4.704	-12970	1.378	0.369	1.489	0.099	0.118	1464	1.717
6500	186.1	6760	-4.325	-13363	1.390	0.378	1.549	0.119	0.129	1387	1.626
7000	349.0	6465	-3.947	-13754	1.401	0.392	1.602	0.146	0.141	1303	1.538
7500	608.9	6148	-3.580	-14164	1.411	0.415	1.661	0.183	0.157	1215	1.460
8000	1006	5803	-3.233	-14613	1.422	0.454	1.743	0.236	0.182	1127	1.395
8500	1602	5418	-2.906	-15126	1.438	0.530	1.875	0.320	0.225	1039	1.348
9000	2504	4971	-2.592	-15741	1.461	0.688	2.121	0.469	0.307	954	1.326
9500	3976	4397	-2.280	-16578	1.505	1.054	2.615	0.772	0.482	878	1.355
9800	5551	3880	-2.075	-17415	1.565	1.511	3.040	1.092	0.662	847	1.448
9900	6487	3598	-1.989	-17932	1.610	1.716	3.084	1.194	0.709	847	1.535
10000	7990	2845	-1.729	-19205	1.748	1.928	2.763	1.353	0.688	847	1.801

Table A.25. Liquid Coexisting with the Saturated Vapour ($(O/U)^{(vap)} = 2.25$)

T K	P bar	ρ $kg\,m^{-3}$	H $10^{3} kJ\,kg^{-1}$	G $kJ\,kg^{-1}$	S $kJ\,kg^{-1}\,K^{-1}$	C_P $kJ\,kg^{-1}\,K^{-1}$	C_P/C_V	β_T $10^{-4} bar^{-1}$	α_T $10^{-3} K^{-1}$	V_S $m\,s^{-1}$	O/U
3000	0.01621	8999	-6.460	-9884	1.141	0.505	1.286	0.038	0.113	1944	1.996
3120	0.03219	8883	-6.397	-10019	1.161	0.508	1.262	0.039	0.109	1899	1.995
3500	0.1970	8553	-6.200	-10457	1.216	0.461	1.240	0.045	0.099	1803	1.989
4000	1.196	8185	-5.956	-11040	1.271	0.394	1.266	0.052	0.094	1724	1.972
4500	4.850	7864	-5.704	-11602	1.311	0.362	1.318	0.061	0.096	1664	1.940
5000	15.26	7572	-5.418	-12119	1.340	0.354	1.380	0.071	0.102	1605	1.887
5500	40.15	7296	-5.090	-12585	1.363	0.358	1.441	0.083	0.110	1539	1.814
6000	92.09	7023	-4.728	-13007	1.380	0.371	1.490	0.099	0.119	1463	1.729
6500	189.0	6743	-4.350	-13402	1.393	0.380	1.551	0.120	0.129	1386	1.638
7000	354.6	6448	-3.971	-13794	1.403	0.395	1.603	0.147	0.142	1302	1.550
7500	619.3	6131	-3.604	-14206	1.414	0.418	1.663	0.184	0.158	1214	1.471
8000	1025	5786	-3.256	-14658	1.425	0.459	1.746	0.238	0.184	1126	1.406
8500	1636	5400	-2.928	-15175	1.441	0.538	1.881	0.323	0.227	1038	1.359
9000	2566	4950	-2.614	-15799	1.465	0.701	2.131	0.473	0.311	954	1.340
9500	4106	4367	-2.302	-16660	1.511	1.085	2.631	0.780	0.491	879	1.374
9800	5824	3824	-2.093	-17564	1.579	1.560	3.027	1.090	0.666	852	1.483
9900	6984	3489	-2.000	-18209	1.637	1.763	2.994	1.162	0.694	859	1.601
10000	8283	2792	-1.719	-19398	1.768	1.899	2.658	1.285	0.652	861	1.849

Table A.26. Saturated Vapour ($(O/U)^{(vap)} = 1.75$)

T K	P bar	ρ $kg\,m^{-3}$	H $10^{3} kJ\,kg^{-1}$	G $kJ\,kg^{-1}$	S $kJ\,kg^{-1}$	C_P K^{-1}	C_P/C_V	β_T $10^{-3} bar^{-1}$	α_T $10^{-3} K^{-1}$	V_S $m\,s^{-1}$	$\mu_O/10kT$
3000	0.01491	0.01586	-4.233	-9258	1.675	0.323	1.121	6.720^{+4}_{10}	0.351	324	-3.380
3120	0.02949	0.03013	-4.198	-9393	1.665	0.340	1.118	3.400^{+4}_{10}	0.343	330	-3.276
3500	0.1788	0.1620	-4.076	-9834	1.645	0.424	1.114	5636.00	0.336	349	-2.996
4000	1.078	0.8410	-3.882	-10442	1.640	0.640	1.123	950.500	0.374	375	-2.720
4500	4.347	2.915	-3.627	-11072	1.655	0.972	1.156	243.300	0.454	404	-2.516
5000	13.59	7.786	-3.300	-11723	1.685	1.325	1.205	81.190	0.534	437	-2.364
5500	35.52	17.39	-2.918	-12393	1.723	1.583	1.256	32.430	0.576	472	-2.248
6000	80.78	34.18	-2.513	-13084	1.762	1.682	1.301	14.820	0.573	507	-2.160
6500	164.0	61.05	-2.115	-13795	1.797	1.684	1.340	7.549	0.550	539	-2.090
7000	303.8	101.5	-1.746	-14526	1.826	1.613	1.376	4.220	0.519	567	-2.034
7500	522.4	160.0	-1.413	-15276	1.848	1.529	1.416	2.549	0.494	589	-1.987
8000	848.5	242.7	-1.121	-16040	1.865	1.458	1.465	1.641	0.480	607	-1.948
8500	1322	359.7	-0.874	-16816	1.876	1.411	1.535	1.110	0.481	620	-1.913
9000	2004	528.6	-0.684	-17600	1.880	1.398	1.643	0.779	0.500	632	-1.880
9500	3012	787.0	-0.572	-18383	1.875	1.441	1.825	0.557	0.548	645	-1.847
9800	3886	1026	-0.568	-18848	1.865	1.518	2.006	0.451	0.600	658	-1.825
9900	4255	1133	-0.587	-18999	1.860	1.559	2.088	0.416	0.622	665	-1.817
10000	4690	1264	-0.621	-19149	1.853	1.612	2.187	0.381	0.649	674	-1.808

Table A.26. Saturated Vapour ($(O/U)^{(vap)} = 1.80$)

T K	P bar	ρ $kg\,m^{-3}$	H $10^{3}kJ\,kg^{-1}$	G $kJ\,kg^{-1}$	S $kJ\,kg^{-1}$	C_P K^{-1}	C_P/C_V	β_T $10^{-3}bar^{-1}$	α_T $10^{-3}K^{-1}$	V_S $m\,s^{-1}$	$\mu_O/10kT$
3000	0.01441	0.01536	-4.347	-9391	1.681	0.337	1.117	6.958^{+4}_{10}	0.354	323	-3.356
3120	0.02857	0.02926	-4.310	-9527	1.672	0.357	1.115	3.512^{+4}_{10}	0.348	329	-3.253
3500	0.1748	0.1587	-4.184	-9970	1.653	0.455	1.112	5773.00	0.346	348	-2.977
4000	1.064	0.8299	-3.980	-10581	1.650	0.695	1.124	965.600	0.393	375	-2.704
4500	4.329	2.889	-3.712	-11216	1.668	1.047	1.161	245.700	0.479	404	-2.503
5000	13.62	7.731	-3.370	-11871	1.700	1.408	1.212	81.540	0.558	439	-2.354
5500	35.76	17.29	-2.973	-12547	1.741	1.663	1.264	32.420	0.595	475	-2.240
6000	81.57	34.00	-2.555	-13244	1.782	1.754	1.308	14.750	0.587	511	-2.153
6500	166.1	60.80	-2.147	-13962	1.818	1.746	1.346	7.489	0.561	544	-2.084
7000	308.2	101.2	-1.770	-14700	1.847	1.666	1.383	4.174	0.527	572	-2.028
7500	531.2	159.8	-1.431	-15456	1.870	1.574	1.422	2.514	0.500	595	-1.982
8000	864.9	242.9	-1.135	-16228	1.887	1.497	1.471	1.612	0.485	613	-1.943
8500	1351	360.9	-0.8869	-17011	1.897	1.446	1.541	1.086	0.484	627	-1.908
9000	2055	532.0	-0.6966	-17801	1.900	1.429	1.649	0.758	0.502	640	-1.875
9500	3104	796.3	-0.5892	-18590	1.895	1.471	1.829	0.537	0.548	654	-1.842
9800	4029	1045	-0.5940	-19055	1.884	1.549	2.008	0.429	0.597	669	-1.819
9900	4426	1158	-0.6172	-19207	1.878	1.591	2.090	0.394	0.618	677	-1.810
10000	4902	1300	-0.6599	-19356	1.870	1.645	2.190	0.357	0.643	688	-1.801

Table A.26. Saturated Vapour ($(O/U)^{(vap)} = 1.85$)

T K	P bar	ρ $kg\,m^{-3}$	H $10\,^3kJ\,kg^{-1}$	G $kJ\,kg^{-1}$	S $kJ\,kg^{-1}$	C_P K^{-1}	C_P/C_V	β_T $10^{-3}bar^{-1}$	α_T $10^{-3}K^{-1}$	V_S $m\,s^{-1}$	$\mu_O/10kT$
3000	0.01399	0.01495	-4.458	-9521	1.688	0.355	1.113	7.167^{+4}_{10}	0.359	322	-3.330
3120	0.02784	0.02857	-4.420	-9658	1.679	0.378	1.111	3.607^{+4}_{10}	0.353	328	-3.227
3500	0.1718	0.1562	-4.289	-10104	1.661	0.490	1.110	5883.00	0.357	348	-2.956
4000	1.057	0.8223	-4.075	-10718	1.661	0.754	1.127	976.600	0.413	375	-2.688
4500	4.328	2.871	-3.793	-11357	1.681	1.124	1.167	247.200	0.504	406	-2.491
5000	13.69	7.691	-3.436	-12017	1.716	1.490	1.220	81.680	0.581	441	-2.344
5500	36.05	17.20	-3.025	-12699	1.759	1.740	1.271	32.340	0.613	478	-2.233
6000	82.45	33.85	-2.595	-13402	1.801	1.822	1.314	14.660	0.600	515	-2.146
6500	168.2	60.59	-2.177	-14127	1.838	1.805	1.353	7.422	0.571	548	-2.078
7000	312.9	101.0	-1.792	-14872	1.869	1.717	1.388	4.125	0.535	577	-2.023
7500	540.3	159.7	-1.448	-15635	1.892	1.619	1.428	2.478	0.506	601	-1.977
8000	881.5	243.2	-1.149	-16413	1.908	1.536	1.477	1.584	0.489	619	-1.938
8500	1380	362.2	-0.899	-17203	1.918	1.480	1.547	1.063	0.487	634	-1.903
9000	2107	535.6	-0.709	-18000	1.921	1.461	1.653	0.737	0.503	647	-1.870
9500	3200	806.2	-0.607	-18793	1.914	1.501	1.833	0.517	0.547	663	-1.836
9800	4179	1065	-0.621	-19261	1.902	1.580	2.011	0.409	0.594	680	-1.813
9900	4608	1186	-0.650	-19412	1.895	1.622	2.092	0.372	0.614	689	-1.804
10000	5132	1339	-0.702	-19560	1.886	1.677	2.192	0.333	0.637	701	-1.793

Table A.26. Saturated Vapour ($(O/U)^{(vap)} = 1.875$)

T K	P bar	ρ $kg\,m^{-3}$	H $10^{3} kJ\,kg^{-1}$	G $kJ\,kg^{-1}$	S $kJ\,kg^{-1}$	C_P K^{-1}	C_P/C_V	β_T $10^{-3} bar^{-1}$	α_T $10^{-3} K^{-1}$	V_S $m\,s^{-1}$	$\mu_O/10kT$
3000	0.01382	0.01479	-4.513	-9586	1.691	0.365	1.111	7.256^{+4}_{10}	0.362	322	-3.315
3120	0.02755	0.02831	-4.474	-9722	1.682	0.390	1.109	3.646^{+4}_{10}	0.357	328	-3.214
3500	0.1708	0.1553	-4.340	-10169	1.666	0.509	1.110	5925.00	0.364	347	-2.945
4000	1.055	0.8199	-4.121	-10785	1.666	0.785	1.129	980.500	0.424	375	-2.680
4500	4.333	2.865	-3.832	-11427	1.688	1.162	1.170	247.600	0.516	406	-2.485
5000	13.73	7.676	-3.468	-12090	1.724	1.529	1.223	81.670	0.592	442	-2.340
5500	36.21	17.17	-3.051	-12774	1.768	1.777	1.275	32.280	0.621	480	-2.229
6000	82.92	33.79	-2.615	-13481	1.811	1.855	1.317	14.610	0.607	517	-2.143
6500	169.4	60.50	-2.192	-14209	1.849	1.834	1.355	7.386	0.575	551	-2.075
7000	315.2	100.9	-1.803	-14957	1.879	1.742	1.391	4.100	0.538	580	-2.020
7500	544.9	159.6	-1.456	-15724	1.902	1.640	1.431	2.460	0.508	604	-1.974
8000	890.0	243.4	-1.155	-16506	1.919	1.555	1.480	1.570	0.491	622	-1.935
8500	1395	362.9	-0.905	-17299	1.929	1.497	1.549	1.051	0.488	637	-1.900
9000	2133	537.5	-0.716	-18098	1.931	1.476	1.656	0.727	0.504	651	-1.867
9500	3249	811.3	-0.616	-18895	1.924	1.516	1.835	0.507	0.547	668	-1.833
9800	4257	1076	-0.634	-19362	1.911	1.595	2.012	0.398	0.592	685	-1.810
9900	4704	1200	-0.667	-19514	1.904	1.637	2.093	0.361	0.612	695	-1.801
10000	5256	1361	-0.725	-19660	1.894	1.693	2.193	0.321	0.634	709	-1.790

Table A.26. Saturated Vapour ($(O/U)^{(vap)} = 1.90$)

T K	P bar	ρ $kg\ m^{-3}$	H $10^{3} kJ\ kg^{-1}$	G $kJ\ kg^{-1}$	S $kJ\ kg^{-1}\ K^{-1}$	C_P $kJ\ kg^{-1}\ K^{-1}$	C_P/C_V	β_T $10^{-3} bar^{-1}$	α_T $10^{-3} K^{-1}$	V_S $m\ s^{-1}$	$\mu_O/10kT$
3000	0.01369	0.01466	-4.567	-9649	1.694	0.377	1.109	7.332^{+4}_{10}	0.365	321	-3.300
3120	0.02731	0.02810	-4.527	-9786	1.686	0.404	1.108	3.679^{+4}_{10}	0.361	327	-3.200
3500	0.1700	0.1546	-4.390	-10234	1.670	0.530	1.109	5957.00	0.371	347	-2.934
4000	1.054	0.8184	-4.166	-10852	1.672	0.816	1.130	983.200	0.435	375	-2.672
4500	4.341	2.861	-3.870	-11496	1.695	1.200	1.173	247.900	0.529	407	-2.479
5000	13.78	7.665	-3.499	-12162	1.732	1.568	1.227	81.600	0.602	443	-2.335
5500	36.39	17.14	-3.075	-12849	1.777	1.813	1.278	32.200	0.629	481	-2.225
6000	83.41	33.74	-2.633	-13559	1.821	1.887	1.320	14.550	0.612	519	-2.140
6500	170.5	60.42	-2.206	-14290	1.859	1.862	1.358	7.349	0.579	553	-2.072
7000	317.6	100.8	-1.814	-15042	1.890	1.766	1.394	4.075	0.541	582	-2.018
7500	549.6	159.6	-1.465	-15812	1.913	1.661	1.433	2.442	0.511	606	-1.972
8000	898.5	243.6	-1.162	-16597	1.929	1.574	1.483	1.557	0.493	625	-1.933
8500	1410	363.6	-0.9104	-17394	1.939	1.514	1.552	1.040	0.489	641	-1.898
9000	2160	539.5	-0.7218	-18196	1.942	1.491	1.658	0.717	0.505	655	-1.865
9500	3299	816.6	-0.6249	-18995	1.934	1.531	1.836	0.498	0.546	672	-1.831
9800	4338	1087	-0.6482	-19463	1.920	1.610	2.013	0.388	0.591	691	-1.807
9900	4804	1216	-0.6840	-19614	1.912	1.653	2.094	0.351	0.610	701	-1.797
10000	5388	1385	-0.7489	-19760	1.901	1.709	2.194	0.309	0.631	716	-1.786

Table A.26. Saturated Vapour ($(O/U)^{(vap)} = 1.92$)

T K	P bar	ρ $kg\,m^{-3}$	H $10^{3} kJ\,kg^{-1}$	G $kJ\,kg^{-1}$	S $kJ\,kg^{-1}$	C_P K^{-1}	C_P/C_V	β_T $10^{-3} bar^{-1}$	α_T $10^{-3} K^{-1}$	V_S $m\,s^{-1}$	$\mu_O/10kT$
3000	0.01360	0.01458	-4.609	-9699	1.697	0.387	1.108	7.380^{+4}_{10}	0.368	321	-3.288
3120	0.02718	0.02797	-4.569	-9837	1.688	0.415	1.107	3.699^{+4}_{10}	0.365	327	-3.189
3500	0.1697	0.1543	-4.429	-10286	1.673	0.547	1.110	5976.00	0.377	347	-2.926
4000	1.055	0.8178	-4.201	-10905	1.676	0.841	1.132	984.500	0.444	375	-2.666
4500	4.351	2.860	-3.900	-11551	1.700	1.231	1.176	247.900	0.538	407	-2.475
5000	13.83	7.658	-3.524	-12219	1.739	1.599	1.230	81.510	0.610	444	-2.332
5500	36.55	17.12	-3.094	-12909	1.784	1.841	1.281	32.130	0.635	482	-2.223
6000	83.82	33.70	-2.648	-13621	1.829	1.912	1.323	14.510	0.617	520	-2.138
6500	171.4	60.36	-2.218	-14355	1.867	1.884	1.360	7.318	0.582	555	-2.070
7000	319.6	100.8	-1.822	-15109	1.898	1.785	1.396	4.054	0.544	584	-2.016
7500	553.3	159.6	-1.471	-15882	1.922	1.678	1.435	2.427	0.513	609	-1.970
8000	905.4	243.7	-1.167	-16670	1.938	1.588	1.485	1.545	0.494	628	-1.931
8500	1423	364.2	-0.915	-17470	1.948	1.527	1.554	1.031	0.490	643	-1.896
9000	2182	541.0	-0.727	-18274	1.950	1.504	1.660	0.709	0.505	658	-1.863
9500	3340	821.0	-0.632	-19075	1.941	1.542	1.837	0.490	0.546	676	-1.829
9800	4404	1096	-0.660	-19543	1.927	1.621	2.014	0.380	0.589	695	-1.804
9900	4886	1229	-0.699	-19694	1.919	1.664	2.095	0.342	0.608	706	-1.795
10000	5497	1405	-0.769	-19839	1.907	1.722	2.195	0.299	0.628	722	-1.783

Table A.26. Saturated Vapour ($(O/U)^{(vap)} = 1.94$)

T	P	ρ	H	G	S	C_P	C_P/C_V	β_T	α_T	V_S	$\mu_O/10kT$
K	bar	$kg\,m^{-3}$	$10^{3}kJ\,kg^{-1}$	$kJ\,kg^{-1}$	$kJ\,kg^{-1}\,K^{-1}$			$10^{-3}bar^{-1}$	$10^{-3}K^{-1}$	$m\,s^{-1}$	
3000	0.01354	0.01453	-4.650	-9749	1.700	0.398	1.107	7.416^{+4}_{10}	0.371	320	-3.275
3120	0.02708	0.02790	-4.609	-9887	1.691	0.427	1.106	3.714^{+4}_{10}	0.369	327	-3.177
3500	0.1695	0.1542	-4.467	-10337	1.677	0.565	1.110	5987.00	0.384	347	-2.917
4000	1.056	0.8179	-4.235	-10958	1.681	0.867	1.134	985.000	0.454	375	-2.659
4500	4.363	2.859	-3.930	-11605	1.706	1.261	1.178	247.800	0.548	408	-2.470
5000	13.88	7.653	-3.548	-12275	1.746	1.629	1.232	81.390	0.618	445	-2.328
5500	36.71	17.11	-3.113	-12968	1.792	1.868	1.283	32.040	0.641	484	-2.220
6000	84.24	33.66	-2.663	-13683	1.837	1.937	1.325	14.450	0.621	522	-2.135
6500	172.4	60.31	-2.229	-14419	1.875	1.905	1.362	7.287	0.585	557	-2.068
7000	321.6	100.7	-1.830	-15176	1.907	1.804	1.398	4.034	0.546	587	-2.014
7500	557.1	159.6	-1.477	-15952	1.930	1.694	1.437	2.412	0.515	611	-1.968
8000	912.3	243.9	-1.172	-16743	1.946	1.603	1.487	1.534	0.496	630	-1.929
8500	1435	364.8	-0.919	-17545	1.956	1.540	1.556	1.022	0.491	646	-1.894
9000	2203	542.7	-0.732	-18352	1.958	1.516	1.661	0.702	0.505	661	-1.861
9500	3381	825.4	-0.640	-19155	1.949	1.554	1.839	0.483	0.545	679	-1.826
9800	4472	1106	-0.671	-19623	1.934	1.633	2.015	0.372	0.588	700	-1.802
9900	4970	1242	-0.714	-19774	1.925	1.676	2.095	0.334	0.606	711	-1.792
10000	5613	1427	-0.791	-19917	1.913	1.734	2.196	0.290	0.625	729	-1.780

Table A.26. Saturated Vapour ($(O/U)^{(vap)} = 1.95$)

T K	P bar	ρ $kg\,m^{-3}$	H $10^{3} kJ\,kg^{-1}$	G $kJ\,kg^{-1}$	S $kJ\,kg^{-1}\,K^{-1}$	C_P	C_P/C_V	β_T $10^{-3} bar^{-1}$	α_T $10^{-3} K^{-1}$	V_S $m\,s^{-1}$	$\mu_O/10kT$
3000	0.01352	0.01451	-4.671	-9774	1.701	0.403	1.106	7.428^{+4}_{10}	0.373	320	-3.268
3120	0.02706	0.02788	-4.629	-9911	1.693	0.433	1.106	3.719^{+4}_{10}	0.371	327	-3.171
3500	0.1695	0.1542	-4.487	-10362	1.679	0.574	1.110	5990.00	0.387	347	-2.912
4000	1.057	0.8181	-4.252	-10984	1.683	0.880	1.135	985.000	0.459	375	-2.656
4500	4.370	2.859	-3.944	-11632	1.709	1.276	1.180	247.700	0.553	408	-2.468
5000	13.91	7.652	-3.559	-12304	1.749	1.643	1.234	81.320	0.622	445	-2.327
5500	36.79	17.10	-3.123	-12997	1.795	1.882	1.284	32.000	0.644	484	-2.219
6000	84.45	33.65	-2.670	-13713	1.841	1.949	1.326	14.430	0.623	523	-2.134
6500	172.9	60.29	-2.234	-14451	1.880	1.916	1.363	7.271	0.587	558	-2.067
7000	322.6	100.7	-1.834	-15210	1.911	1.813	1.399	4.023	0.547	588	-2.013
7500	559.0	159.6	-1.480	-15987	1.934	1.702	1.438	2.405	0.515	612	-1.967
8000	915.7	244.0	-1.174	-16780	1.951	1.610	1.488	1.529	0.496	631	-1.928
8500	1441	365.1	-0.922	-17583	1.960	1.546	1.557	1.018	0.492	647	-1.893
9000	2214	543.5	-0.734	-18391	1.962	1.522	1.662	0.698	0.505	662	-1.860
9500	3402	827.7	-0.643	-19194	1.953	1.560	1.839	0.479	0.545	681	-1.825
9800	4506	1111	-0.677	-19663	1.937	1.639	2.015	0.368	0.587	702	-1.801
9900	5014	1249	-0.721	-19813	1.928	1.682	2.096	0.329	0.605	714	-1.791
10000	5674	1438	-0.802	-19956	1.915	1.740	2.197	0.285	0.623	732	-1.779

Table A.26. Saturated Vapour ($(O/U)^{(vap)} = 1.96$)

T K	P bar	ρ $kg\,m^{-3}$	H $10^{3} kJ\,kg^{-1}$	G $kJ\,kg^{-1}$	S $kJ\,kg^{-1}\,K^{-1}$	C_P	C_P/C_V	β_T $10^{-3} bar^{-1}$	α_T $10^{-3} K^{-1}$	V_S $m\,s^{-1}$	$\mu_O/10kT$
3000	0.01351	0.01451	-4.691	-9798	1.702	0.409	1.106	7.437^{+4}_{10}	0.375	320	-3.261
3120	0.02704	0.02787	-4.649	-9936	1.694	0.440	1.105	3.722^{+4}_{10}	0.373	326	-3.165
3500	0.1696	0.1543	-4.505	-10388	1.681	0.583	1.110	5991.00	0.391	347	-2.908
4000	1.058	0.8185	-4.269	-11010	1.685	0.893	1.136	984.800	0.463	375	-2.653
4500	4.377	2.860	-3.958	-11660	1.711	1.290	1.181	247.500	0.558	408	-2.466
5000	13.94	7.651	-3.571	-12332	1.752	1.658	1.235	81.240	0.626	446	-2.325
5500	36.88	17.09	-3.132	-13027	1.799	1.895	1.285	31.950	0.647	485	-2.217
6000	84.67	33.63	-2.677	-13744	1.844	1.961	1.327	14.400	0.625	523	-2.133
6500	173.4	60.27	-2.239	-14483	1.884	1.926	1.364	7.255	0.588	559	-2.066
7000	323.6	100.7	-1.838	-15243	1.915	1.822	1.400	4.013	0.548	589	-2.012
7500	560.9	159.6	-1.483	-16022	1.938	1.710	1.439	2.398	0.516	613	-1.966
8000	919.2	244.1	-1.177	-16816	1.955	1.617	1.489	1.523	0.497	633	-1.927
8500	1447	365.4	-0.924	-17620	1.964	1.553	1.558	1.013	0.492	649	-1.892
9000	2225	544.3	-0.737	-18430	1.966	1.527	1.663	0.694	0.506	664	-1.859
9500	3423	829.9	-0.647	-19234	1.957	1.565	1.840	0.476	0.545	683	-1.824
9800	4541	1116	-0.683	-19703	1.941	1.645	2.015	0.364	0.586	704	-1.800
9900	5058	1256	-0.729	-19853	1.932	1.688	2.096	0.325	0.603	716	-1.790
10000	5736	1450	-0.814	-19995	1.918	1.746	2.197	0.280	0.622	735	-1.777

Table A.26. Saturated Vapour ($(O/U)^{(vap)} = 1.97$)

T K	P bar	ρ $kg\,m^{-3}$	H $10^{3} kJ\,kg^{-1}$	G $kJ\,kg^{-1}$	S $kJ\,kg^{-1}\,K^{-1}$	C_P	C_P/C_V	β_T $10^{-3}bar^{-1}$	α_T $10^{-3}K^{-1}$	V_S $m\,s^{-1}$	$\mu_O/10kT$
3000	0.01350	0.01450	-4.711	-9823	1.704	0.415	1.105	7.443^{+4}_{10}	0.377	320	-3.254
3120	0.02704	0.02787	-4.669	-9961	1.696	0.446	1.105	3.724^{+4}_{10}	0.376	326	-3.159
3500	0.1697	0.1544	-4.524	-10413	1.682	0.592	1.110	5990.00	0.395	347	-2.903
4000	1.060	0.8190	-4.286	-11036	1.688	0.906	1.136	984.300	0.468	375	-2.650
4500	4.386	2.861	-3.973	-11687	1.714	1.305	1.182	247.400	0.563	409	-2.463
5000	13.97	7.650	-3.583	-12360	1.755	1.672	1.236	81.150	0.630	446	-2.323
5500	36.97	17.09	-3.141	-13056	1.803	1.908	1.287	31.910	0.649	486	-2.216
6000	84.89	33.62	-2.684	-13775	1.848	1.972	1.328	14.370	0.626	524	-2.132
6500	173.9	60.24	-2.245	-14515	1.888	1.936	1.365	7.238	0.589	559	-2.065
7000	324.6	100.7	-1.842	-15277	1.919	1.831	1.400	4.002	0.549	590	-2.011
7500	562.9	159.7	-1.486	-16057	1.943	1.718	1.440	2.390	0.517	614	-1.966
8000	922.7	244.2	-1.179	-16852	1.959	1.624	1.490	1.518	0.497	634	-1.927
8500	1453	365.7	-0.926	-17658	1.968	1.559	1.558	1.009	0.492	650	-1.891
9000	2236	545.2	-0.739	-18468	1.970	1.533	1.664	0.690	0.506	665	-1.858
9500	3444	832.3	-0.651	-19273	1.960	1.571	1.840	0.472	0.545	685	-1.823
9800	4577	1121	-0.690	-19742	1.944	1.650	2.015	0.360	0.585	706	-1.799
9900	5103	1264	-0.737	-19892	1.935	1.694	2.096	0.321	0.602	719	-1.788
10000	5800	1462	-0.825	-20034	1.921	1.752	2.198	0.275	0.620	739	-1.776

Table A.26. Saturated Vapour ($(O/U)^{(vap)} = 1.98$)

T K	P bar	ρ $kg\,m^{-3}$	H $10^{3} kJ\,kg^{-1}$	G $kJ\,kg^{-1}$	S $kJ\,kg^{-1}$	C_P K^{-1}	C_P/C_V	β_T $10^{-3} bar^{-1}$	α_T $10^{-3} K^{-1}$	V_S $m\,s^{-1}$	$\mu_O/10kT$
3000	0.01350	0.01451	-4.731	-9847	1.705	0.420	1.105	7.444^{+4}_{10}	0.379	320	-3.248
3120	0.02705	0.02789	-4.689	-9985	1.698	0.452	1.105	3.723^{+4}_{10}	0.378	326	-3.153
3500	0.1699	0.1545	-4.542	-10438	1.684	0.602	1.111	5988.00	0.398	347	-2.899
4000	1.061	0.8197	-4.302	-11062	1.690	0.919	1.137	983.700	0.473	376	-2.647
4500	4.394	2.862	-3.987	-11713	1.717	1.320	1.184	247.200	0.567	409	-2.461
5000	14.00	7.650	-3.594	-12388	1.759	1.687	1.238	81.060	0.634	447	-2.321
5500	37.06	17.08	-3.150	-13085	1.806	1.921	1.288	31.860	0.652	486	-2.214
6000	85.11	33.60	-2.691	-13805	1.852	1.984	1.329	14.350	0.628	525	-2.131
6500	174.4	60.22	-2.250	-14547	1.892	1.946	1.366	7.222	0.591	560	-2.064
7000	325.6	100.6	-1.846	-15310	1.923	1.840	1.401	3.991	0.550	591	-2.010
7500	564.8	159.7	-1.489	-16091	1.947	1.726	1.441	2.383	0.518	615	-1.965
8000	926.2	244.3	-1.181	-16888	1.963	1.631	1.490	1.512	0.498	635	-1.926
8500	1460	366.0	-0.928	-17695	1.973	1.566	1.559	1.004	0.493	651	-1.891
9000	2247	546.0	-0.741	-18507	1.974	1.539	1.665	0.686	0.506	666	-1.857
9500	3465	834.6	-0.654	-19313	1.964	1.577	1.841	0.468	0.544	686	-1.822
9800	4612	1126	-0.696	-19782	1.948	1.656	2.016	0.356	0.585	709	-1.797
9900	5149	1271	-0.745	-19931	1.938	1.700	2.096	0.317	0.601	722	-1.787
10000	5866	1475	-0.838	-20072	1.923	1.758	2.198	0.270	0.618	742	-1.774

Table A.26. Saturated Vapour ($(O/U)^{(vap)} = 1.99$)

T K	P bar	ρ $kg\,m^{-3}$	H $10^{3}kJ\,kg^{-1}$	G $kJ\,kg^{-1}$	S $kJ\,kg^{-1}$	C_P K^{-1}	C_P/C_V	β_T $10^{-3}bar^{-1}$	α_T $10^{-3}K^{-1}$	V_S $m\,s^{-1}$	$\mu_O/10kT$
3000	0.01351	0.01452	-4.750	-9871	1.707	0.426	1.105	7.441^{+4}_{10}	0.381	320	-3.241
3120	0.02707	0.02792	-4.708	-10009	1.699	0.459	1.105	3.721^{+4}_{10}	0.381	326	-3.147
3500	0.1701	0.1547	-4.561	-10463	1.686	0.611	1.111	5983.00	0.402	346	-2.894
4000	1.063	0.8206	-4.319	-11088	1.692	0.932	1.138	982.900	0.478	376	-2.644
4500	4.404	2.863	-4.000	-11740	1.720	1.334	1.185	246.900	0.572	409	-2.459
5000	14.03	7.650	-3.605	-12416	1.762	1.701	1.239	80.950	0.637	447	-2.320
5500	37.15	17.08	-3.159	-13115	1.810	1.934	1.289	31.800	0.654	487	-2.213
6000	85.33	33.59	-2.698	-13836	1.856	1.995	1.330	14.320	0.630	526	-2.130
6500	174.9	60.20	-2.255	-14579	1.896	1.957	1.367	7.205	0.592	561	-2.063
7000	326.6	100.6	-1.850	-15343	1.928	1.849	1.402	3.981	0.551	592	-2.009
7500	566.7	159.7	-1.492	-16126	1.951	1.734	1.442	2.376	0.519	617	-1.964
8000	929.8	244.4	-1.184	-16924	1.968	1.639	1.491	1.507	0.499	636	-1.925
8500	1466	366.4	-0.930	-17732	1.977	1.572	1.560	1.000	0.493	653	-1.890
9000	2259	546.9	-0.744	-18545	1.978	1.545	1.665	0.683	0.506	668	-1.856
9500	3486	837.0	-0.658	-19352	1.968	1.582	1.841	0.465	0.544	688	-1.821
9800	4649	1131	-0.702	-19821	1.951	1.662	2.016	0.353	0.584	711	-1.796
9900	5196	1279	-0.753	-19970	1.941	1.706	2.097	0.313	0.600	724	-1.786
10000	5934	1489	-0.851	-20110	1.926	1.764	2.199	0.265	0.616	746	-1.772

Table A.26. Saturated Vapour ($(O/U)^{(vap)} = 2.00$)

T K	P bar	ρ $kg\,m^{-3}$	H $10^{3} kJ\,kg^{-1}$	G $kJ\,kg^{-1}$	S $kJ\,kg^{-1}\,K^{-1}$	C_P	C_P/C_V	β_T $10^{-3} bar^{-1}$	α_T $10^{-3} K^{-1}$	V_S $m\,s^{-1}$	$\mu_O/10kT$
3000	0.01353	0.01455	-4.769	-9895	1.709	0.432	1.104	7.434^{+4}_{10}	0.384	320	-3.234
3120	0.02711	0.02797	-4.727	-10033	1.701	0.466	1.105	3.718^{+4}_{10}	0.383	326	-3.140
3500	0.1704	0.1550	-4.579	-10487	1.688	0.621	1.112	5976.00	0.406	346	-2.890
4000	1.066	0.8216	-4.335	-11114	1.695	0.945	1.140	981.900	0.483	376	-2.641
4500	4.414	2.865	-4.014	-11767	1.723	1.349	1.186	246.700	0.577	410	-2.457
5000	14.06	7.651	-3.617	-12444	1.765	1.715	1.240	80.840	0.641	448	-2.318
5500	37.24	17.07	-3.168	-13144	1.814	1.947	1.290	31.750	0.657	488	-2.212
6000	85.56	33.58	-2.705	-13866	1.860	2.007	1.331	14.290	0.632	527	-2.129
6500	175.4	60.19	-2.260	-14611	1.900	1.967	1.368	7.188	0.593	562	-2.062
7000	327.6	100.6	-1.854	-15377	1.932	1.858	1.403	3.970	0.552	593	-2.008
7500	568.7	159.7	-1.495	-16161	1.955	1.742	1.443	2.368	0.519	618	-1.963
8000	933.3	244.5	-1.186	-16960	1.972	1.645	1.492	1.501	0.499	638	-1.924
8500	1472	366.7	-0.932	-17770	1.981	1.578	1.561	0.996	0.494	654	-1.889
9000	2270	547.7	-0.746	-18584	1.982	1.551	1.666	0.679	0.506	669	-1.855
9500	3508	839.3	-0.662	-19391	1.972	1.588	1.842	0.461	0.544	690	-1.820
9800	4685	1137	-0.708	-19860	1.954	1.668	2.016	0.349	0.583	713	-1.795
9900	5243	1287	-0.762	-20009	1.944	1.711	2.097	0.308	0.599	727	-1.784
10000	6005	1503	-0.864	-20148	1.928	1.770	2.199	0.260	0.615	750	-1.771

Table A.26. Saturated Vapour ($(O/U)^{(vap)} = 2.01$)

T K	P bar	ρ $kg\,m^{-3}$	H $10^{3} kJ\,kg^{-1}$	G $kJ\,kg^{-1}$	S $kJ\,kg^{-1}$	C_P K^{-1}	C_P/C_V	β_T $10^{-3} bar^{-1}$	α_T $10^{-3} K^{-1}$	V_S $m\,s^{-1}$	$\mu_O/10kT$
3000	0.01355	0.01458	-4.788	-9919	1.710	0.438	1.104	7.423_{10}^{+4}	0.386	319	-3.227
3120	0.02716	0.02802	-4.746	-10057	1.702	0.472	1.105	3.712_{10}^{+4}	0.386	326	-3.134
3500	0.1708	0.1552	-4.597	-10512	1.690	0.630	1.112	5968.00	0.410	346	-2.885
4000	1.068	0.8227	-4.351	-11139	1.697	0.958	1.141	980.600	0.488	376	-2.638
4500	4.424	2.867	-4.028	-11794	1.726	1.363	1.188	246.300	0.581	410	-2.454
5000	14.10	7.652	-3.628	-12472	1.769	1.728	1.241	80.730	0.644	448	-2.316
5500	37.34	17.07	-3.177	-13173	1.817	1.959	1.291	31.690	0.659	488	-2.211
6000	85.79	33.57	-2.712	-13897	1.864	2.018	1.332	14.260	0.633	527	-2.128
6500	175.9	60.17	-2.265	-14643	1.904	1.976	1.368	7.171	0.594	563	-2.061
7000	328.6	100.6	-1.858	-15410	1.936	1.866	1.404	3.959	0.553	594	-2.007
7500	570.6	159.7	-1.498	-16195	1.960	1.750	1.444	2.361	0.520	619	-1.962
8000	936.8	244.7	-1.188	-16996	1.976	1.652	1.493	1.496	0.500	639	-1.923
8500	1478	367.0	-0.935	-17807	1.985	1.585	1.562	0.991	0.494	655	-1.888
9000	2281	548.6	-0.749	-18622	1.986	1.557	1.667	0.675	0.506	671	-1.854
9500	3530	841.8	-0.666	-19430	1.975	1.594	1.842	0.458	0.543	692	-1.819
9800	4723	1142	-0.715	-19899	1.958	1.673	2.016	0.345	0.582	716	-1.794
9900	5291	1295	-0.771	-20048	1.947	1.717	2.097	0.304	0.598	730	-1.783
10000	6078	1518	-0.878	-20186	1.931	1.776	2.200	0.255	0.613	753	-1.769

Table A.26. Saturated Vapour ($(O/U)^{(vap)} = 2.02$)

T K	P bar	ρ $kg\,m^{-3}$	H $10^{3} kJ\,kg^{-1}$	G $kJ\,kg^{-1}$	S $kJ\,kg^{-1}$	C_P K^{-1}	C_P/C_V	β_T $10^{-3} bar^{-1}$	α_T $10^{-3} K^{-1}$	V_S $m\,s^{-1}$	$\mu_O/10kT$
3000	0.01359	0.01462	-4.807	-9942	1.712	0.443	1.104	7.407^{+4}_{10}	0.389	319	-3.220
3120	0.02722	0.02809	-4.764	-10081	1.704	0.479	1.105	3.705^{+4}_{10}	0.389	326	-3.128
3500	0.1712	0.1556	-4.614	-10537	1.692	0.640	1.113	5957.00	0.414	346	-2.881
4000	1.071	0.8240	-4.367	-11165	1.700	0.971	1.142	979.200	0.493	376	-2.635
4500	4.435	2.870	-4.041	-11820	1.729	1.377	1.189	246.000	0.586	410	-2.452
5000	14.13	7.654	-3.639	-12500	1.772	1.742	1.243	80.600	0.648	449	-2.315
5500	37.43	17.07	-3.185	-13202	1.821	1.971	1.292	31.640	0.662	489	-2.209
6000	86.03	33.56	-2.719	-13927	1.868	2.029	1.333	14.230	0.635	528	-2.126
6500	176.4	60.15	-2.271	-14674	1.908	1.986	1.369	7.153	0.596	564	-2.060
7000	329.7	100.6	-1.862	-15443	1.940	1.875	1.405	3.948	0.554	595	-2.006
7500	572.6	159.7	-1.501	-16230	1.964	1.758	1.444	2.353	0.521	620	-1.961
8000	940.4	244.8	-1.191	-17032	1.980	1.659	1.494	1.490	0.500	640	-1.922
8500	1485	367.4	-0.937	-17844	1.989	1.591	1.563	0.987	0.494	657	-1.887
9000	2292	549.5	-0.751	-18660	1.990	1.563	1.667	0.671	0.506	672	-1.854
9500	3551	844.2	-0.669	-19469	1.979	1.599	1.843	0.454	0.543	693	-1.818
9800	4760	1148	-0.721	-19938	1.961	1.679	2.017	0.341	0.581	718	-1.793
9900	5341	1303	-0.779	-20086	1.950	1.722	2.097	0.300	0.596	733	-1.782
10000	6154	1533	-0.893	-20223	1.933	1.781	2.200	0.250	0.611	757	-1.767

Table A.26. Saturated Vapour ($(O/U)^{(vap)} = 2.03$)

T K	P bar	ρ $kg\,m^{-3}$	H $10^{3} kJ\,kg^{-1}$	G $kJ\,kg^{-1}$	S $kJ\,kg^{-1}\,K^{-1}$	C_P	C_P/C_V	β_T $10^{-3} bar^{-1}$	α_T $10^{-3} K^{-1}$	V_S $m\,s^{-1}$	$\mu_O/10kT$
3000	0.01363	0.01466	-4.826	-9966	1.713	0.449	1.104	7.388^{+4}_{10}	0.392	319	-3.213
3120	0.02730	0.02818	-4.783	-10105	1.706	0.486	1.105	3.696^{+4}_{10}	0.392	326	-3.122
3500	0.1717	0.1560	-4.632	-10561	1.694	0.650	1.113	5945.00	0.418	346	-2.877
4000	1.074	0.8255	-4.382	-11190	1.702	0.984	1.143	977.600	0.498	376	-2.632
4500	4.447	2.872	-4.055	-11847	1.732	1.392	1.190	245.600	0.591	411	-2.450
5000	14.17	7.656	-3.650	-12527	1.776	1.755	1.244	80.470	0.651	449	-2.313
5500	37.53	17.07	-3.194	-13231	1.825	1.984	1.293	31.580	0.664	490	-2.208
6000	86.26	33.55	-2.725	-13957	1.872	2.040	1.334	14.200	0.637	529	-2.125
6500	176.9	60.14	-2.276	-14706	1.912	1.996	1.370	7.136	0.597	565	-2.059
7000	330.7	100.6	-1.866	-15476	1.944	1.884	1.405	3.937	0.554	596	-2.005
7500	574.5	159.8	-1.504	-16264	1.968	1.765	1.445	2.346	0.521	621	-1.960
8000	943.9	244.9	-1.193	-17067	1.984	1.666	1.495	1.485	0.501	641	-1.921
8500	1491	367.7	-0.939	-17881	1.993	1.597	1.563	0.983	0.495	658	-1.886
9000	2304	550.4	-0.753	-18699	1.994	1.568	1.668	0.668	0.506	674	-1.853
9500	3573	846.7	-0.673	-19508	1.983	1.605	1.843	0.450	0.543	695	-1.817
9800	4799	1153	-0.728	-19977	1.964	1.684	2.017	0.337	0.580	720	-1.791
9900	5391	1311	-0.788	-20125	1.953	1.728	2.097	0.296	0.595	736	-1.780
10000	6234	1550	-0.908	-20260	1.935	1.787	2.201	0.245	0.608	761	-1.766

Table A.26. Saturated Vapour ($(O/U)^{(vap)} = 2.04$)

T K	P bar	ρ $kg\,m^{-3}$	H $10^{3} kJ\,kg^{-1}$	G $kJ\,kg^{-1}$	S $kJ\,kg^{-1}\,K^{-1}$	C_P $kJ\,kg^{-1}\,K^{-1}$	C_P/C_V	β_T $10^{-3}bar^{-1}$	α_T $10^{-3}K^{-1}$	V_S $m\,s^{-1}$	$\mu_O/10kT$
3000	0.01368	0.01472	-4.844	-9989	1.715	0.455	1.104	7.365^{+4}_{10}	0.394	319	-3.206
3120	0.02739	0.02828	-4.801	-10129	1.708	0.492	1.105	3.685^{+4}_{10}	0.395	326	-3.116
3500	0.1722	0.1564	-4.649	-10585	1.696	0.660	1.114	5931.00	0.423	346	-2.872
4000	1.077	0.8271	-4.398	-11216	1.704	0.998	1.144	975.800	0.503	376	-2.628
4500	4.459	2.875	-4.068	-11873	1.735	1.406	1.192	245.200	0.595	411	-2.448
5000	14.21	7.658	-3.660	-12555	1.779	1.768	1.245	80.330	0.655	450	-2.312
5500	37.63	17.07	-3.203	-13260	1.829	1.996	1.294	31.520	0.666	491	-2.207
6000	86.50	33.54	-2.732	-13987	1.876	2.050	1.335	14.170	0.638	530	-2.124
6500	177.4	60.13	-2.281	-14738	1.916	2.005	1.371	7.119	0.598	566	-2.058
7000	331.7	100.6	-1.869	-15509	1.948	1.892	1.406	3.926	0.555	597	-2.005
7500	576.5	159.8	-1.507	-16298	1.972	1.773	1.446	2.339	0.522	622	-1.959
8000	947.5	245.0	-1.195	-17103	1.988	1.673	1.496	1.479	0.501	642	-1.921
8500	1497	368.1	-0.941	-17918	1.997	1.603	1.564	0.978	0.495	659	-1.885
9000	2315	551.3	-0.756	-18737	1.998	1.574	1.669	0.664	0.507	675	-1.852
9500	3596	849.2	-0.677	-19547	1.986	1.610	1.843	0.447	0.543	697	-1.816
9800	4837	1159	-0.735	-20016	1.967	1.690	2.017	0.333	0.579	723	-1.790
9900	5443	1320	-0.797	-20163	1.956	1.734	2.097	0.291	0.594	738	-1.779
10000	6318	1567	-0.924	-20297	1.937	1.793	2.202	0.240	0.606	766	-1.764

Table A.26. Saturated Vapour ($(O/U)^{(vap)} = 2.05$)

T K	P bar	ρ $kg\,m^{-3}$	H $10^3 kJ\,kg^{-1}$	G $kJ\,kg^{-1}$	S $kJ\,kg^{-1}$	C_P K^{-1}	C_P/C_V	β_T $10^{-3}bar^{-1}$	α_T $10^{-3}K^{-1}$	V_S $m\,s^{-1}$	$\mu_O/10kT$
3000	0.01373	0.01478	-4.862	-10012	1.717	0.461	1.105	7.338^{+4}_{10}	0.397	319	-3.200
3120	0.02750	0.02839	-4.819	-10152	1.709	0.499	1.105	3.672^{+4}_{10}	0.399	326	-3.110
3500	0.1728	0.1569	-4.666	-10610	1.698	0.670	1.114	5915.00	0.427	346	-2.868
4000	1.080	0.8288	-4.413	-11241	1.707	1.011	1.145	973.800	0.508	377	-2.625
4500	4.472	2.878	-4.081	-11900	1.738	1.419	1.193	244.800	0.600	411	-2.446
5000	14.25	7.661	-3.671	-12582	1.782	1.781	1.246	80.180	0.658	450	-2.310
5500	37.74	17.07	-3.211	-13289	1.832	2.007	1.295	31.450	0.668	491	-2.205
6000	86.74	33.54	-2.738	-14018	1.880	2.061	1.336	14.140	0.640	531	-2.123
6500	177.9	60.11	-2.286	-14769	1.921	2.015	1.372	7.101	0.599	567	-2.057
7000	332.8	100.6	-1.873	-15542	1.953	1.900	1.407	3.916	0.556	598	-2.004
7500	578.5	159.8	-1.509	-16333	1.976	1.780	1.447	2.331	0.523	623	-1.959
8000	951.1	245.1	-1.198	-17138	1.993	1.680	1.496	1.474	0.502	644	-1.920
8500	1504	368.4	-0.943	-17955	2.001	1.609	1.565	0.974	0.495	660	-1.885
9000	2326	552.2	-0.758	-18775	2.002	1.580	1.669	0.660	0.507	677	-1.851
9500	3618	851.7	-0.681	-19586	1.990	1.616	1.844	0.443	0.542	699	-1.815
9800	4877	1165	-0.742	-20054	1.971	1.695	2.017	0.329	0.579	725	-1.789
9900	5495	1329	-0.807	-20201	1.959	1.739	2.098	0.287	0.592	741	-1.778
10000	6406	1586	-0.941	-20334	1.939	1.798	2.202	0.234	0.604	770	-1.762

Table A.26. Saturated Vapour ($(O/U)^{(vap)} = 2.06$)

T K	P bar	ρ $kg\,m^{-3}$	H $10^{3} kJ\,kg^{-1}$	G $kJ\,kg^{-1}$	S $kJ\,kg^{-1}$	C_P K^{-1}	C_P/C_V	β_T $10^{-3} bar^{-1}$	α_T $10^{-3} K^{-1}$	V_S $m\,s^{-1}$	$\mu_O/10kT$
3000	0.01379	0.01485	-4.880	-10036	1.719	0.467	1.105	7.307^{+4}_{10}	0.400	319	-3.193
3120	0.02762	0.02852	-4.837	-10175	1.711	0.506	1.106	3.658^{+4}_{10}	0.402	326	-3.104
3500	0.1734	0.1575	-4.683	-10634	1.700	0.680	1.115	5897.00	0.431	347	-2.863
4000	1.084	0.8307	-4.429	-11266	1.709	1.024	1.146	971.600	0.513	377	-2.622
4500	4.485	2.882	-4.094	-11926	1.740	1.433	1.194	244.300	0.604	412	-2.444
5000	14.29	7.664	-3.681	-12610	1.786	1.794	1.248	80.030	0.661	451	-2.308
5500	37.84	17.07	-3.219	-13317	1.836	2.019	1.296	31.390	0.670	492	-2.204
6000	86.99	33.53	-2.745	-14048	1.884	2.071	1.336	14.100	0.641	532	-2.122
6500	178.4	60.10	-2.291	-14801	1.925	2.024	1.372	7.083	0.600	568	-2.056
7000	333.8	100.5	-1.877	-15574	1.957	1.909	1.408	3.905	0.557	599	-2.003
7500	580.4	159.8	-1.512	-16367	1.981	1.788	1.448	2.324	0.523	624	-1.958
8000	954.7	245.2	-1.200	-17174	1.997	1.686	1.497	1.469	0.502	645	-1.919
8500	1510	368.8	-0.945	-17992	2.005	1.616	1.566	0.970	0.496	662	-1.884
9000	2338	553.1	-0.761	-18813	2.006	1.585	1.670	0.657	0.507	678	-1.850
9500	3641	854.3	-0.685	-19625	1.994	1.621	1.844	0.440	0.542	701	-1.814
9800	4917	1171	-0.748	-20093	1.974	1.700	2.017	0.326	0.578	727	-1.788
9900	5549	1338	-0.817	-20239	1.962	1.744	2.098	0.283	0.591	744	-1.776
10000	6499	1606	-0.959	-20370	1.941	1.804	2.203	0.229	0.601	775	-1.760

Table A.26. Saturated Vapour ($(O/U)^{(vap)} = 2.08$)

T K	P bar	ρ $kg\,m^{-3}$	H $10^{3} kJ\,kg^{-1}$	G $kJ\,kg^{-1}$	S $kJ\,kg^{-1}\,K^{-1}$	C_P	C_P/C_V	β_T $10^{-3} bar^{-1}$	α_T $10^{-3} K^{-1}$	V_S $m\,s^{-1}$	$\mu_O/10kT$
3000	0.01394	0.01502	-4.915	-10081	1.722	0.479	1.105	7.235^{+4}_{10}	0.407	319	-3.180
3120	0.02790	0.02881	-4.871	-10222	1.715	0.520	1.106	3.625^{+4}_{10}	0.409	325	-3.092
3500	0.1749	0.1587	-4.717	-10681	1.704	0.700	1.117	5855.00	0.441	347	-2.855
4000	1.091	0.8349	-4.458	-11316	1.714	1.050	1.149	966.600	0.523	377	-2.616
4500	4.514	2.890	-4.119	-11978	1.746	1.460	1.197	243.300	0.613	413	-2.439
5000	14.37	7.672	-3.702	-12665	1.792	1.819	1.250	79.710	0.667	452	-2.305
5500	38.05	17.07	-3.236	-13375	1.843	2.041	1.298	31.250	0.674	493	-2.202
6000	87.48	33.52	-2.758	-14108	1.892	2.092	1.338	14.040	0.644	533	-2.120
6500	179.5	60.08	-2.300	-14863	1.933	2.042	1.374	7.047	0.602	570	-2.055
7000	335.9	100.5	-1.884	-15640	1.965	1.925	1.409	3.883	0.558	601	-2.001
7500	584.4	159.9	-1.518	-16435	1.989	1.802	1.449	2.309	0.524	626	-1.956
8000	961.9	245.5	-1.204	-17245	2.005	1.700	1.499	1.458	0.503	647	-1.917
8500	1523	369.5	-0.949	-18065	2.014	1.628	1.567	0.961	0.496	664	-1.882
9000	2361	555.0	-0.765	-18888	2.014	1.597	1.671	0.649	0.507	681	-1.848
9500	3686	859.5	-0.693	-19702	2.001	1.632	1.845	0.433	0.541	704	-1.812
9800	4998	1183	-0.763	-20170	1.980	1.711	2.017	0.318	0.576	732	-1.785
9900	5660	1358	-0.837	-20315	1.968	1.755	2.098	0.275	0.588	750	-1.774
10000	6705	1652	-1.000	-20441	1.944	1.814	2.205	0.217	0.596	785	-1.756

Table A.26. Saturated Vapour ($(O/U)^{(vap)} = 2.10$)

T K	P bar	ρ $kg\,m^{-3}$	H $10^{3} kJ\,kg^{-1}$	G $kJ\,kg^{-1}$	S $kJ\,kg^{-1}$	C_P K^{-1}	C_P/C_V	β_T $10^{-3} bar^{-1}$	α_T $10^{-3} K^{-1}$	V_S $m\,s^{-1}$	$\mu_O/10kT$
3000	0.01412	0.01521	-4.950	-10127	1.726	0.492	1.106	7.151^{+4}_{10}	0.413	319	-3.168
3120	0.02823	0.02915	-4.906	-10268	1.719	0.535	1.107	3.586^{+4}_{10}	0.417	325	-3.081
3500	0.1767	0.1602	-4.750	-10729	1.708	0.720	1.118	5807.00	0.450	347	-2.846
4000	1.100	0.8398	-4.488	-11366	1.719	1.075	1.151	960.900	0.534	378	-2.610
4500	4.545	2.899	-4.144	-12030	1.753	1.486	1.199	242.200	0.621	413	-2.435
5000	14.46	7.682	-3.722	-12719	1.799	1.844	1.252	79.360	0.673	453	-2.302
5500	38.28	17.07	-3.252	-13432	1.851	2.063	1.300	31.110	0.678	495	-2.199
6000	87.99	33.52	-2.770	-14167	1.900	2.112	1.340	13.970	0.646	535	-2.118
6500	180.6	60.07	-2.310	-14926	1.941	2.060	1.375	7.011	0.604	571	-2.053
7000	338.1	100.5	-1.891	-15705	1.973	1.941	1.411	3.860	0.560	603	-1.999
7500	588.4	160.0	-1.523	-16503	1.997	1.817	1.450	2.294	0.526	629	-1.955
8000	969.2	245.8	-1.209	-17315	2.013	1.713	1.500	1.447	0.504	649	-1.916
8500	1536	370.2	-0.953	-18138	2.022	1.640	1.569	0.953	0.497	667	-1.880
9000	2384	556.9	-0.770	-18964	2.021	1.608	1.673	0.642	0.507	684	-1.847
9500	3733	864.8	-0.701	-19779	2.008	1.643	1.846	0.426	0.540	708	-1.810
9800	5083	1196	-0.777	-20246	1.987	1.722	2.017	0.310	0.574	737	-1.783
9900	5777	1378	-0.858	-20390	1.973	1.765	2.098	0.266	0.585	756	-1.771
10000	6951	1710	-1.050	-20510	1.946	1.825	2.209	0.204	0.589	797	-1.751

Table A.26. Saturated Vapour ($(O/U)^{(vap)} = 2.125$)

T K	P bar	ρ $kg\,m^{-3}$	H $kJ\,kg^{-1}$	G $kJ\,kg^{-1}$	S $kJ\,kg^{-1}$	C_P K^{-1}	C_P/C_V	β_T $10^{-3} bar^{-1}$	α_T K^{-1}	V_S $m\,s^{-1}$	$\mu_O/10kT$
3000	0.01438	0.01549	-4.992	-10182	1.730	0.508	1.107	7.031^{+4}_{10}	0.422	319	-3.153
3120	0.02871	0.02965	-4.948	-10324	1.723	0.553	1.108	3.530^{+4}_{10}	0.426	325	-3.067
3500	0.1792	0.1623	-4.790	-10788	1.714	0.747	1.120	5738.0	0.462	347	-2.836
4000	1.113	0.8466	-4.523	-11427	1.726	1.108	1.154	952.80	0.546	378	-2.603
4500	4.587	2.911	-4.174	-12095	1.760	1.518	1.203	240.60	0.631	414	-2.430
5000	14.58	7.696	-3.747	-12787	1.808	1.873	1.255	78.89	0.680	455	-2.298
5500	38.56	17.08	-3.272	-13503	1.860	2.090	1.302	30.93	0.683	496	-2.196
6000	88.64	33.51	-2.786	-14242	1.909	2.136	1.341	13.89	0.649	537	-2.116
6500	181.9	60.05	-2.322	-15003	1.951	2.082	1.377	6.960	0.606	574	-2.051
7000	340.8	100.5	-1.900	-15786	1.984	1.961	1.412	3.833	0.561	605	-1.997
7500	593.5	160.1	-1.530	-16587	2.008	1.835	1.452	2.276	0.527	631	-1.953
8000	978.3	246.1	-1.214	-17403	2.024	1.729	1.502	1.434	0.505	652	-1.914
8500	1552	371.2	-0.9581	-18229	2.032	1.654	1.570	0.942	0.497	670	-1.878
9000	2414	559.3	-0.7762	-19057	2.031	1.622	1.674	0.633	0.507	688	-1.844
9500	3792	871.6	-0.7106	-19874	2.017	1.656	1.846	0.418	0.539	712	-1.808
9800	5192	1213	-0.7962	-20340	1.994	1.734	2.017	0.301	0.571	743	-1.780
9900	5933	1406	-0.8858	-20482	1.979	1.777	2.098	0.255	0.581	765	-1.767
10000	7386	1819	-1.141	-20588	1.945	1.838	2.217	0.182	0.579	817	-1.744

Table A.26. Saturated Vapour ($(O/U)^{(vap)} = 2.15$)

T	P	ρ	H	G	S	C_P	C_P/C_V	β_T	α_T	V_S	$\mu_O/10kT$
K	bar	$kg\,m^{-3}$	$10^{3} kJ\,kg^{-1}$	$kJ\,kg^{-1}$	$kJ\,kg^{-1}$	K^{-1}		$10^{-3} bar^{-1}$	$10^{-3} K^{-1}$	$m\,s^{-1}$	
3000	0.01467	0.01581	-5.033	-10238	1.735	0.526	1.108	6.897^{+4}_{10}	0.432	319	-3.139
3120	0.02927	0.03022	-4.988	-10380	1.728	0.573	1.110	3.468^{+4}_{10}	0.436	325	-3.054
3500	0.1820	0.1646	-4.829	-10846	1.719	0.773	1.123	5660.00	0.474	347	-2.825
4000	1.126	0.8544	-4.558	-11488	1.732	1.139	1.157	943.700	0.559	379	-2.596
4500	4.632	2.925	-4.203	-12159	1.768	1.549	1.206	238.800	0.641	415	-2.425
5000	14.70	7.713	-3.771	-12854	1.817	1.901	1.257	78.380	0.686	456	-2.294
5500	38.86	17.09	-3.291	-13573	1.869	2.115	1.304	30.740	0.686	498	-2.193
6000	89.30	33.51	-2.801	-14316	1.919	2.159	1.343	13.800	0.652	539	-2.113
6500	183.3	60.05	-2.333	-15081	1.961	2.103	1.378	6.918	0.608	576	-2.048
7000	343.5	100.6	-1.909	-15867	1.994	1.980	1.414	3.805	0.563	608	-1.995
7500	598.6	160.2	-1.536	-16671	2.018	1.852	1.454	2.257	0.528	634	-1.951
8000	987.5	246.5	-1.219	-17491	2.034	1.745	1.504	1.420	0.506	655	-1.912
8500	1569	372.1	-0.963	-18320	2.042	1.669	1.572	0.932	0.498	673	-1.876
9000	2444	561.7	-0.782	-19151	2.041	1.635	1.675	0.624	0.507	691	-1.842
9500	3852	878.7	-0.721	-19969	2.026	1.669	1.847	0.409	0.538	717	-1.806
9800	5306	1231	-0.816	-20434	2.002	1.747	2.017	0.291	0.568	750	-1.777
9900	6101	1437	-0.917	-20574	1.986	1.789	2.098	0.244	0.576	773	-1.763
10000	8434	2107	-1.361	-20637	1.928	1.835	2.230	0.140	0.547	868	-1.728

Table A.26. Saturated Vapour ($(O/U)^{(vap)} = 2.20$)

T K	P bar	ρ $kg\,m^{-3}$	H $10^{3} kJ\,kg^{-1}$	G $kJ\,kg^{-1}$	S $kJ\,kg^{-1}\,K^{-1}$	C_P	C_P/C_V	β_T $10^{-4} bar^{-1}$	α_T $10^{-3} K^{-1}$	V_S $m\,s^{-1}$	$\mu_O/10kT$
3000	0.01537	0.01657	-5.114	-10346	1.744	0.564	1.110	6.600^{+5}_{10}	0.451	319	-3.113
3120	0.03059	0.03158	-5.068	-10490	1.738	0.616	1.112	3.328^{+5}_{10}	0.458	325	-3.030
3500	0.1888	0.1703	-4.904	-10960	1.730	0.829	1.128	5.480^{+4}_{10}	0.500	348	-2.806
4000	1.158	0.8727	-4.625	-11608	1.746	1.202	1.164	9226.00	0.584	380	-2.582
4500	4.734	2.959	-4.260	-12285	1.783	1.608	1.212	2348.00	0.659	418	-2.415
5000	14.96	7.754	-3.818	-12987	1.834	1.953	1.262	772.600	0.697	459	-2.287
5500	39.48	17.12	-3.328	-13713	1.888	2.162	1.308	303.300	0.693	502	-2.188
6000	90.67	33.53	-2.830	-14462	1.939	2.203	1.346	136.100	0.657	543	-2.108
6500	186.1	60.05	-2.355	-15234	1.981	2.144	1.381	68.220	0.611	581	-2.044
7000	349.0	100.6	-1.925	-16027	2.015	2.017	1.417	37.490	0.565	613	-1.991
7500	608.9	160.4	-1.549	-16839	2.039	1.886	1.457	22.210	0.530	639	-1.947
8000	1006	247.2	-1.230	-17665	2.054	1.775	1.507	13.940	0.507	661	-1.908
8500	1602	374.2	-0.973	-18500	2.062	1.697	1.575	9.113	0.499	680	-1.873
9000	2504	566.8	-0.794	-19336	2.060	1.662	1.678	6.072	0.507	698	-1.838
9500	3976	893.4	-0.742	-20157	2.044	1.694	1.848	3.928	0.536	726	-1.801
9800	5551	1270	-0.859	-20619	2.016	1.770	2.017	2.725	0.561	763	-1.771
9900	6487	1511	-0.988	-20753	1.996	1.811	2.097	2.210	0.565	792	-1.756
10000	7992	1894	-1.207	-20862	1.966	1.816	2.157	1.563	0.537	854	-1.734

Table A.26. Saturated Vapour ($(O/U)^{(vap)} = 2.25$)

T K	P bar	ρ $kg\,m^{-3}$	H $10^{3} kJ\,kg^{-1}$	G $kJ\,kg^{-1}$	S $kJ\,kg^{-1}\,K^{-1}$	C_P	C_P/C_V	β_T $10^{-4} bar^{-1}$	α_T $10^{-3} K^{-1}$	V_S $m\,s^{-1}$	$\mu_O/10kT$
3000	0.01621	0.01748	-5.192	-10452	1.753	0.605	1.112	6.274^{+5}_{10}	0.472	318	-3.089
3120	0.03219	0.03321	-5.146	-10597	1.747	0.661	1.115	3.173^{+5}_{10}	0.480	325	-3.007
3500	0.1970	0.1772	-4.977	-11071	1.741	0.887	1.133	5.275^{+4}_{10}	0.527	348	-2.787
4000	1.196	0.8947	-4.689	-11726	1.759	1.264	1.171	8982.00	0.608	382	-2.568
4500	4.850	2.999	-4.313	-12410	1.799	1.663	1.218	2302.00	0.676	420	-2.405
5000	15.26	7.803	-3.862	-13119	1.851	2.000	1.267	760.200	0.707	462	-2.280
5500	40.15	17.17	-3.364	-13851	1.907	2.206	1.311	298.900	0.699	505	-2.182
6000	92.09	33.56	-2.858	-14607	1.958	2.243	1.349	134.200	0.660	547	-2.104
6500	189.0	60.08	-2.377	-15386	2.001	2.182	1.384	67.250	0.613	585	-2.040
7000	354.6	100.7	-1.941	-16186	2.035	2.052	1.419	36.920	0.567	618	-1.988
7500	619.3	160.7	-1.562	-17004	2.059	1.918	1.459	21.840	0.532	645	-1.943
8000	1025	248.1	-1.240	-17837	2.075	1.805	1.509	13.680	0.508	667	-1.904
8500	1636	376.3	-0.983	-18678	2.082	1.724	1.577	8.912	0.499	686	-1.869
9000	2566	572.1	-0.806	-19519	2.079	1.687	1.680	5.903	0.506	705	-1.834
9500	4106	909.0	-0.764	-20343	2.061	1.718	1.848	3.767	0.533	735	-1.796
9800	5824	1315	-0.907	-20801	2.030	1.791	2.015	2.534	0.554	778	-1.765
9900	6984	1612	-1.081	-20923	2.004	1.829	2.097	1.953	0.552	816	-1.746
10000	8277	1902	-1.217	-21049	1.983	1.800	2.114	1.459	0.513	873	-1.729

Table A.27. Boiling Liquid ($(O/U)^{(liq)} = 1.75$)

T K	P bar	ρ $kg\,m^{-3}$	H $10^{3}kJ\,kg^{-1}$	G $kJ\,kg^{-1}$	S $kJ\,kg^{-1}\,K^{-1}$	C_P	C_P/C_V	β_T $10^{-4}bar^{-1}$	α_T $10^{-3}K^{-1}$	V_S $m\,s^{-1}$	$\mu_O/10kT$
3000	0.02008	9358	-5.882	-9248	1.122	0.493	1.277	0.037	0.111	1932	-3.504
3120	0.03838	9237	-5.823	-9383	1.141	0.488	1.256	0.038	0.106	1890	-3.391
3500	0.2125	8896	-5.648	-9827	1.194	0.426	1.234	0.043	0.094	1801	-3.087
4000	1.171	8506	-5.454	-10438	1.246	0.356	1.255	0.049	0.087	1729	-2.779
4500	4.437	8142	-5.285	-11071	1.286	0.328	1.304	0.057	0.089	1673	-2.543
5000	13.60	7776	-5.122	-11723	1.320	0.329	1.366	0.067	0.096	1615	-2.359
5500	37.20	7393	-4.953	-12391	1.352	0.347	1.433	0.081	0.107	1546	-2.212
6000	95.07	6990	-4.773	-13074	1.383	0.375	1.493	0.100	0.120	1460	-2.095
6500	229.3	6566	-4.580	-13771	1.414	0.402	1.564	0.127	0.136	1371	-2.000
7000	518.2	6130	-4.369	-14482	1.445	0.443	1.632	0.165	0.157	1271	-1.922
7500	1084	5687	-4.135	-15202	1.476	0.514	1.718	0.220	0.189	1173	-1.858
8000	2071	5244	-3.860	-15930	1.509	0.650	1.854	0.300	0.243	1085	-1.807
8500	3549	4801	-3.521	-16664	1.546	0.883	2.070	0.416	0.327	1018	-1.768
9000	5380	4341	-3.095	-17408	1.590	1.211	2.349	0.571	0.438	974	-1.741
9500	7149	3814	-2.555	-18172	1.644	1.564	2.611	0.782	0.551	935	-1.728
9800	7889	3413	-2.150	-18651	1.684	1.754	2.731	0.983	0.617	902	-1.728
9900	8025	3250	-1.993	-18816	1.699	1.811	2.761	1.080	0.640	887	-1.730
9950	8063	3160	-1.908	-18900	1.708	1.838	2.775	1.140	0.652	878	-1.731
10000	8075	3062	-1.818	-18985	1.717	1.863	2.786	1.209	0.665	867	-1.733

Table A.27. Boiling Liquid ($(O/U)^{(liq)} = 1.80$)

T K	P bar	ρ $kg\,m^{-3}$	H $10\,^3kJ\;kg^{-1}$	G $kJ\;kg^{-1}$	S $kJ\;kg^{-1}\;K^{-1}$	C_P	C_P/C_V	β_T $10^{-4}bar^{-1}$	α_T $10^{-3}K^{-1}$	V_S $m\;s^{-1}$	$\mu_O/10kT$
3000	0.01841	9283	-6.001	-9383	1.127	0.494	1.278	0.037	0.111	1935	-3.475
3120	0.03536	9163	-5.942	-9519	1.147	0.489	1.256	0.038	0.106	1892	-3.362
3500	0.1987	8824	-5.767	-9966	1.200	0.429	1.233	0.043	0.094	1802	-3.058
4000	1.117	8436	-5.571	-10579	1.252	0.360	1.254	0.050	0.088	1729	-2.750
4500	4.346	8073	-5.399	-11216	1.292	0.333	1.305	0.058	0.090	1672	-2.515
5000	13.81	7705	-5.233	-11871	1.328	0.337	1.370	0.068	0.098	1612	-2.333
5500	39.33	7318	-5.060	-12542	1.360	0.356	1.439	0.083	0.109	1541	-2.189
6000	104.1	6910	-4.876	-13230	1.392	0.384	1.499	0.103	0.123	1453	-2.074
6500	257.0	6483	-4.677	-13931	1.424	0.412	1.570	0.131	0.140	1362	-1.981
7000	587.8	6045	-4.461	-14646	1.455	0.457	1.638	0.170	0.162	1261	-1.905
7500	1232	5603	-4.218	-15370	1.487	0.537	1.727	0.228	0.196	1164	-1.843
8000	2333	5165	-3.931	-16101	1.521	0.688	1.868	0.310	0.253	1080	-1.794
8500	3934	4730	-3.576	-16839	1.560	0.937	2.083	0.423	0.338	1021	-1.757
9000	5842	4277	-3.131	-17588	1.606	1.263	2.341	0.567	0.442	982	-1.732
9500	7593	3750	-2.569	-18361	1.662	1.592	2.564	0.762	0.541	947	-1.721
9800	8250	3340	-2.147	-18848	1.704	1.764	2.659	0.956	0.599	913	-1.723
9900	8339	3171	-1.982	-19016	1.721	1.815	2.683	1.055	0.620	896	-1.725
9950	8348	3076	-1.892	-19102	1.730	1.839	2.693	1.116	0.632	886	-1.727
10000	8325	2971	-1.795	-19190	1.739	1.862	2.701	1.189	0.644	874	-1.729

Table A.27. Boiling Liquid ($(O/U)^{(liq)} = 1.85$)

T K	P bar	ρ $kg\,m^{-3}$	H $10^{3} kJ\,kg^{-1}$	G $kJ\,kg^{-1}$	S $kJ\,kg^{-1}$	C_P K^{-1}	C_P/C_V	β_T $10^{-4} bar^{-1}$	α_T $10^{-3} K^{-1}$	V_S $m\,s^{-1}$	$\mu_O/10kT$
3000	0.01678	9209	-6.120	-9516	1.132	0.494	1.280	0.037	0.111	1938	-3.439
3120	0.03243	9090	-6.061	-9653	1.151	0.491	1.257	0.039	0.106	1895	-3.326
3500	0.1857	8753	-5.884	-10101	1.205	0.432	1.233	0.043	0.094	1804	-3.022
4000	1.073	8367	-5.687	-10717	1.258	0.365	1.255	0.050	0.088	1729	-2.715
4500	4.336	8003	-5.512	-11357	1.299	0.341	1.308	0.059	0.091	1671	-2.483
5000	14.41	7630	-5.341	-12015	1.335	0.346	1.375	0.070	0.100	1609	-2.304
5500	42.81	7238	-5.163	-12691	1.369	0.365	1.445	0.085	0.112	1535	-2.164
6000	116.8	6826	-4.974	-13382	1.401	0.394	1.505	0.106	0.126	1445	-2.052
6500	293.7	6396	-4.771	-14088	1.433	0.423	1.575	0.135	0.143	1351	-1.962
7000	675.5	5958	-4.548	-14807	1.465	0.472	1.643	0.176	0.167	1250	-1.888
7500	1410	5519	-4.297	-15535	1.498	0.563	1.734	0.236	0.203	1155	-1.829
8000	2635	5088	-3.996	-16270	1.534	0.729	1.879	0.319	0.263	1077	-1.782
8500	4355	4661	-3.625	-17011	1.575	0.988	2.089	0.427	0.347	1025	-1.747
9000	6327	4216	-3.162	-17766	1.623	1.307	2.324	0.559	0.442	993	-1.724
9500	8038	3688	-2.580	-18548	1.681	1.613	2.512	0.740	0.528	959	-1.715
9800	8595	3268	-2.140	-19043	1.725	1.769	2.589	0.931	0.581	923	-1.718
9900	8630	3091	-1.966	-19215	1.742	1.816	2.608	1.032	0.601	904	-1.721
9950	8604	2990	-1.870	-19303	1.752	1.838	2.615	1.097	0.612	893	-1.723
10000	8539	2878	-1.766	-19393	1.763	1.858	2.621	1.176	0.624	880	-1.726

Table A.27. Boiling Liquid ($(O/U)^{(liq)} = 1.875$)

T K	P bar	ρ $kg\,m^{-3}$	H $10^{3} kJ\,kg^{-1}$	G $kJ\,kg^{-1}$	S $kJ\,kg^{-1}$	C_P K^{-1}	C_P/C_V	β_T $10^{-4} bar^{-1}$	α_T $10^{-3} K^{-1}$	V_S $m\,s^{-1}$	$\mu_O/10kT$
3000	0.01598	9173	-6.179	-9582	1.134	0.494	1.280	0.037	0.111	1939	-3.418
3120	0.03102	9054	-6.120	-9719	1.154	0.492	1.257	0.039	0.106	1896	-3.305
3500	0.1798	8718	-5.943	-10168	1.207	0.434	1.233	0.043	0.094	1804	-3.001
4000	1.059	8333	-5.744	-10785	1.260	0.368	1.255	0.050	0.089	1729	-2.695
4500	4.383	7966	-5.567	-11426	1.302	0.346	1.310	0.059	0.092	1670	-2.464
5000	14.94	7591	-5.394	-12086	1.339	0.351	1.378	0.070	0.102	1606	-2.288
5500	45.27	7196	-5.213	-12764	1.373	0.370	1.448	0.086	0.113	1531	-2.150
6000	125.1	6782	-5.022	-13457	1.406	0.398	1.508	0.107	0.128	1440	-2.041
6500	316.4	6351	-4.816	-14165	1.438	0.428	1.576	0.137	0.145	1345	-1.952
7000	727.8	5913	-4.590	-14886	1.471	0.480	1.645	0.180	0.169	1244	-1.880
7500	1512	5477	-4.334	-15616	1.504	0.576	1.737	0.240	0.207	1150	-1.822
8000	2801	5050	-4.027	-16352	1.541	0.750	1.883	0.322	0.267	1076	-1.776
8500	4579	4628	-3.648	-17096	1.582	1.013	2.090	0.428	0.351	1027	-1.742
9000	6577	4187	-3.176	-17854	1.631	1.326	2.312	0.554	0.440	998	-1.720
9500	8261	3658	-2.584	-18641	1.690	1.620	2.485	0.729	0.521	966	-1.712
9800	8761	3232	-2.135	-19140	1.735	1.770	2.555	0.919	0.571	928	-1.716
9900	8765	3050	-1.956	-19314	1.753	1.815	2.572	1.022	0.591	908	-1.719
9950	8719	2946	-1.857	-19404	1.763	1.836	2.578	1.089	0.602	896	-1.722
10000	8631	2830	-1.749	-19495	1.775	1.856	2.583	1.172	0.614	882	-1.725

Table A.27. Boiling Liquid ($(O/U)^{(liq)} = 1.90$)

T K	P bar	ρ $kg\,m^{-3}$	H $10^{3} kJ\,kg^{-1}$	G $kJ\,kg^{-1}$	S $kJ\,kg^{-1}\,K^{-1}$	C_P $kJ\,kg^{-1}\,K^{-1}$	C_P/C_V	β_T $10^{-4} bar^{-1}$	α_T $10^{-3} K^{-1}$	V_S $m\,s^{-1}$	$\mu_O/10kT$
3000	0.01521	9137	-6.238	-9646	1.136	0.494	1.281	0.037	0.111	1940	-3.392
3120	0.02966	9019	-6.178	-9784	1.156	0.492	1.257	0.039	0.106	1897	-3.279
3500	0.1746	8683	-6.001	-10234	1.209	0.436	1.233	0.044	0.094	1805	-2.976
4000	1.054	8297	-5.801	-10852	1.263	0.372	1.257	0.051	0.089	1729	-2.671
4500	4.487	7929	-5.621	-11495	1.305	0.351	1.313	0.060	0.094	1668	-2.443
5000	15.69	7550	-5.445	-12157	1.342	0.357	1.382	0.071	0.103	1603	-2.271
5500	48.39	7153	-5.262	-12836	1.377	0.375	1.450	0.087	0.115	1526	-2.136
6000	135.1	6736	-5.068	-13532	1.411	0.403	1.510	0.109	0.129	1435	-2.029
6500	342.5	6305	-4.860	-14242	1.443	0.434	1.578	0.140	0.147	1339	-1.942
7000	786.3	5868	-4.631	-14965	1.476	0.488	1.646	0.183	0.172	1238	-1.872
7500	1623	5434	-4.370	-15696	1.510	0.590	1.740	0.244	0.211	1146	-1.814
8000	2978	5012	-4.057	-16434	1.547	0.770	1.887	0.326	0.272	1075	-1.770
8500	4812	4596	-3.669	-17180	1.589	1.036	2.089	0.428	0.354	1031	-1.737
9000	6831	4158	-3.189	-17941	1.639	1.343	2.299	0.548	0.438	1004	-1.717
9500	8482	3629	-2.587	-18733	1.700	1.627	2.458	0.717	0.514	972	-1.709
9800	8921	3196	-2.129	-19236	1.746	1.771	2.522	0.907	0.562	933	-1.714
9900	8892	3009	-1.945	-19413	1.764	1.815	2.537	1.014	0.582	912	-1.718
9950	8824	2902	-1.843	-19504	1.775	1.835	2.543	1.083	0.593	899	-1.720
10000	8709	2780	-1.730	-19597	1.787	1.855	2.546	1.171	0.605	884	-1.724

Table A.27. Boiling Liquid ($(O/U)^{(liq)} = 1.92$)

T	P	ρ	H	G	S	C_P	C_P/C_V	β_T	α_T	V_S	$\mu_O/10kT$
K	bar	$kg\,m^{-3}$	$10^{3} kJ\,kg^{-1}$	$kJ\,kg^{-1}$	$kJ\,kg^{-1}\,K^{-1}$			$10^{-4} bar^{-1}$	$10^{-3} K^{-1}$	$m\,s^{-1}$	
3000	0.01463	9109	-6.285	-9698	1.138	0.494	1.281	0.037	0.111	1942	-3.367
3120	0.02865	8991	-6.225	-9835	1.157	0.493	1.257	0.039	0.106	1898	-3.255
3500	0.1714	8656	-6.047	-10286	1.211	0.439	1.233	0.044	0.095	1805	-2.951
4000	1.061	8269	-5.845	-10905	1.265	0.377	1.258	0.051	0.090	1728	-2.649
4500	4.630	7897	-5.663	-11549	1.308	0.357	1.315	0.060	0.095	1666	-2.425
5000	16.50	7516	-5.485	-12212	1.345	0.361	1.384	0.072	0.104	1600	-2.256
5500	51.50	7116	-5.300	-12893	1.381	0.378	1.452	0.088	0.116	1522	-2.125
6000	144.5	6698	-5.104	-13591	1.414	0.406	1.512	0.110	0.130	1430	-2.020
6500	366.6	6267	-4.894	-14303	1.447	0.438	1.579	0.142	0.148	1333	-1.934
7000	838.8	5831	-4.663	-15027	1.481	0.495	1.647	0.186	0.174	1233	-1.865
7500	1720	5400	-4.398	-15760	1.515	0.602	1.741	0.247	0.213	1143	-1.809
8000	3129	4982	-4.079	-16499	1.552	0.787	1.888	0.328	0.275	1075	-1.765
8500	5007	4571	-3.685	-17247	1.595	1.054	2.086	0.427	0.355	1034	-1.733
9000	7041	4137	-3.199	-18010	1.646	1.355	2.286	0.542	0.436	1009	-1.714
9500	8661	3606	-2.589	-18806	1.707	1.630	2.436	0.707	0.508	977	-1.707
9800	9048	3168	-2.124	-19313	1.754	1.771	2.496	0.898	0.555	937	-1.712
9900	8989	2977	-1.936	-19492	1.773	1.813	2.510	1.007	0.575	915	-1.716
9950	8903	2866	-1.831	-19584	1.784	1.834	2.515	1.079	0.586	902	-1.719
10000	8764	2739	-1.714	-19678	1.796	1.853	2.517	1.171	0.599	886	-1.723

Table A.27. Boiling Liquid ($(O/U)^{(liq)} = 1.94$)

T K	P bar	ρ $kg\,m^{-3}$	H $10^{3} kJ\,kg^{-1}$	G $kJ\,kg^{-1}$	S $kJ\,kg^{-1}\,K^{-1}$	C_P	C_P/C_V	β_T $10^{-4} bar^{-1}$	α_T $10^{-3} K^{-1}$	V_S $m\,s^{-1}$	$\mu_O/10kT$
3000	0.01408	9080	-6.332	-9748	1.139	0.495	1.282	0.037	0.111	1943	-3.336
3120	0.02774	8963	-6.272	-9886	1.158	0.494	1.258	0.039	0.106	1899	-3.224
3500	0.1696	8628	-6.093	-10337	1.212	0.442	1.234	0.044	0.095	1805	-2.921
4000	1.084	8239	-5.889	-10957	1.267	0.383	1.260	0.051	0.091	1727	-2.622
4500	4.852	7864	-5.704	-11602	1.311	0.362	1.318	0.061	0.096	1664	-2.405
5000	17.56	7479	-5.523	-12267	1.349	0.365	1.387	0.073	0.105	1596	-2.241
5500	55.25	7078	-5.337	-12950	1.384	0.382	1.454	0.089	0.117	1517	-2.112
6000	155.4	6659	-5.140	-13649	1.418	0.409	1.513	0.112	0.131	1424	-2.010
6500	393.6	6228	-4.927	-14363	1.452	0.442	1.579	0.144	0.150	1328	-1.926
7000	896.3	5794	-4.694	-15088	1.485	0.502	1.647	0.189	0.176	1228	-1.858
7500	1825	5366	-4.425	-15823	1.520	0.613	1.742	0.250	0.216	1139	-1.803
8000	3287	4953	-4.101	-16564	1.558	0.803	1.889	0.330	0.278	1074	-1.760
8500	5206	4547	-3.701	-17313	1.601	1.070	2.083	0.426	0.356	1037	-1.730
9000	7252	4115	-3.208	-18078	1.652	1.366	2.273	0.536	0.433	1015	-1.711
9500	8838	3583	-2.591	-18879	1.714	1.633	2.414	0.697	0.502	983	-1.705
9800	9169	3139	-2.118	-19390	1.762	1.770	2.470	0.890	0.548	940	-1.711
9900	9080	2944	-1.926	-19570	1.782	1.812	2.483	1.002	0.568	918	-1.715
9950	8974	2829	-1.817	-19663	1.794	1.832	2.488	1.077	0.579	904	-1.718
10000	8809	2697	-1.696	-19759	1.806	1.851	2.490	1.174	0.592	887	-1.723

Table A.27. Boiling Liquid ($(O/U)^{(liq)} = 1.95$)

T K	P bar	ρ $kg\ m^{-3}$	H $10^{3} kJ\ kg^{-1}$	G $kJ\ kg^{-1}$	S $kJ\ kg^{-1}\ K^{-1}$	C_P	C_P/C_V	β_T $10^{-4} bar^{-1}$	α_T $10^{-3} K^{-1}$	V_S $m\ s^{-1}$	$\mu_O/10kT$
3000	0.01384	9066	-6.355	-9773	1.139	0.495	1.282	0.037	0.111	1943	-3.317
3120	0.02738	8949	-6.296	-9911	1.159	0.495	1.258	0.039	0.107	1899	-3.204
3500	0.1697	8614	-6.116	-10362	1.213	0.444	1.234	0.044	0.096	1805	-2.903
4000	1.105	8223	-5.910	-10983	1.268	0.386	1.262	0.052	0.092	1727	-2.607
4500	5.004	7846	-5.724	-11629	1.312	0.364	1.320	0.061	0.097	1662	-2.394
5000	18.20	7460	-5.542	-12294	1.350	0.367	1.388	0.073	0.106	1594	-2.232
5500	57.41	7058	-5.355	-12978	1.386	0.383	1.454	0.090	0.118	1514	-2.106
6000	161.5	6640	-5.157	-13678	1.420	0.411	1.514	0.113	0.132	1422	-2.005
6500	408.4	6209	-4.944	-14393	1.454	0.444	1.579	0.145	0.150	1325	-1.922
7000	927.2	5775	-4.709	-15119	1.487	0.506	1.647	0.190	0.176	1225	-1.855
7500	1880	5349	-4.438	-15854	1.522	0.619	1.743	0.252	0.218	1138	-1.800
8000	3369	4939	-4.111	-16596	1.561	0.811	1.889	0.331	0.280	1075	-1.758
8500	5308	4535	-3.708	-17345	1.604	1.078	2.080	0.425	0.356	1039	-1.728
9000	7359	4105	-3.213	-18113	1.656	1.371	2.266	0.533	0.432	1018	-1.709
9500	8927	3572	-2.592	-18915	1.718	1.635	2.403	0.693	0.499	985	-1.704
9800	9228	3125	-2.115	-19429	1.767	1.770	2.457	0.886	0.545	942	-1.710
9900	9123	2927	-1.920	-19610	1.787	1.812	2.470	0.999	0.564	919	-1.715
9950	9007	2811	-1.810	-19703	1.798	1.832	2.475	1.076	0.576	905	-1.718
10000	8828	2675	-1.687	-19800	1.811	1.850	2.476	1.176	0.589	887	-1.722

Table A.27. Boiling Liquid ($(O/U)^{(liq)} = 1.96$)

T	P	ρ	H	G	S	C_P	C_P/C_V	β_T	α_T	V_S	$\mu_O/10kT$
K	bar	$kg\ m^{-3}$	$10^{3} kJ\ kg^{-1}$	$kJ\ kg^{-1}$	$kJ\ kg^{-1}$	K^{-1}		$10^{-4} bar^{-1}$	$10^{-3} K^{-1}$	$m\ s^{-1}$	
3000	0.01364	9052	-6.378	-9798	1.140	0.495	1.282	0.038	0.111	1944	-3.293
3120	0.02711	8935	-6.319	-9936	1.159	0.496	1.258	0.039	0.107	1900	-3.181
3500	0.1712	8599	-6.139	-10387	1.214	0.447	1.235	0.044	0.096	1805	-2.881
4000	1.137	8207	-5.931	-11009	1.269	0.390	1.263	0.052	0.093	1726	-2.591
4500	5.190	7827	-5.743	-11655	1.314	0.367	1.321	0.061	0.098	1660	-2.382
5000	18.94	7440	-5.560	-12321	1.352	0.368	1.388	0.074	0.106	1592	-2.224
5500	59.80	7038	-5.372	-13006	1.388	0.384	1.455	0.091	0.118	1512	-2.100
6000	168.1	6619	-5.174	-13707	1.422	0.412	1.514	0.114	0.132	1419	-2.000
6500	424.0	6189	-4.960	-14422	1.456	0.446	1.579	0.146	0.151	1321	-1.918
7000	959.5	5756	-4.724	-15150	1.489	0.509	1.647	0.191	0.177	1223	-1.851
7500	1937	5332	-4.451	-15885	1.525	0.625	1.743	0.253	0.219	1136	-1.797
8000	3453	4924	-4.122	-16627	1.563	0.819	1.889	0.332	0.281	1075	-1.756
8500	5412	4523	-3.716	-17378	1.607	1.086	2.078	0.424	0.357	1040	-1.726
9000	7466	4095	-3.217	-18147	1.659	1.376	2.259	0.530	0.430	1020	-1.708
9500	9015	3561	-2.592	-18951	1.722	1.636	2.392	0.688	0.495	988	-1.703
9800	9286	3110	-2.112	-19467	1.771	1.770	2.445	0.882	0.541	944	-1.709
9900	9165	2910	-1.914	-19649	1.791	1.811	2.458	0.997	0.561	920	-1.714
9950	9038	2792	-1.803	-19743	1.803	1.831	2.462	1.075	0.573	905	-1.718
10000	8844	2653	-1.677	-19841	1.816	1.850	2.463	1.179	0.586	888	-1.722

Table A.27. Boiling Liquid ($(O/U)^{(liq)} = 1.97$)

T K	P bar	ρ $kg\ m^{-3}$	H $10^{3} kJ\ kg^{-1}$	G $kJ\ kg^{-1}$	S $kJ\ kg^{-1}\ K^{-1}$	C_P $kJ\ kg^{-1}\ K^{-1}$	C_P/C_V	β_T $10^{-4} bar^{-1}$	α_T $10^{-3} K^{-1}$	V_S $m\ s^{-1}$	$\mu_O/10kT$
3000	0.01351	9038	-6.402	-9823	1.140	0.496	1.283	0.038	0.111	1944	-3.264
3120	0.02705	8921	-6.342	-9961	1.160	0.497	1.259	0.039	0.107	1900	-3.152
3500	0.1749	8585	-6.161	-10412	1.215	0.450	1.236	0.044	0.097	1805	-2.855
4000	1.182	8189	-5.951	-11034	1.271	0.393	1.265	0.052	0.094	1724	-2.573
4500	5.417	7808	-5.762	-11681	1.315	0.369	1.323	0.062	0.098	1658	-2.370
5000	19.78	7420	-5.578	-12348	1.354	0.370	1.389	0.074	0.107	1589	-2.216
5500	62.43	7017	-5.390	-13034	1.390	0.386	1.455	0.091	0.118	1509	-2.094
6000	175.2	6599	-5.190	-13736	1.424	0.413	1.514	0.115	0.133	1416	-1.995
6500	440.7	6169	-4.976	-14452	1.458	0.449	1.579	0.147	0.152	1318	-1.914
7000	993.4	5737	-4.738	-15180	1.492	0.513	1.647	0.193	0.178	1220	-1.848
7500	1996	5315	-4.464	-15916	1.527	0.631	1.743	0.255	0.220	1135	-1.795
8000	3539	4910	-4.132	-16659	1.566	0.827	1.889	0.333	0.282	1075	-1.754
8500	5517	4512	-3.723	-17411	1.610	1.093	2.075	0.423	0.357	1042	-1.724
9000	7574	4085	-3.221	-18181	1.662	1.380	2.252	0.527	0.428	1023	-1.707
9500	9103	3550	-2.593	-18987	1.726	1.637	2.381	0.683	0.492	991	-1.702
9800	9344	3096	-2.108	-19505	1.775	1.770	2.433	0.878	0.538	946	-1.709
9900	9205	2893	-1.909	-19688	1.796	1.811	2.445	0.995	0.558	922	-1.714
9950	9067	2773	-1.795	-19783	1.808	1.830	2.449	1.075	0.570	906	-1.717
10000	8856	2631	-1.666	-19882	1.822	1.849	2.450	1.182	0.583	888	-1.722

Table A.27. Boiling Liquid ($(O/U)^{(liq)} = 1.98$)

T K	P bar	ρ $kg\,m^{-3}$	H $10^{3} kJ\,kg^{-1}$	G $kJ\,kg^{-1}$	S $kJ\,kg^{-1}$	C_P K^{-1}	C_P/C_V	β_T $10^{-4} bar^{-1}$	α_T $10^{-3} K^{-1}$	V_S $m\,s^{-1}$	$\mu_O/10kT$
3000	0.01357	9024	-6.425	-9847	1.141	0.497	1.283	0.038	0.111	1945	-3.223
3120	0.02746	8907	-6.365	-9985	1.160	0.499	1.259	0.039	0.107	1900	-3.112
3500	0.1829	8569	-6.182	-10437	1.215	0.456	1.238	0.044	0.098	1804	-2.822
4000	1.248	8170	-5.971	-11059	1.272	0.397	1.267	0.052	0.094	1723	-2.552
4500	5.693	7787	-5.780	-11707	1.317	0.370	1.324	0.062	0.099	1656	-2.357
5000	20.74	7399	-5.596	-12375	1.356	0.371	1.390	0.075	0.107	1586	-2.207
5500	65.33	6996	-5.407	-13061	1.392	0.387	1.455	0.092	0.119	1506	-2.087
6000	182.9	6578	-5.207	-13764	1.426	0.414	1.514	0.115	0.134	1412	-1.990
6500	458.3	6148	-4.991	-14481	1.460	0.451	1.578	0.149	0.152	1315	-1.910
7000	1029	5718	-4.753	-15210	1.494	0.516	1.646	0.194	0.179	1217	-1.845
7500	2057	5299	-4.476	-15947	1.529	0.637	1.743	0.256	0.222	1133	-1.792
8000	3627	4896	-4.141	-16691	1.569	0.835	1.888	0.334	0.283	1075	-1.751
8500	5623	4500	-3.730	-17443	1.613	1.100	2.071	0.422	0.357	1044	-1.723
9000	7683	4075	-3.225	-18215	1.665	1.384	2.245	0.523	0.426	1026	-1.705
9500	9191	3539	-2.593	-19023	1.729	1.637	2.370	0.678	0.489	994	-1.701
9800	9400	3082	-2.105	-19543	1.779	1.769	2.421	0.875	0.534	948	-1.708
9900	9244	2876	-1.903	-19728	1.801	1.810	2.433	0.993	0.555	923	-1.713
9950	9093	2753	-1.787	-19823	1.813	1.830	2.436	1.076	0.567	907	-1.717
10000	8865	2607	-1.655	-19923	1.827	1.848	2.437	1.186	0.581	888	-1.722

Table A.27. Boiling Liquid ($(O/U)^{(liq)} = 1.99$)

T K	P bar	ρ $kg\,m^{-3}$	H $10^{3} kJ\,kg^{-1}$	G $kJ\,kg^{-1}$	S $kJ\,kg^{-1}\,K^{-1}$	C_P $kJ\,kg^{-1}\,K^{-1}$	C_P/C_V	β_T $10^{-4} bar^{-1}$	α_T $10^{-3} K^{-1}$	V_S $m\,s^{-1}$	$\mu_O/10kT$
3000	0.01429	9010	-6.448	-9870	1.141	0.500	1.284	0.038	0.112	1945	-3.158
3120	0.02944	8891	-6.387	-10009	1.161	0.504	1.261	0.039	0.108	1900	-3.051
3500	0.2002	8550	-6.202	-10461	1.217	0.461	1.240	0.045	0.099	1803	-2.781
4000	1.339	8149	-5.989	-11084	1.274	0.399	1.268	0.053	0.095	1720	-2.531
4500	6.028	7766	-5.797	-11732	1.319	0.371	1.324	0.062	0.099	1653	-2.344
5000	21.83	7377	-5.613	-12402	1.358	0.371	1.390	0.075	0.107	1583	-2.198
5500	68.52	6974	-5.423	-13089	1.394	0.387	1.455	0.093	0.119	1502	-2.081
6000	191.1	6556	-5.223	-13793	1.428	0.415	1.514	0.116	0.134	1409	-1.985
6500	477.1	6128	-5.007	-14511	1.462	0.452	1.578	0.150	0.153	1312	-1.906
7000	1066	5699	-4.767	-15240	1.496	0.520	1.646	0.196	0.180	1215	-1.841
7500	2120	5282	-4.488	-15978	1.532	0.643	1.742	0.257	0.223	1132	-1.789
8000	3717	4882	-4.151	-16722	1.571	0.843	1.888	0.334	0.284	1076	-1.749
8500	5731	4489	-3.737	-17476	1.616	1.107	2.068	0.421	0.357	1046	-1.721
9000	7793	4065	-3.229	-18248	1.669	1.388	2.237	0.520	0.425	1029	-1.704
9500	9279	3529	-2.594	-19059	1.733	1.638	2.359	0.673	0.486	996	-1.700
9800	9455	3067	-2.101	-19582	1.784	1.769	2.409	0.871	0.531	949	-1.707
9900	9281	2859	-1.896	-19767	1.805	1.810	2.420	0.992	0.552	924	-1.713
9950	9118	2734	-1.779	-19863	1.818	1.829	2.424	1.076	0.564	908	-1.717
10000	8870	2583	-1.643	-19964	1.832	1.848	2.425	1.192	0.578	887	-1.722

Table A.27. Boiling Liquid ($(O/U)^{(liq)} = 2.00$)

T K	P bar	ρ $kg\,m^{-3}$	H $10^{3} kJ\,kg^{-1}$	G $kJ\,kg^{-1}$	S $kJ\,kg^{-1}\,K^{-1}$	C_P	C_P/C_V	β_T $10^{-4} bar^{-1}$	α_T $10^{-3} K^{-1}$	V_S $m\,s^{-1}$	$\mu_O/10kT$
3000	0.02024	8990	-6.467	-9893	1.142	0.508	1.286	0.038	0.113	1943	-3.021
3120	0.03960	8871	-6.406	-10032	1.162	0.511	1.263	0.040	0.109	1898	-2.943
3500	0.2360	8528	-6.220	-10484	1.218	0.464	1.241	0.045	0.099	1800	-2.730
4000	1.464	8126	-6.005	-11108	1.276	0.399	1.268	0.053	0.095	1717	-2.508
4500	6.431	7743	-5.814	-11758	1.321	0.371	1.324	0.063	0.099	1650	-2.331
5000	23.08	7355	-5.629	-12428	1.360	0.372	1.390	0.076	0.108	1580	-2.189
5500	72.04	6952	-5.440	-13116	1.396	0.388	1.455	0.093	0.120	1499	-2.074
6000	200.0	6535	-5.239	-13821	1.430	0.416	1.514	0.117	0.134	1406	-1.979
6500	497.0	6107	-5.022	-14540	1.464	0.454	1.577	0.151	0.154	1308	-1.902
7000	1105	5679	-4.781	-15270	1.498	0.523	1.645	0.197	0.181	1212	-1.838
7500	2185	5265	-4.500	-16008	1.534	0.649	1.742	0.259	0.224	1131	-1.786
8000	3809	4869	-4.160	-16753	1.574	0.850	1.887	0.334	0.285	1076	-1.747
8500	5841	4478	-3.743	-17508	1.619	1.114	2.064	0.419	0.356	1049	-1.719
9000	7904	4055	-3.233	-18282	1.672	1.392	2.229	0.516	0.422	1032	-1.703
9500	9366	3518	-2.594	-19095	1.737	1.638	2.348	0.669	0.483	999	-1.699
9800	9509	3053	-2.097	-19620	1.788	1.768	2.397	0.868	0.528	951	-1.707
9900	9316	2842	-1.890	-19806	1.810	1.809	2.408	0.991	0.549	925	-1.712
9950	9140	2714	-1.770	-19903	1.822	1.829	2.412	1.077	0.561	908	-1.716
10000	8867	2557	-1.630	-20005	1.838	1.848	2.412	1.199	0.576	887	-1.722
10015 [a]	*8758*	*2501*	*-1.581*	*-20038*	*1.843*	*1.853*	*2.411*	*1.247*	*0.581*	*879*	*-1.724*

[a] critical point

Table A.27. Boiling Liquid ($(O/U)^{(liq)} = 2.01$)

T K	P bar	ρ $kg\,m^{-3}$	H $10^{3} kJ\,kg^{-1}$	G $kJ\,kg^{-1}$	S $kJ\,kg^{-1}\,K^{-1}$	C_P $kJ\,kg^{-1}\,K^{-1}$	C_P/C_V	β_T $10^{-4} bar^{-1}$	α_T $10^{-3} K^{-1}$	V_S $m\,s^{-1}$	$\mu_O/10kT$
3000	0.04524	8955	-6.477	-9915	1.146	0.502	1.280	0.038	0.112	1934	-2.882
3120	0.07140	8838	-6.417	-10053	1.165	0.504	1.258	0.040	0.108	1890	-2.833
3500	0.2985	8499	-6.233	-10507	1.221	0.459	1.239	0.045	0.099	1795	-2.679
4000	1.629	8101	-6.020	-11133	1.278	0.397	1.268	0.053	0.095	1714	-2.484
4500	6.912	7719	-5.830	-11783	1.323	0.370	1.324	0.063	0.099	1647	-2.317
5000	24.49	7331	-5.645	-12454	1.362	0.372	1.389	0.076	0.108	1576	-2.180
5500	75.91	6930	-5.455	-13143	1.398	0.388	1.455	0.094	0.120	1495	-2.067
6000	209.6	6512	-5.254	-13849	1.432	0.417	1.514	0.118	0.135	1402	-1.974
6500	518.1	6086	-5.037	-14569	1.466	0.456	1.576	0.152	0.154	1305	-1.898
7000	1146	5660	-4.794	-15300	1.501	0.527	1.644	0.199	0.182	1209	-1.834
7500	2253	5248	-4.512	-16039	1.537	0.655	1.741	0.260	0.225	1129	-1.784
8000	3904	4855	-4.169	-16784	1.577	0.858	1.885	0.335	0.286	1077	-1.745
8500	5952	4467	-3.750	-17540	1.622	1.120	2.060	0.418	0.356	1051	-1.717
9000	8015	4046	-3.237	-18316	1.675	1.395	2.221	0.512	0.420	1035	-1.701
9500	9454	3507	-2.594	-19131	1.741	1.639	2.337	0.664	0.479	1002	-1.698
9800	9562	3039	-2.093	-19658	1.792	1.768	2.385	0.864	0.525	953	-1.706
9900	9350	2824	-1.883	-19845	1.814	1.809	2.397	0.990	0.546	926	-1.712
9950	9160	2694	-1.761	-19943	1.827	1.828	2.400	1.079	0.558	909	-1.716
10000	8852	2528	-1.614	-20047	1.843	1.849	2.401	1.211	0.575	886	-1.722

Table A.27. Boiling Liquid ($(O/U)^{(liq)} = 2.02$)

T K	P bar	ρ $kg\,m^{-3}$	H $10^{3} kJ\,kg^{-1}$	G $kJ\,kg^{-1}$	S $kJ\,kg^{-1}\,K^{-1}$	C_P	C_P/C_V	β_T $10^{-4} bar^{-1}$	α_T $10^{-3} K^{-1}$	V_S $m\,s^{-1}$	$\mu_O/10kT$
3000	0.07853	8916	-6.484	-9935	1.150	0.500	1.275	0.039	0.111	1923	-2.813
3120	0.1151	8800	-6.424	-10074	1.170	0.498	1.254	0.040	0.107	1881	-2.769
3500	0.3858	8466	-6.243	-10529	1.225	0.450	1.236	0.046	0.097	1789	-2.635
4000	1.839	8073	-6.034	-11156	1.281	0.393	1.266	0.054	0.095	1709	-2.461
4500	7.482	7693	-5.845	-11808	1.325	0.369	1.323	0.064	0.099	1643	-2.304
5000	26.09	7307	-5.661	-12480	1.364	0.371	1.389	0.077	0.108	1573	-2.171
5500	80.18	6906	-5.471	-13171	1.400	0.389	1.454	0.095	0.120	1491	-2.060
6000	219.9	6490	-5.269	-13877	1.435	0.418	1.513	0.119	0.135	1398	-1.969
6500	540.6	6064	-5.051	-14598	1.469	0.458	1.575	0.153	0.155	1301	-1.893
7000	1189	5641	-4.808	-15330	1.503	0.531	1.643	0.200	0.183	1206	-1.831
7500	2323	5232	-4.523	-16069	1.539	0.661	1.740	0.261	0.226	1128	-1.781
8000	4000	4842	-4.178	-16815	1.580	0.865	1.884	0.335	0.287	1078	-1.743
8500	6064	4456	-3.756	-17572	1.625	1.126	2.056	0.416	0.355	1053	-1.716
9000	8127	4037	-3.240	-18349	1.679	1.398	2.213	0.509	0.418	1038	-1.700
9500	9541	3497	-2.594	-19167	1.745	1.639	2.327	0.659	0.476	1005	-1.697
9700	9691	3204	-2.274	-19515	1.777	1.725	2.360	0.774	0.504	976	-1.702
9800	9614	3024	-2.089	-19696	1.797	1.767	2.374	0.861	0.521	955	-1.706
9900	9382	2806	-1.876	-19885	1.819	1.808	2.385	0.989	0.543	927	-1.712
9950	9178	2673	-1.752	-19983	1.832	1.828	2.388	1.081	0.556	909	-1.716
10000	8849	2503	-1.601	-20088	1.849	1.848	2.388	1.218	0.572	885	-1.722

Table A.27. Boiling Liquid ($(O/U)^{(liq)} = 2.03$)

T K	P bar	ρ $kg\,m^{-3}$	H $10^{3} kJ\,kg^{-1}$	G $kJ\,kg^{-1}$	S $kJ\,kg^{-1}\,K^{-1}$	C_P $kJ\,kg^{-1}\,K^{-1}$	C_P/C_V	β_T $10^{-4} bar^{-1}$	α_T $10^{-3} K^{-1}$	V_S $m\,s^{-1}$	$\mu_O/10kT$
3000	0.1167	8876	-6.491	-9955	1.155	0.500	1.271	0.039	0.111	1912	-2.768
3120	0.1655	8761	-6.431	-10095	1.174	0.495	1.251	0.041	0.107	1872	-2.725
3500	0.4921	8431	-6.252	-10551	1.228	0.443	1.234	0.046	0.097	1782	-2.599
4000	2.097	8043	-6.046	-11180	1.283	0.387	1.265	0.054	0.094	1704	-2.439
4500	8.150	7667	-5.859	-11833	1.327	0.366	1.322	0.064	0.099	1639	-2.290
5000	27.89	7282	-5.676	-12506	1.366	0.370	1.388	0.078	0.108	1569	-2.161
5500	84.87	6882	-5.486	-13197	1.402	0.389	1.453	0.095	0.120	1487	-2.054
6000	231.1	6467	-5.284	-13905	1.437	0.419	1.513	0.120	0.136	1394	-1.964
6500	564.3	6043	-5.066	-14627	1.471	0.460	1.574	0.155	0.155	1297	-1.889
7000	1234	5621	-4.821	-15359	1.505	0.534	1.642	0.202	0.184	1204	-1.828
7500	2395	5215	-4.534	-16099	1.542	0.666	1.739	0.263	0.227	1127	-1.778
8000	4098	4828	-4.187	-16846	1.582	0.872	1.882	0.335	0.287	1079	-1.741
8500	6179	4446	-3.762	-17604	1.628	1.132	2.051	0.414	0.354	1056	-1.714
9000	8240	4028	-3.244	-18382	1.682	1.400	2.205	0.505	0.416	1042	-1.699
9500	9628	3487	-2.594	-19203	1.748	1.639	2.316	0.654	0.473	1008	-1.696
9700	9757	3191	-2.271	-19552	1.782	1.725	2.349	0.770	0.501	978	-1.701
9800	9664	3010	-2.085	-19734	1.801	1.767	2.363	0.858	0.518	956	-1.705
9900	9412	2789	-1.869	-19924	1.824	1.808	2.373	0.989	0.540	928	-1.711
9950	9192	2652	-1.742	-20024	1.837	1.828	2.376	1.083	0.553	909	-1.716
10000	8970	2515	-1.611	-20124	1.851	1.838	2.370	1.183	0.562	892	-1.720

Table A.27. Boiling Liquid ($(O/U)^{(liq)} = 2.04$)

T K	P bar	ρ $kg\,m^{-3}$	H $10^{3} kJ\,kg^{-1}$	G $kJ\,kg^{-1}$	S $kJ\,kg^{-1}\,K^{-1}$	C_P $kJ\,kg^{-1}\,K^{-1}$	C_P/C_V	β_T $10^{-4} bar^{-1}$	α_T $10^{-3} K^{-1}$	V_S $m\,s^{-1}$	$\mu_O/10kT$
3000	0.1602	8837	-6.497	-9974	1.159	0.500	1.267	0.040	0.111	1902	-2.734
3120	0.2227	8722	-6.438	-10115	1.178	0.493	1.248	0.041	0.106	1862	-2.692
3500	0.6152	8396	-6.261	-10573	1.232	0.437	1.232	0.047	0.096	1775	-2.570
4000	2.402	8012	-6.058	-11203	1.286	0.382	1.263	0.055	0.093	1699	-2.418
4500	8.928	7639	-5.873	-11857	1.330	0.364	1.321	0.065	0.099	1634	-2.276
5000	29.93	7256	-5.690	-12532	1.368	0.369	1.387	0.078	0.108	1564	-2.152
5500	90.02	6858	-5.501	-13224	1.404	0.389	1.452	0.096	0.121	1483	-2.047
6000	243.1	6444	-5.299	-13933	1.439	0.419	1.512	0.121	0.136	1390	-1.959
6500	589.5	6021	-5.080	-14655	1.473	0.462	1.573	0.156	0.156	1294	-1.885
7000	1280	5602	-4.834	-15388	1.508	0.538	1.641	0.203	0.185	1201	-1.824
7500	2470	5199	-4.545	-16129	1.545	0.672	1.738	0.264	0.228	1126	-1.776
8000	4199	4815	-4.195	-16877	1.585	0.879	1.880	0.335	0.288	1080	-1.738
8500	6294	4436	-3.768	-17635	1.631	1.137	2.047	0.412	0.354	1058	-1.712
9000	8354	4019	-3.247	-18416	1.685	1.403	2.197	0.501	0.414	1045	-1.697
9500	9715	3476	-2.594	-19238	1.752	1.639	2.306	0.650	0.470	1010	-1.695
9700	9822	3179	-2.269	-19589	1.786	1.725	2.338	0.766	0.498	980	-1.700
9800	9714	2995	-2.080	-19772	1.805	1.766	2.352	0.856	0.515	958	-1.704
9900	9439	2771	-1.861	-19963	1.828	1.807	2.362	0.989	0.537	929	-1.711
9950	9203	2631	-1.732	-20064	1.842	1.827	2.365	1.087	0.551	910	-1.716
10000	9098	2529	-1.622	-20160	1.854	1.827	2.353	1.149	0.553	900	-1.719

Table A.27. Boiling Liquid ($(O/U)^{(liq)} = 2.05$)

T K	P bar	ρ $kg\,m^{-3}$	H $10^{3} kJ\,kg^{-1}$	G $kJ\,kg^{-1}$	S $kJ\,kg^{-1}$	C_P K^{-1}	C_P/C_V	β_T $10^{-4} bar^{-1}$	α_T $10^{-3} K^{-1}$	V_S $m\,s^{-1}$	$\mu_O/10kT$
3000	0.2096	8797	-6.503	-9993	1.163	0.500	1.264	0.040	0.111	1891	-2.705
3120	0.2871	8683	-6.444	-10134	1.183	0.491	1.246	0.042	0.106	1853	-2.664
3500	0.7544	8359	-6.268	-10594	1.236	0.431	1.231	0.047	0.095	1767	-2.544
4000	2.757	7979	-6.068	-11226	1.289	0.376	1.261	0.055	0.093	1693	-2.398
4500	9.823	7610	-5.886	-11882	1.332	0.360	1.319	0.065	0.098	1629	-2.263
5000	32.22	7230	-5.704	-12557	1.371	0.368	1.386	0.079	0.108	1560	-2.143
5500	95.67	6833	-5.515	-13251	1.406	0.388	1.451	0.097	0.121	1479	-2.040
6000	256.0	6420	-5.313	-13960	1.441	0.420	1.511	0.122	0.136	1386	-1.953
6500	616.3	5999	-5.093	-14684	1.475	0.463	1.571	0.157	0.157	1290	-1.881
7000	1329	5582	-4.846	-15418	1.510	0.542	1.639	0.205	0.186	1198	-1.821
7500	2547	5182	-4.556	-16159	1.547	0.678	1.737	0.265	0.229	1125	-1.773
8000	4301	4802	-4.204	-16907	1.588	0.885	1.877	0.335	0.288	1081	-1.736
8500	6411	4425	-3.774	-17667	1.635	1.142	2.042	0.410	0.353	1061	-1.711
9000	8469	4010	-3.250	-18449	1.689	1.405	2.189	0.497	0.411	1048	-1.696
9500	9802	3466	-2.594	-19274	1.756	1.639	2.295	0.645	0.467	1013	-1.694
9700	9886	3166	-2.266	-19626	1.790	1.724	2.327	0.762	0.495	982	-1.699
9800	9762	2981	-2.076	-19810	1.810	1.766	2.341	0.853	0.512	960	-1.704
9900	9465	2752	-1.854	-20002	1.833	1.807	2.351	0.989	0.534	929	-1.711
9950	9211	2609	-1.721	-20104	1.847	1.827	2.354	1.091	0.548	910	-1.716

Table A.27. Boiling Liquid ($(O/U)^{(liq)} = 2.06$)

T K	P bar	ρ $kg\,m^{-3}$	H $10^{3} kJ\,kg^{-1}$	G $kJ\,kg^{-1}$	S $kJ\,kg^{-1}\,K^{-1}$	C_P	C_P/C_V	β_T $10^{-4} bar^{-1}$	α_T $10^{-3} K^{-1}$	V_S $m\,s^{-1}$	$\mu_O/10kT$
3000	0.2650	8758	-6.509	-10012	1.168	0.499	1.260	0.041	0.111	1881	-2.681
3120	0.3589	8645	-6.450	-10154	1.187	0.488	1.244	0.042	0.106	1843	-2.640
3500	0.9099	8323	-6.276	-10615	1.240	0.427	1.230	0.048	0.095	1760	-2.522
4000	3.161	7946	-6.078	-11249	1.293	0.371	1.260	0.056	0.092	1687	-2.380
4500	10.84	7580	-5.898	-11906	1.335	0.357	1.318	0.066	0.098	1624	-2.250
5000	34.78	7203	-5.718	-12583	1.373	0.366	1.384	0.080	0.108	1555	-2.133
5500	101.9	6807	-5.529	-13277	1.409	0.388	1.449	0.098	0.121	1474	-2.033
6000	269.9	6396	-5.327	-13988	1.443	0.420	1.510	0.124	0.137	1382	-1.948
6500	644.6	5977	-5.107	-14712	1.478	0.465	1.570	0.159	0.157	1286	-1.877
7000	1380	5563	-4.858	-15447	1.513	0.545	1.638	0.206	0.186	1195	-1.818
7500	2626	5166	-4.566	-16189	1.550	0.684	1.735	0.266	0.230	1124	-1.770
8000	4406	4790	-4.212	-16938	1.591	0.892	1.875	0.334	0.289	1082	-1.734
8500	6530	4416	-3.779	-17699	1.638	1.147	2.036	0.408	0.352	1064	-1.709
9000	8584	4001	-3.254	-18482	1.692	1.407	2.181	0.493	0.409	1051	-1.695
9500	9888	3456	-2.594	-19309	1.760	1.639	2.285	0.641	0.463	1016	-1.694
9700	9949	3154	-2.263	-19663	1.794	1.723	2.316	0.758	0.491	984	-1.699
9800	9809	2966	-2.071	-19848	1.814	1.765	2.330	0.851	0.509	961	-1.703
9900	9489	2734	-1.846	-20042	1.838	1.807	2.340	0.990	0.532	930	-1.710
9950	9215	2586	-1.710	-20144	1.853	1.827	2.343	1.096	0.546	909	-1.715

Table A.27. Boiling Liquid ($(O/U)^{(liq)} = 2.08$)

T K	P bar	ρ $kg\,m^{-3}$	H $10^{3} kJ\,kg^{-1}$	G $kJ\,kg^{-1}$	S $kJ\,kg^{-1}\,K^{-1}$	C_P $kJ\,kg^{-1}\,K^{-1}$	C_P/C_V	β_T $10^{-4} bar^{-1}$	α_T $10^{-3} K^{-1}$	V_S $m\,s^{-1}$	$\mu_O/10kT$
3000	0.3916	8680	-6.521	-10049	1.176	0.497	1.254	0.042	0.110	1860	-2.642
3120	0.5227	8568	-6.462	-10192	1.195	0.484	1.240	0.044	0.106	1825	-2.602
3500	1.268	8250	-6.291	-10656	1.247	0.419	1.228	0.049	0.095	1744	-2.485
4000	4.120	7879	-6.097	-11293	1.299	0.363	1.259	0.057	0.092	1674	-2.348
4500	13.29	7518	-5.921	-11953	1.341	0.350	1.315	0.067	0.097	1613	-2.225
5000	40.80	7146	-5.743	-12633	1.378	0.362	1.381	0.081	0.108	1544	-2.115
5500	116.0	6755	-5.556	-13330	1.413	0.386	1.446	0.100	0.121	1464	-2.019
6000	300.8	6347	-5.354	-14042	1.448	0.420	1.507	0.126	0.137	1373	-1.938
6500	706.3	5932	-5.133	-14768	1.482	0.468	1.566	0.162	0.158	1278	-1.868
7000	1490	5524	-4.882	-15504	1.517	0.552	1.634	0.209	0.188	1190	-1.811
7500	2793	5134	-4.586	-16248	1.555	0.695	1.731	0.267	0.232	1123	-1.765
8000	4622	4765	-4.227	-16998	1.596	0.904	1.869	0.333	0.289	1085	-1.730
8500	6772	4396	-3.790	-17761	1.644	1.156	2.025	0.403	0.349	1069	-1.706
9000	8817	3984	-3.260	-18548	1.699	1.410	2.164	0.485	0.403	1058	-1.692
9500	10060	3436	-2.593	-19380	1.767	1.638	2.265	0.631	0.457	1022	-1.692
9700	10070	3129	-2.257	-19737	1.802	1.722	2.295	0.751	0.485	988	-1.697
9800	9898	2936	-2.061	-19924	1.823	1.764	2.309	0.846	0.504	964	-1.702
9900	9529	2696	-1.829	-20120	1.848	1.806	2.319	0.992	0.527	931	-1.710
9950	9194	2534	-1.682	-20226	1.864	1.829	2.322	1.113	0.543	908	-1.716

Table A.27. Boiling Liquid ($(O/U)^{(liq)} = 2.10$)

T K	P bar	ρ $kg\,m^{-3}$	H $10^{3} kJ\,kg^{-1}$	G $kJ\,kg^{-1}$	S $kJ\,kg^{-1}\,K^{-1}$	C_P	C_P/C_V	β_T $10^{-4} bar^{-1}$	α_T $10^{-3} K^{-1}$	V_S $m\,s^{-1}$	$\mu_O/10kT$
3000	0.5322	8603	-6.532	-10086	1.184	0.495	1.249	0.043	0.110	1840	-2.613
3120	0.7059	8493	-6.474	-10229	1.204	0.479	1.236	0.045	0.105	1807	-2.572
3500	1.681	8178	-6.305	-10697	1.255	0.411	1.228	0.050	0.095	1729	-2.456
4000	5.273	7810	-6.115	-11337	1.306	0.356	1.259	0.058	0.091	1661	-2.320
4500	16.29	7454	-5.942	-12000	1.346	0.344	1.313	0.069	0.097	1601	-2.202
5000	48.12	7087	-5.767	-12682	1.383	0.358	1.378	0.083	0.107	1533	-2.097
5500	132.7	6700	-5.581	-13382	1.418	0.384	1.442	0.102	0.121	1454	-2.006
6000	336.3	6297	-5.380	-14096	1.453	0.420	1.503	0.128	0.138	1364	-1.927
6500	775.3	5887	-5.158	-14824	1.487	0.472	1.561	0.164	0.159	1270	-1.860
7000	1609	5485	-4.905	-15561	1.522	0.560	1.629	0.212	0.189	1185	-1.805
7500	2970	5103	-4.605	-16306	1.560	0.706	1.727	0.269	0.233	1122	-1.760
8000	4847	4741	-4.241	-17058	1.602	0.915	1.862	0.332	0.289	1088	-1.726
8500	7020	4378	-3.800	-17823	1.650	1.163	2.013	0.398	0.346	1075	-1.703
9000	9053	3968	-3.265	-18613	1.705	1.412	2.146	0.477	0.398	1065	-1.690
9500	10230	3416	-2.592	-19451	1.775	1.637	2.245	0.622	0.451	1027	-1.690
9700	10190	3104	-2.251	-19810	1.810	1.721	2.275	0.744	0.479	992	-1.696
9800	9982	2907	-2.050	-20000	1.832	1.763	2.288	0.842	0.498	967	-1.702
9900	9560	2657	-1.810	-20199	1.857	1.806	2.299	0.996	0.522	932	-1.710
9950	9331	2527	-1.682	-20301	1.871	1.818	2.295	1.084	0.531	915	-1.714

Table A.27. Boiling Liquid ($(O/U)^{(liq)} = 2.125$)

T K	P bar	ρ $kg\,m^{-3}$	H $10^{3} kJ\,kg^{-1}$	G $kJ\,kg^{-1}$	S $kJ\,kg^{-1}\,K^{-1}$	C_P	C_P/C_V	β_T $10^{-4} bar^{-1}$	α_T $10^{-3} K^{-1}$	V_S $m\,s^{-1}$	$\mu_O/10kT$
3000	0.7126	8509	-6.547	-10131	1.195	0.491	1.244	0.044	0.110	1817	-2.585
3120	0.9453	8400	-6.489	-10275	1.214	0.472	1.233	0.046	0.105	1785	-2.545
3500	2.253	8089	-6.323	-10746	1.264	0.403	1.228	0.052	0.095	1711	-2.427
4000	6.966	7724	-6.137	-11391	1.314	0.349	1.259	0.060	0.091	1644	-2.292
4500	20.86	7373	-5.967	-12058	1.354	0.338	1.312	0.071	0.097	1585	-2.176
5000	59.26	7010	-5.795	-12743	1.390	0.353	1.374	0.085	0.107	1518	-2.076
5500	157.7	6629	-5.611	-13445	1.424	0.381	1.437	0.105	0.121	1440	-1.990
6000	387.9	6232	-5.410	-14163	1.459	0.419	1.498	0.132	0.138	1352	-1.914
6500	872.8	5829	-5.187	-14892	1.493	0.475	1.554	0.168	0.160	1260	-1.850
7000	1773	5436	-4.931	-15632	1.529	0.569	1.622	0.215	0.191	1179	-1.796
7500	3207	5065	-4.627	-16378	1.567	0.719	1.720	0.270	0.234	1121	-1.753
8000	5141	4712	-4.258	-17132	1.609	0.928	1.853	0.329	0.288	1093	-1.721
8500	7339	4356	-3.812	-17900	1.657	1.171	1.998	0.391	0.342	1083	-1.699
9000	9353	3949	-3.272	-18694	1.714	1.414	2.125	0.466	0.391	1075	-1.686
9500	10440	3392	-2.590	-19539	1.784	1.635	2.220	0.612	0.443	1035	-1.688
9700	10340	3072	-2.242	-19902	1.821	1.719	2.250	0.736	0.472	997	-1.695
9800	10080	2869	-2.036	-20094	1.843	1.762	2.264	0.839	0.491	970	-1.700
9900	9581	2605	-1.785	-20298	1.870	1.806	2.274	1.004	0.517	932	-1.709

Table A.27. Boiling Liquid ($(O/U)^{(liq)} = 2.15$)

T K	P bar	ρ $kg\,m^{-3}$	H $10^{3} kJ\,kg^{-1}$	G $kJ\,kg^{-1}$	S	C_P $kJ\,kg^{-1}\,K^{-1}$	C_P/C_V	β_T $10^{-4} bar^{-1}$	α_T $10^{-3} K^{-1}$	V_S $m\,s^{-1}$	$\mu_O/10kT$
3000	0.8843	8416	-6.561	-10175	1.204	0.486	1.239	0.046	0.110	1794	-2.566
3120	1.180	8309	-6.504	-10320	1.223	0.465	1.230	0.048	0.105	1763	-2.524
3500	2.860	8001	-6.341	-10795	1.273	0.396	1.228	0.054	0.095	1693	-2.404
4000	8.899	7639	-6.158	-11444	1.321	0.343	1.261	0.062	0.092	1628	-2.268
4500	26.30	7290	-5.990	-12114	1.361	0.333	1.312	0.073	0.097	1569	-2.154
5000	72.73	6932	-5.821	-12803	1.397	0.348	1.370	0.088	0.107	1503	-2.057
5500	187.5	6556	-5.639	-13508	1.431	0.378	1.431	0.108	0.121	1425	-1.974
6000	448.3	6166	-5.439	-14228	1.465	0.418	1.492	0.135	0.138	1339	-1.902
6500	984.0	5771	-5.215	-14960	1.499	0.479	1.546	0.172	0.161	1250	-1.840
7000	1954	5388	-4.956	-15701	1.535	0.578	1.614	0.218	0.192	1173	-1.788
7500	3461	5028	-4.648	-16449	1.574	0.732	1.712	0.271	0.235	1122	-1.747
8000	5449	4685	-4.274	-17205	1.616	0.939	1.842	0.326	0.286	1098	-1.716
8500	7668	4336	-3.823	-17976	1.665	1.177	1.981	0.383	0.338	1092	-1.695
9000	9659	3931	-3.278	-18775	1.722	1.414	2.103	0.455	0.384	1084	-1.683
9500	10650	3368	-2.588	-19626	1.794	1.633	2.196	0.601	0.435	1042	-1.686
9700	10480	3041	-2.233	-19994	1.831	1.718	2.227	0.729	0.465	1002	-1.693
9800	10160	2830	-2.021	-20189	1.854	1.761	2.240	0.836	0.485	973	-1.700
9900	9561	2546	-1.753	-20398	1.883	1.808	2.251	1.021	0.514	931	-1.710

Table A.27. Boiling Liquid ($(O/U)^{(liq)} = 2.20$)

T K	P bar	ρ $kg\,m^{-3}$	H $10\,^3 kJ\,kg^{-1}$	G $kJ\,kg^{-1}$	S $kJ\,kg^{-1}\,K^{-1}$	C_P	C_P/C_V	β_T $10^{-4} bar^{-1}$	α_T $10^{-3} K^{-1}$	V_S $m\,s^{-1}$	$\mu_O/10kT$
3000	1.177	8237	-6.592	-10261	1.223	0.474	1.232	0.049	0.109	1751	-2.541
3120	1.602	8132	-6.536	-10409	1.241	0.451	1.226	0.051	0.105	1723	-2.497
3500	4.105	7830	-6.379	-10890	1.289	0.383	1.230	0.057	0.096	1658	-2.371
4000	13.34	7471	-6.201	-11547	1.337	0.336	1.265	0.067	0.094	1595	-2.232
4500	39.68	7124	-6.036	-12225	1.375	0.327	1.313	0.078	0.098	1536	-2.118
5000	107.0	6772	-5.869	-12920	1.410	0.343	1.366	0.093	0.108	1470	-2.024
5500	263.7	6407	-5.690	-13632	1.444	0.374	1.420	0.114	0.121	1393	-1.946
6000	599.5	6031	-5.491	-14357	1.478	0.417	1.478	0.142	0.139	1312	-1.878
6500	1253	5655	-5.265	-15093	1.512	0.487	1.528	0.179	0.162	1229	-1.821
7000	2372	5296	-5.001	-15837	1.548	0.595	1.596	0.222	0.193	1164	-1.773
7500	4023	4958	-4.684	-16588	1.587	0.753	1.693	0.270	0.235	1125	-1.735
8000	6110	4635	-4.302	-17349	1.631	0.957	1.816	0.317	0.281	1111	-1.707
8500	8359	4299	-3.844	-18126	1.680	1.184	1.946	0.367	0.327	1111	-1.687
9000	10290	3898	-3.289	-18934	1.738	1.412	2.060	0.434	0.370	1104	-1.678
9500	11070	3322	-2.582	-19799	1.812	1.628	2.151	0.580	0.421	1056	-1.682
9700	10730	2978	-2.212	-20177	1.852	1.715	2.182	0.717	0.452	1011	-1.691
9800	10300	2750	-1.986	-20378	1.877	1.760	2.196	0.835	0.474	978	-1.698
9900	9882	2523	-1.748	-20583	1.902	1.787	2.194	0.967	0.490	948	-1.706

Table A.27. Boiling Liquid ($(O/U)^{(liq)} = 2.25$)

T K	P bar	ρ $kg\,m^{-3}$	H $10^{3} kJ\ kg^{-1}$	G $kJ\ kg^{-1}$	S $kJ\ kg^{-1}\ K^{-1}$	C_P $kJ\ kg^{-1}\ K^{-1}$	C_P/C_V	β_T $10^{-4} bar^{-1}$	α_T $10^{-3} K^{-1}$	V_S $m\ s^{-1}$	$\mu_O/10kT$
3000	1.402	8064	-6.625	-10347	1.241	0.461	1.226	0.052	0.109	1711	-2.526
3120	1.952	7962	-6.571	-10497	1.258	0.436	1.223	0.054	0.105	1686	-2.480
3500	5.343	7665	-6.418	-10984	1.304	0.372	1.233	0.061	0.097	1624	-2.349
4000	18.41	7308	-6.245	-11648	1.351	0.330	1.270	0.071	0.096	1563	-2.205
4500	56.29	6962	-6.082	-12333	1.389	0.325	1.316	0.084	0.101	1503	-2.090
5000	151.5	6612	-5.915	-13035	1.424	0.341	1.363	0.100	0.110	1436	-1.997
5500	364.6	6256	-5.736	-13752	1.457	0.372	1.410	0.122	0.122	1361	-1.921
6000	797.6	5896	-5.537	-14481	1.491	0.417	1.463	0.150	0.140	1286	-1.856
6500	1592	5543	-5.309	-15221	1.525	0.496	1.507	0.185	0.162	1211	-1.803
7000	2874	5209	-5.039	-15969	1.561	0.611	1.574	0.225	0.193	1159	-1.758
7500	4663	4896	-4.715	-16724	1.601	0.770	1.670	0.266	0.232	1132	-1.723
8000	6833	4592	-4.326	-17489	1.645	0.967	1.787	0.306	0.274	1128	-1.697
8500	9098	4268	-3.861	-18273	1.696	1.186	1.909	0.349	0.314	1133	-1.680
9000	10940	3870	-3.299	-19090	1.755	1.406	2.018	0.412	0.354	1125	-1.672
9500	11480	3277	-2.575	-19971	1.831	1.622	2.109	0.561	0.406	1071	-1.678
9700	10960	2913	-2.189	-20359	1.873	1.712	2.141	0.707	0.440	1019	-1.689
9800	10390	2665	-1.945	-20568	1.900	1.761	2.156	0.841	0.465	981	-1.697

Table A.28. Vapour Coexisting with the Boiling Liquid ($(O/U)^{(liq)} = 1.75$)

T K	P bar	ρ $kg\ m^{-3}$	H $10^{3} kJ\ kg^{-1}$	G $kJ\ kg^{-1}$	S $kJ\ kg^{-1}$	C_P K^{-1}	C_P/C_V	β_T $10^{-3} bar^{-1}$	α_T $10^{-3} K^{-1}$	V_S $m\ s^{-1}$	O/U
3000	0.02008	0.02098	-3.501	-8386	1.628	0.278	1.138	4.986_{10}^{+4}	0.342	330	1.440
3120	0.03838	0.03858	-3.509	-8565	1.620	0.287	1.135	2.609_{10}^{+4}	0.331	336	1.457
3500	0.2125	0.1905	-3.517	-9133	1.605	0.331	1.126	4723.00	0.309	354	1.505
4000	1.171	0.9132	-3.492	-9906	1.603	0.479	1.122	866.400	0.321	377	1.565
4500	4.437	3.000	-3.436	-10762	1.628	0.820	1.146	235.700	0.405	402	1.644
5000	13.60	7.757	-3.334	-11795	1.692	1.366	1.208	81.390	0.546	438	1.774
5500	37.20	17.08	-3.164	-13131	1.812	1.941	1.290	31.770	0.656	488	1.996
6000	95.07	33.66	-2.909	-14890	1.997	2.314	1.353	13.030	0.665	555	2.349
6500	229.3	61.73	-2.577	-17117	2.237	2.478	1.393	5.539	0.607	638	2.853
7000	518.2	108.6	-2.203	-19667	2.495	2.490	1.424	2.483	0.534	726	3.466
7500	1084	187.6	-1.846	-22133	2.705	2.429	1.455	1.195	0.477	806	4.042
8000	2071	320.9	-1.567	-23921	2.794	2.322	1.495	0.626	0.439	863	4.346
8500	3549	538.3	-1.394	-24559	2.725	2.179	1.552	0.361	0.421	894	4.203
9000	5380	867.3	-1.308	-24034	2.525	2.025	1.643	0.231	0.420	905	3.670
9500	7149	1330	-1.274	-22663	2.251	1.890	1.810	0.169	0.448	897	2.942
9800	7889	1701	-1.278	-21523	2.066	1.842	2.003	0.154	0.496	875	2.469
9900	8025	1851	-1.286	-21080	1.999	1.839	2.104	0.152	0.524	864	2.304
9950	8063	1935	-1.292	-20842	1.965	1.841	2.167	0.153	0.543	857	2.220
10000	8075	2026	-1.300	-20590	1.929	1.846	2.242	0.154	0.564	849	2.134

Table A.28. Vapour Coexisting with the Boiling Liquid ($(O/U)^{(liq)} = 1.80$)

T K	P bar	ρ $kg\,m^{-3}$	H $10^{3} kJ\,kg^{-1}$	G $kJ\,kg^{-1}$	S $kJ\,kg^{-1}$	C_P K^{-1}	C_P/C_V	β_T $10^{-3} bar^{-1}$	α_T $10^{-3} K^{-1}$	V_S $m\,s^{-1}$	O/U
3000	0.0184[illegible]	0.01932	-3.688	-8612	1.641	0.285	1.134	5.440^{+4}_{10}	0.344	329	1.518
3120	0.03536	0.03570	-3.697	-8793	1.633	0.296	1.131	2.833^{+4}_{10}	0.333	334	1.535
3500	0.1987	0.1788	-3.707	-9373	1.619	0.354	1.122	5057.00	0.315	352	1.586
4000	1.117	0.8718	-3.687	-10172	1.621	0.550	1.121	912.300	0.344	375	1.655
4500	4.346	2.913	-3.631	-11080	1.655	0.976	1.156	243.400	0.455	404	1.752
5000	13.81	7.660	-3.517	-12203	1.737	1.591	1.229	81.540	0.608	444	1.915
5500	39.33	17.12	-3.319	-13679	1.884	2.151	1.307	30.430	0.692	501	2.188
6000	104.1	34.19	-3.028	-15626	2.100	2.450	1.359	11.920	0.663	577	2.615
6500	257.0	63.53	-2.660	-18059	2.369	2.555	1.390	4.906	0.586	668	3.209
7000	587.8	113.3	-2.262	-20763	2.643	2.534	1.416	2.163	0.511	760	3.902
7500	1232	198.3	-1.899	-23238	2.845	2.455	1.446	1.036	0.455	839	4.500
8000	2333	342.2	-1.628	-24841	2.902	2.330	1.485	0.545	0.421	892	4.730
8500	3934	575.0	-1.466	-25175	2.789	2.175	1.543	0.317	0.405	920	4.451
9000	5842	920.4	-1.383	-24367	2.554	2.016	1.634	0.206	0.406	928	3.795
9500	7593	1397	-1.345	-22793	2.258	1.883	1.802	0.154	0.435	917	2.989
9800	8250	1778	-1.345	-21553	2.062	1.838	2.002	0.142	0.487	891	2.482
9900	8339	1935	-1.353	-21074	1.992	1.836	2.110	0.142	0.517	877	2.307
9950	8348	2023	-1.359	-20816	1.955	1.839	2.178	0.143	0.537	869	2.217
10000	8325	2120	-1.367	-20540	1.917	1.847	2.260	0.144	0.561	860	2.125

Table A.28. Vapour Coexisting with the Boiling Liquid ($(O/U)^{(liq)} = 1.85$)

T K	P bar	ρ $kg\,m^{-3}$	H $10^{3} kJ\,kg^{-1}$	G $kJ\,kg^{-1}$	S $kJ\,kg^{-1}\,K^{-1}$	C_P $kJ\,kg^{-1}\,K^{-1}$	C_P/C_V	β_T $10^{-3} bar^{-1}$	α_T $10^{-3} K^{-1}$	V_S $m\,s^{-1}$	O/U
3000	0.01678	0.01770	-3.908	-8874	1.656	0.297	1.129	5.970_{10}^{+4}	0.346	327	1.610
3120	0.03243	0.03291	-3.916	-9057	1.648	0.310	1.126	3.090_{10}^{+4}	0.337	333	1.628
3500	0.1857	0.1679	-3.927	-9648	1.634	0.390	1.117	5417.00	0.326	350	1.683
4000	1.073	0.8375	-3.910	-10482	1.643	0.655	1.123	955.100	0.379	375	1.764
4500	4.336	2.863	-3.849	-11457	1.691	1.179	1.171	247.800	0.522	406	1.886
5000	14.41	7.677	-3.712	-12692	1.796	1.832	1.251	79.540	0.670	453	2.090
5500	42.81	17.42	-3.476	-14329	1.973	2.326	1.320	28.170	0.709	519	2.427
6000	116.8	35.23	-3.142	-16484	2.224	2.536	1.359	10.570	0.645	604	2.942
6500	293.7	66.24	-2.739	-19126	2.521	2.596	1.383	4.248	0.558	701	3.637
7000	675.5	119.6	-2.321	-21956	2.805	2.556	1.407	1.856	0.484	796	4.408
7500	1410	211.5	-1.955	-24377	2.990	2.463	1.437	0.891	0.434	873	5.004
8000	2635	367.3	-1.695	-25730	3.004	2.326	1.476	0.473	0.404	922	5.124
8500	4355	615.8	-1.542	-25731	2.846	2.162	1.533	0.279	0.390	945	4.686
9000	6327	976.2	-1.459	-24650	2.577	2.001	1.624	0.184	0.392	952	3.906
9500	8038	1465	-1.413	-22892	2.261	1.872	1.795	0.140	0.423	936	3.026
9800	8595	1857	-1.410	-21557	2.056	1.831	2.004	0.132	0.478	906	2.488
9900	8630	2020	-1.417	-21042	1.982	1.832	2.118	0.132	0.511	890	2.303
9950	8604	2113	-1.423	-20763	1.944	1.837	2.192	0.134	0.533	881	2.207
10000	8539	2217	-1.432	-20462	1.903	1.846	2.282	0.136	0.559	870	2.108
10070	8358	2390	-1.450	-19987	1.841	1.870	2.453	0.141	0.609	852	1.960

Table A.28. Vapour Coexisting with the Boiling Liquid ($(O/U)^{(liq)} = 1.875$)

T K	P bar	ρ $kg\,m^{-3}$	H $10^{3} kJ\,kg^{-1}$	G $kJ\,kg^{-1}$	S $kJ\,kg^{-1}$	C_P K^{-1}	C_P/C_V	β_T $10^{-3}bar^{-1}$	α_T $10^{-3}K^{-1}$	V_S $m\,s^{-1}$	O/U
3000	0.01598	0.01691	-4.034	-9023	1.663	0.305	1.126	6.267^{+4}_{10}	0.348	326	1.664
3120	0.03102	0.03157	-4.041	-9206	1.656	0.321	1.123	3.232^{+4}_{10}	0.339	332	1.682
3500	0.1798	0.1629	-4.051	-9803	1.643	0.418	1.114	5601.00	0.334	349	1.739
4000	1.059	0.8251	-4.035	-10660	1.656	0.729	1.126	972.500	0.404	375	1.829
4500	4.383	2.861	-3.968	-11678	1.713	1.301	1.182	247.400	0.561	409	1.967
5000	14.94	7.750	-3.814	-12976	1.832	1.949	1.262	77.360	0.696	459	2.196
5500	45.27	17.70	-3.554	-14703	2.027	2.392	1.323	26.670	0.708	529	2.570
6000	125.1	35.99	-3.197	-16967	2.295	2.559	1.357	9.830	0.630	619	3.135
6500	316.4	68.03	-2.777	-19712	2.605	2.605	1.379	3.918	0.542	719	3.885
7000	727.8	123.5	-2.351	-22587	2.891	2.560	1.402	1.711	0.471	815	4.692
7500	1512	219.3	-1.986	-24954	3.062	2.462	1.432	0.824	0.423	890	5.273
8000	2801	381.3	-1.731	-26157	3.053	2.320	1.472	0.440	0.395	936	5.321
8500	4579	637.7	-1.581	-25985	2.871	2.152	1.529	0.261	0.382	958	4.797
9000	6577	1005	-1.496	-24775	2.587	1.991	1.620	0.174	0.384	963	3.957
9500	8261	1500	-1.447	-22932	2.262	1.865	1.792	0.134	0.417	946	3.042
9800	8761	1896	-1.442	-21551	2.052	1.827	2.005	0.127	0.474	913	2.489
9900	8765	2064	-1.449	-21017	1.977	1.829	2.124	0.128	0.509	896	2.298
9950	8719	2159	-1.455	-20727	1.937	1.835	2.201	0.130	0.531	886	2.199
10000	8631	2267	-1.464	-20413	1.895	1.846	2.296	0.133	0.560	874	2.097
10070	8400	2452	-1.484	-19906	1.829	1.872	2.481	0.139	0.614	855	1.941

Table A.28. Vapour Coexisting with the Boiling Liquid ($(O/U)^{(liq)} = 1.90$)

T K	P bar	ρ $kg\,m^{-3}$	H $10^{3} kJ\,kg^{-1}$	G $kJ\,kg^{-1}$	S $kJ\,kg^{-1}$	C_P K^{-1}	C_P/C_V	β_T $10^{-3} bar^{-1}$	α_T $10^{-3} K^{-1}$	V_S $m\,s^{-1}$	O/U
3000	0.01521	0.01615	-4.173	-9187	1.671	0.317	1.122	6.585^{+4}_{10}	0.350	325	1.723
3120	0.02966	0.03029	-4.179	-9370	1.664	0.337	1.119	3.381^{+4}_{10}	0.343	331	1.742
3500	0.1746	0.1585	-4.188	-9975	1.654	0.456	1.111	5778.00	0.346	348	1.802
4000	1.054	0.8182	-4.172	-10862	1.673	0.821	1.131	983.500	0.437	375	1.904
4500	4.487	2.882	-4.095	-11928	1.741	1.434	1.194	244.300	0.604	412	2.061
5000	15.69	7.884	-3.917	-13294	1.875	2.056	1.272	74.190	0.717	466	2.317
5500	48.39	18.10	-3.631	-15113	2.088	2.439	1.325	24.930	0.701	542	2.732
6000	135.1	36.96	-3.251	-17489	2.373	2.569	1.354	9.060	0.613	636	3.350
6500	342.5	70.15	-2.814	-20330	2.695	2.606	1.374	3.595	0.525	738	4.155
7000	786.3	127.9	-2.380	-23235	2.979	2.560	1.398	1.572	0.457	834	4.994
7500	1623	227.9	-2.017	-25526	3.135	2.459	1.428	0.761	0.413	907	5.549
8000	2978	396.4	-1.767	-26565	3.100	2.311	1.467	0.410	0.386	951	5.516
8500	4812	660.4	-1.620	-26220	2.894	2.141	1.524	0.245	0.374	970	4.902
9000	6831	1034	-1.533	-24886	2.595	1.981	1.615	0.164	0.377	975	4.003
9500	8482	1534	-1.480	-22963	2.261	1.858	1.789	0.128	0.411	955	3.056
9800	8921	1936	-1.474	-21538	2.047	1.824	2.007	0.123	0.471	920	2.487
9900	8892	2107	-1.480	-20985	1.970	1.827	2.130	0.125	0.507	901	2.291
9950	8824	2206	-1.487	-20683	1.929	1.834	2.211	0.127	0.531	890	2.190
10000	8709	2318	-1.496	-20354	1.886	1.846	2.312	0.130	0.561	878	2.084

Table A.28. Vapour Coexisting with the Boiling Liquid ($((O/U)^{(liq)} = 1.92)$)

T K	P bar	ρ $kg\,m^{-3}$	H $10^{3} kJ\,kg^{-1}$	G $kJ\,kg^{-1}$	S $kJ\,kg^{-1}\,K^{-1}$	C_P $kJ\,kg^{-1}\,K^{-1}$	C_P/C_V	β_T $10^{-3} bar^{-1}$	α_T $10^{-3} K^{-1}$	V_S $m\,s^{-1}$	O/U
3000	0.01463	0.01557	-4.295	-9331	1.678	0.330	1.119	6.852_{10}^{+4}	0.353	324	1.777
3120	0.02865	0.02933	-4.300	-9515	1.671	0.355	1.115	3.502_{10}^{+4}	0.347	329	1.796
3500	0.1714	0.1558	-4.309	-10131	1.663	0.498	1.110	5901.00	0.360	347	1.860
4000	1.061	0.8193	-4.294	-11048	1.689	0.912	1.137	984.100	0.471	376	1.975
4500	4.630	2.925	-4.202	-12155	1.767	1.548	1.206	238.900	0.640	415	2.149
5000	16.50	8.047	-4.001	-13577	1.915	2.131	1.279	70.840	0.728	474	2.429
5500	51.50	18.52	-3.692	-15475	2.142	2.463	1.325	23.390	0.690	553	2.878
6000	144.5	37.90	-3.293	-17942	2.442	2.569	1.351	8.429	0.596	650	3.543
6500	366.6	72.13	-2.844	-20857	2.771	2.604	1.371	3.340	0.511	754	4.395
7000	838.8	131.9	-2.406	-23777	3.053	2.557	1.395	1.465	0.447	850	5.256
7500	1720	235.5	-2.045	-25993	3.193	2.454	1.424	0.713	0.405	921	5.782
8000	3129	409.4	-1.799	-26889	3.136	2.302	1.464	0.386	0.380	962	5.674
8500	5007	679.5	-1.653	-26403	2.912	2.130	1.520	0.233	0.368	981	4.986
9000	7041	1058	-1.563	-24974	2.601	1.972	1.611	0.157	0.372	985	4.040
9500	8661	1561	-1.506	-22988	2.261	1.852	1.786	0.123	0.406	963	3.066
9800	9048	1968	-1.498	-21526	2.044	1.820	2.008	0.119	0.468	925	2.486
9900	8989	2142	-1.505	-20957	1.965	1.825	2.135	0.122	0.505	905	2.286
9950	8903	2243	-1.511	-20645	1.923	1.832	2.219	0.124	0.530	894	2.181
10000	8764	2360	-1.521	-20303	1.878	1.845	2.326	0.127	0.562	881	2.072

Table A.28. Vapour Coexisting with the Boiling Liquid ($(O/U)^{(liq)} = 1.94$)

T K	P bar	ρ $kg\,m^{-3}$	H $10^3 kJ\,kg^{-1}$	G $kJ\,kg^{-1}$	S $kJ\,kg^{-1}$	C_P K^{-1}	C_P/C_V	β_T $10^{-3} bar^{-1}$	α_T $10^{-3} K^{-1}$	V_S $m\,s^{-1}$	O/U
3000	0.01408	0.01504	-4.432	-9490	1.686	0.350	1.114	7.121^{+4}_{10}	0.358	322	1.838
3120	0.02774	0.02849	-4.436	-9677	1.680	0.381	1.111	3.619^{+4}_{10}	0.354	328	1.857
3500	0.1696	0.1542	-4.447	-10310	1.675	0.555	1.110	5982.00	0.380	347	1.930
4000	1.084	0.8307	-4.428	-11266	1.709	1.023	1.146	971.600	0.513	377	2.060
4500	4.852	3.000	-4.314	-12413	1.800	1.664	1.218	230.100	0.676	420	2.251
5000	17.56	8.275	-4.085	-13889	1.961	2.191	1.284	66.750	0.732	482	2.555
5500	55.25	19.05	-3.751	-15866	2.203	2.474	1.324	21.730	0.675	565	3.042
6000	155.4	39.02	-3.333	-18424	2.515	2.562	1.348	7.796	0.579	666	3.755
6500	393.6	74.38	-2.873	-21406	2.851	2.599	1.367	3.093	0.497	771	4.653
7000	896.3	136.3	-2.432	-24325	3.128	2.553	1.391	1.363	0.437	866	5.530
7500	1825	243.6	-2.074	-26451	3.250	2.447	1.421	0.668	0.397	935	6.018
8000	3287	423.0	-1.831	-27197	3.171	2.292	1.460	0.364	0.373	974	5.828
8500	5206	699.0	-1.685	-26572	2.928	2.119	1.517	0.221	0.362	991	5.065
9000	7252	1082	-1.592	-25052	2.607	1.962	1.607	0.150	0.366	994	4.074
9500	8838	1589	-1.532	-23007	2.261	1.846	1.783	0.119	0.402	971	3.075
9800	9169	2000	-1.523	-21509	2.039	1.816	2.010	0.116	0.465	930	2.483
9900	9080	2178	-1.529	-20925	1.959	1.822	2.141	0.119	0.504	909	2.278
9950	8974	2282	-1.536	-20602	1.916	1.831	2.229	0.121	0.530	897	2.171
10000	8809	2403	-1.546	-20246	1.870	1.845	2.342	0.125	0.564	883	2.058

Table A.28. Vapour Coexisting with the Boiling Liquid ($(O/U)^{(liq)} = 1.95$)

T K	P bar	ρ $kg\ m^{-3}$	H $10^{3} kJ\ kg^{-1}$	G $kJ\ kg^{-1}$	S $kJ\ kg^{-1}$	C_P K^{-1}	C_P/C_V	β_T $10^{-3} bar^{-1}$	α_T $10^{-3} K^{-1}$	V_S $m\ s^{-1}$	O/U
3000	0.01384	0.01481	-4.507	-9579	1.691	0.364	1.111	7.248^{+4}_{10}	0.361	322	1.872
3120	0.02738	0.02815	-4.512	-9768	1.685	0.399	1.108	3.670^{+4}_{10}	0.360	328	1.893
3500	0.1697	0.1544	-4.526	-10415	1.683	0.593	1.110	5990.00	0.395	347	1.971
4000	1.105	0.8425	-4.502	-11391	1.722	1.089	1.152	957.700	0.539	378	2.110
4500	5.004	3.054	-4.372	-12555	1.818	1.721	1.224	224.200	0.693	423	2.309
5000	18.20	8.418	-4.127	-14057	1.986	2.215	1.287	64.440	0.732	487	2.625
5500	57.41	19.37	-3.780	-16075	2.235	2.476	1.323	20.88	0.667	572	3.131
6000	161.5	39.65	-3.353	-18677	2.554	2.557	1.347	7.482	0.571	674	3.869
6500	408.4	75.62	-2.888	-21689	2.892	2.596	1.366	2.973	0.490	779	4.789
7000	927.2	138.7	-2.446	-24603	3.165	2.551	1.390	1.313	0.431	874	5.673
7500	1880	247.9	-2.088	-26678	3.279	2.443	1.420	0.646	0.393	942	6.138
8000	3369	430.1	-1.847	-27347	3.187	2.286	1.459	0.354	0.370	979	5.905
8500	5308	709.0	-1.702	-26653	2.935	2.113	1.515	0.215	0.359	996	5.103
9000	7359	1094	-1.607	-25089	2.609	1.957	1.606	0.147	0.363	999	4.090
9500	8927	1603	-1.545	-23015	2.260	1.842	1.782	0.117	0.400	974	3.079
9800	9228	2016	-1.535	-21500	2.037	1.814	2.011	0.115	0.464	933	2.481
9900	9123	2195	-1.542	-20907	1.956	1.821	2.145	0.118	0.504	911	2.274
9950	9007	2301	-1.548	-20579	1.913	1.831	2.234	0.120	0.530	898	2.166
10000	8828	2425	-1.559	-20216	1.866	1.845	2.350	0.124	0.565	884	2.051

Table A.28. Vapour Coexisting with the Boiling Liquid ($(O/U)^{(liq)} = 1.96$)

T K	P bar	ρ $kg\,m^{-3}$	H $10^{3} kJ\,kg^{-1}$	G $kJ\,kg^{-1}$	S $kJ\,kg^{-1}\,K^{-1}$	C_P $kJ\,kg^{-1}\,K^{-1}$	C_P/C_V	β_T $10^{-3} bar^{-1}$	α_T $10^{-3} K^{-1}$	V_S $m\,s^{-1}$	O/U
3000	0.01364	0.01462	-4.590	-9677	1.696	0.383	1.108	7.359^{+4}_{10}	0.366	321	1.911
3120	0.02711	0.02792	-4.595	-9869	1.690	0.423	1.106	3.709^{+4}_{10}	0.367	327	1.933
3500	0.1712	0.1556	-4.613	-10535	1.692	0.639	1.113	5958.00	0.414	346	2.019
4000	1.137	0.8604	-4.582	-11530	1.737	1.161	1.160	936.800	0.568	379	2.167
4500	5.190	3.121	-4.432	-12706	1.839	1.777	1.230	217.100	0.708	426	2.372
5000	18.94	8.583	-4.169	-14234	2.013	2.235	1.288	61.950	0.729	492	2.699
5500	59.80	19.71	-3.809	-16292	2.270	2.475	1.322	20.000	0.657	579	3.225
6000	168.1	40.33	-3.373	-18937	2.594	2.551	1.345	7.169	0.562	682	3.988
6500	424.0	76.95	-2.903	-21978	2.935	2.593	1.364	2.855	0.484	788	4.931
7000	959.5	141.2	-2.459	-24883	3.203	2.548	1.388	1.265	0.426	882	5.820
7500	1937	252.4	-2.104	-26905	3.307	2.439	1.418	0.624	0.389	948	6.259
8000	3453	437.3	-1.864	-27494	3.204	2.280	1.457	0.343	0.366	985	5.981
8500	5412	719.1	-1.718	-26732	2.943	2.107	1.513	0.210	0.356	1002	5.141
9000	7466	1106	-1.621	-25125	2.612	1.952	1.604	0.144	0.360	1004	4.106
9500	9015	1616	-1.557	-23023	2.260	1.839	1.781	0.115	0.398	978	3.083
9800	9286	2032	-1.547	-21490	2.035	1.813	2.012	0.113	0.463	935	2.479
9900	9165	2213	-1.554	-20889	1.953	1.820	2.148	0.116	0.503	913	2.270
9950	9038	2321	-1.561	-20555	1.909	1.830	2.240	0.119	0.531	900	2.160
10000	8844	2447	-1.571	-20184	1.861	1.846	2.359	0.123	0.566	885	2.043

Table A.28. Vapour Coexisting with the Boiling Liquid ($(O/U)^{(liq)} = 1.97$)

T K	P bar	ρ $kg\,m^{-3}$	H $10^{3} kJ\,kg^{-1}$	G $kJ\,kg^{-1}$	S $kJ\,kg^{-1}\,K^{-1}$	C_P $kJ\,kg^{-1}\,K^{-1}$	C_P/C_V	β_T $10^{-3} bar^{-1}$	α_T $10^{-3} K^{-1}$	V_S $m\,s^{-1}$	O/U
3000	0.01351	0.01451	-4.683	-9789	1.702	0.407	1.106	7.434^{+4}_{10}	0.374	320	1.956
3120	0.02705	0.02789	-4.691	-9988	1.698	0.453	1.105	3.723^{+4}_{10}	0.378	326	1.981
3500	0.1749	0.1587	-4.715	-10679	1.704	0.699	1.117	5858.00	0.440	347	2.079
4000	1.182	0.8867	-4.667	-11686	1.755	1.243	1.168	906.900	0.600	381	2.233
4500	5.417	3.205	-4.492	-12868	1.861	1.831	1.237	208.800	0.722	430	2.440
5000	19.78	8.774	-4.211	-14420	2.042	2.250	1.289	59.310	0.725	498	2.779
5500	62.43	20.09	-3.838	-16518	2.306	2.471	1.321	19.120	0.647	586	3.325
6000	175.2	41.07	-3.393	-19207	2.636	2.545	1.344	6.860	0.553	691	4.114
6500	440.7	78.35	-2.918	-22273	2.978	2.589	1.362	2.740	0.477	797	5.079
7000	993.4	143.8	-2.474	-25166	3.242	2.545	1.387	1.218	0.422	890	5.970
7500	1996	257.1	-2.119	-27129	3.335	2.435	1.417	0.604	0.385	955	6.382
8000	3539	444.8	-1.881	-27638	3.220	2.274	1.455	0.333	0.363	991	6.056
8500	5517	729.4	-1.734	-26808	2.950	2.100	1.511	0.204	0.353	1007	5.177
9000	7574	1119	-1.635	-25160	2.614	1.947	1.602	0.141	0.358	1009	4.121
9500	9103	1630	-1.570	-23029	2.259	1.836	1.780	0.113	0.395	982	3.086
9800	9344	2048	-1.559	-21479	2.033	1.811	2.013	0.112	0.462	937	2.477
9900	9205	2231	-1.566	-20869	1.950	1.819	2.152	0.115	0.503	915	2.266
9950	9067	2340	-1.573	-20530	1.905	1.829	2.246	0.118	0.531	901	2.154
10000	8856	2467	-1.582	-20155	1.857	1.846	2.368	0.122	0.567	885	2.036

Table A.28. Vapour Coexisting with the Boiling Liquid ($(O/U)^{(liq)} = 1.98$)

T	P	ρ	H	G	S	C_P	C_P/C_V	β_T	α_T	V_S	O/U
K	bar	$kg\,m^{-3}$	$10^{3} kJ\,kg^{-1}$	$kJ\,kg^{-1}$	$kJ\,kg^{-1}\,K^{-1}$			$10^{-3} bar^{-1}$	$10^{-3} K^{-1}$	$m\,s^{-1}$	
3000	0.01357	0.01460	-4.799	-9932	1.711	0.441	1.104	7.415^{+4}_{10}	0.388	319	2.016
3120	0.02746	0.02835	-4.812	-10144	1.709	0.497	1.105	3.677^{+4}_{10}	0.398	326	2.046
3500	0.1829	0.1654	-4.840	-10863	1.721	0.782	1.123	5635.00	0.478	347	2.158
4000	1.248	0.9246	-4.759	-11861	1.776	1.334	1.178	866.200	0.635	384	2.308
4500	5.693	3.307	-4.553	-13041	1.886	1.881	1.242	199.400	0.733	434	2.515
5000	20.74	8.991	-4.252	-14616	2.073	2.261	1.290	56.540	0.719	504	2.864
5500	65.33	20.52	-3.866	-16754	2.343	2.466	1.319	18.220	0.637	594	3.431
6000	182.9	41.87	-3.413	-19485	2.679	2.538	1.342	6.554	0.544	699	4.246
6500	458.3	79.85	-2.934	-22574	3.022	2.585	1.361	2.626	0.470	806	5.232
7000	1029	146.6	-2.488	-25450	3.280	2.542	1.385	1.173	0.417	898	6.124
7500	2057	261.9	-2.135	-27352	3.362	2.430	1.415	0.584	0.381	962	6.505
8000	3627	452.4	-1.898	-27778	3.235	2.268	1.454	0.324	0.360	997	6.130
8500	5623	739.7	-1.750	-26882	2.957	2.094	1.510	0.199	0.350	1012	5.214
9000	7683	1131	-1.650	-25194	2.616	1.942	1.600	0.138	0.355	1014	4.136
9500	9191	1644	-1.582	-23036	2.258	1.833	1.779	0.111	0.393	986	3.090
9800	9400	2064	-1.571	-21467	2.030	1.809	2.015	0.111	0.461	940	2.475
9900	9244	2249	-1.578	-20849	1.947	1.818	2.156	0.114	0.503	916	2.261
9950	9093	2360	-1.585	-20503	1.901	1.829	2.252	0.117	0.532	902	2.148
10000	8865	2485	-1.592	-20127	1.853	1.846	2.375	0.122	0.569	886	2.030

Table A.28. Vapour Coexisting with the Boiling Liquid ($(O/U)^{(liq)} = 1.99$)

T K	P bar	ρ $kg\,m^{-3}$	H $10\,^{3}kJ\,kg^{-1}$	G $kJ\,kg^{-1}$	S $kJ\,kg^{-1}$	C_P K^{-1}	C_P/C_V	β_T $10^{-3}bar^{-1}$	α_T $10^{-3}K^{-1}$	V_S $m\,s^{-1}$	O/U
3000	0.01429	0.01539	-4.978	-10164	1.729	0.503	1.107	7.071^{+4}_{10}	0.419	319	2.117
3120	0.02944	0.03039	-5.000	-10395	1.729	0.579	1.110	3.450^{+4}_{10}	0.439	325	2.157
3500	0.2002	0.1798	-5.002	-11110	1.745	0.907	1.134	5199.00	0.536	348	2.267
4000	1.339	0.9780	-4.856	-12056	1.800	1.433	1.188	813.700	0.672	386	2.394
4500	6.028	3.432	-4.614	-13225	1.913	1.926	1.247	188.900	0.741	439	2.595
5000	21.83	9.240	-4.293	-14822	2.106	2.267	1.291	53.660	0.711	510	2.955
5500	68.52	20.98	-3.894	-17000	2.383	2.458	1.318	17.330	0.626	602	3.543
6000	191.1	42.73	-3.433	-19771	2.723	2.530	1.341	6.253	0.535	708	4.385
6500	477.1	81.44	-2.949	-22880	3.066	2.582	1.360	2.516	0.464	815	5.392
7000	1066	149.5	-2.503	-25735	3.319	2.539	1.384	1.128	0.412	906	6.282
7500	2120	267.0	-2.151	-27572	3.390	2.425	1.414	0.564	0.377	969	6.629
8000	3717	460.2	-1.915	-27916	3.250	2.261	1.452	0.314	0.357	1002	6.204
8500	5731	750.2	-1.766	-26954	2.963	2.087	1.508	0.194	0.347	1018	5.249
9000	7793	1143	-1.664	-25226	2.618	1.937	1.598	0.135	0.352	1019	4.151
9500	9279	1657	-1.594	-23041	2.258	1.829	1.777	0.110	0.391	989	3.093
9800	9455	2080	-1.583	-21454	2.028	1.807	2.016	0.109	0.460	942	2.472
9900	9281	2268	-1.589	-20828	1.943	1.817	2.160	0.113	0.503	918	2.256
9950	9118	2381	-1.597	-20476	1.897	1.828	2.259	0.116	0.532	904	2.141
10000	8870	2495	-1.598	-20112	1.851	1.846	2.380	0.121	0.569	886	2.026

Table A.28. Vapour Coexisting with the Boiling Liquid ($(O/U)^{(liq)} = 2.00$)

T K	P bar	ρ $kg\,m^{-3}$	H $10^{3} kJ\,kg^{-1}$	G $kJ\,kg^{-1}$	S $kJ\,kg^{-1}\,K^{-1}$	C_P	C_P/C_V	β_T $10^{-3} bar^{-1}$	α_T $10^{-3} K^{-1}$	V_S $m\,s^{-1}$	O/U
3000	0.02024	0.02181	-5.448	-10795	1.782	0.759	1.118	5.069^{+4}_{10}	0.544	318	2.419
3120	0.03960	0.04074	-5.390	-10935	1.777	0.827	1.124	2.605^{+4}_{10}	0.557	325	2.414
3500	0.2360	0.2095	-5.206	-11431	1.779	1.092	1.150	4471.00	0.616	350	2.418
4000	1.464	1.051	-4.958	-12267	1.827	1.535	1.199	750.000	0.708	390	2.491
4500	6.431	3.581	-4.676	-13419	1.943	1.964	1.251	177.500	0.746	444	2.682
5000	23.08	9.521	-4.334	-15038	2.141	2.269	1.290	50.700	0.702	517	3.053
5500	72.04	21.50	-3.922	-17255	2.424	2.450	1.316	16.440	0.615	610	3.663
6000	200.0	43.66	-3.452	-20066	2.769	2.522	1.339	5.958	0.526	718	4.530
6500	497.0	83.13	-2.965	-23192	3.112	2.578	1.358	2.409	0.458	824	5.558
7000	1105	152.5	-2.519	-26022	3.358	2.535	1.383	1.085	0.407	914	6.444
7500	2185	272.2	-2.168	-27791	3.416	2.419	1.412	0.545	0.374	976	6.753
8000	3809	468.2	-1.932	-28050	3.265	2.254	1.451	0.305	0.354	1008	6.277
8500	5841	760.9	-1.783	-27024	2.970	2.080	1.506	0.189	0.344	1023	5.283
9000	7904	1155	-1.678	-25257	2.620	1.931	1.596	0.132	0.349	1024	4.165
9500	9366	1671	-1.606	-23046	2.257	1.826	1.776	0.108	0.389	993	3.096
9800	9509	2096	-1.594	-21441	2.025	1.805	2.018	0.108	0.459	944	2.469
9900	9316	2286	-1.601	-20806	1.940	1.816	2.164	0.112	0.503	920	2.251
9950	9140	2401	-1.609	-20447	1.893	1.828	2.265	0.115	0.533	905	2.134
10000	8867	2490	-1.595	-20118	1.852	1.846	2.378	0.122	0.569	886	2.027

Table A.28. Vapour Coexisting with the Boiling Liquid ($(O/U)^{(liq)} = 2.01$)

T K	P bar	ρ $kg\,m^{-3}$	H $10^{3} kJ\,kg^{-1}$	G $kJ\,kg^{-1}$	S $kJ\,kg^{-1}\,K^{-1}$	C_P $kJ\,kg^{-1}\,K^{-1}$	C_P/C_V	β_T $10^{-3} bar^{-1}$	α_T $10^{-3} K^{-1}$	V_S $m\,s^{-1}$	O/U
3000	0.04524	0.04839	-5.973	-11465	1.831	1.077	1.130	2.305^{+4}_{10}	0.679	318	2.792
3120	0.07140	0.07269	-5.821	-11514	1.825	1.126	1.138	1.469^{+4}_{10}	0.684	326	2.726
3500	0.2985	0.2608	-5.418	-11771	1.815	1.289	1.164	3580.00	0.696	353	2.588
4000	1.629	1.146	-5.059	-12490	1.858	1.629	1.208	678.400	0.738	394	2.595
4500	6.912	3.757	-4.737	-13624	1.975	1.995	1.254	165.300	0.747	449	2.776
5000	24.49	9.840	-4.374	-15264	2.178	2.268	1.290	47.700	0.692	524	3.157
5500	75.91	22.06	-3.949	-17522	2.468	2.441	1.315	15.560	0.604	619	3.789
6000	209.6	44.66	-3.472	-20370	2.816	2.515	1.338	5.670	0.518	727	4.683
6500	518.1	84.93	-2.981	-23509	3.158	2.574	1.357	2.304	0.452	833	5.730
7000	1146	155.7	-2.534	-26310	3.396	2.531	1.381	1.043	0.403	922	6.610
7500	2253	277.6	-2.185	-28007	3.443	2.413	1.411	0.527	0.370	982	6.878
8000	3904	476.4	-1.949	-28181	3.279	2.247	1.449	0.296	0.350	1014	6.349
8500	5952	771.6	-1.799	-27091	2.976	2.073	1.504	0.184	0.341	1028	5.317
9000	8015	1168	-1.691	-25287	2.622	1.926	1.595	0.129	0.347	1029	4.179
9500	9454	1685	-1.618	-23050	2.256	1.822	1.775	0.106	0.387	997	3.099
9800	9562	2112	-1.606	-21427	2.023	1.804	2.019	0.107	0.458	946	2.466
9900	9350	2304	-1.613	-20782	1.936	1.815	2.168	0.111	0.503	921	2.245
9950	9160	2422	-1.621	-20417	1.889	1.828	2.273	0.114	0.534	906	2.127
10000	8852	2474	-1.586	-20140	1.855	1.846	2.372	0.122	0.569	885	2.033

Table A.28. Vapour Coexisting with the Boiling Liquid ($(O/U)^{(liq)} = 2.02$)

T K	P bar	ρ $kg\,m^{-3}$	H $10^{3} kJ\,kg^{-1}$	G $kJ\,kg^{-1}$	S $kJ\,kg^{-1}\,K^{-1}$	C_P	C_P/C_V	β_T $10^{-3} bar^{-1}$	α_T $10^{-3} K^{-1}$	V_S $m\,s^{-1}$	O/U
3000	0.07853	0.08315	-6.154	-11685	1.844	1.079	1.129	1.327^{+4}_{10}	0.673	320	2.939
3120	0.1151	0.1158	-6.023	-11778	1.845	1.183	1.140	9132.00	0.701	328	2.890
3500	0.3858	0.3312	-5.593	-12056	1.847	1.417	1.172	2790.00	0.741	356	2.739
4000	1.839	1.266	-5.158	-12714	1.889	1.706	1.214	603.700	0.759	399	2.705
4500	7.482	3.964	-4.797	-13837	2.009	2.017	1.256	152.800	0.744	455	2.876
5000	26.09	10.20	-4.414	-15501	2.218	2.263	1.289	44.680	0.680	532	3.268
5500	80.18	22.68	-3.976	-17798	2.513	2.432	1.313	14.690	0.593	628	3.923
6000	219.9	45.73	-3.492	-20682	2.865	2.508	1.337	5.388	0.509	737	4.844
6500	540.6	86.83	-2.998	-23830	3.205	2.571	1.356	2.202	0.446	842	5.908
7000	1189	159.0	-2.551	-26598	3.435	2.527	1.380	1.003	0.398	930	6.779
7500	2323	283.1	-2.202	-28220	3.469	2.407	1.409	0.509	0.367	989	7.004
8000	4000	484.8	-1.967	-28309	3.293	2.239	1.448	0.287	0.347	1020	6.420
8500	6064	782.4	-1.815	-27157	2.981	2.066	1.502	0.180	0.338	1034	5.350
9000	8127	1180	-1.705	-25317	2.623	1.920	1.593	0.126	0.344	1034	4.193
9500	9541	1698	-1.630	-23054	2.255	1.819	1.774	0.104	0.385	1000	3.101
9700	9691	1967	-1.617	-21998	2.101	1.801	1.915	0.104	0.425	970	2.678
9800	9614	2128	-1.617	-21413	2.020	1.802	2.021	0.106	0.457	949	2.463
9900	9382	2323	-1.625	-20758	1.933	1.814	2.173	0.110	0.503	922	2.239
9950	9178	2442	-1.633	-20386	1.885	1.827	2.280	0.114	0.535	907	2.119
10000	8849	2487	-1.593	-20116	1.852	1.847	2.380	0.122	0.570	885	2.027

Table A.28. Vapour Coexisting with the Boiling Liquid ($(O/U)^{(liq)} = 2.03$)

T K	P bar	ρ $kg\,m^{-3}$	H $10^{3} kJ\,kg^{-1}$	G $kJ\,kg^{-1}$	S $kJ\,kg^{-1}\,K^{-1}$	C_P $kJ\,kg^{-1}\,K^{-1}$	C_P/C_V	β_T $10^{-3} bar^{-1}$	α_T $10^{-3} K^{-1}$	V_S $m\,s^{-1}$	O/U
3000	0.1167	0.1221	-6.239	-11796	1.852	0.995	1.125	8887.00	0.633	322	3.023
3120	0.1655	0.1646	-6.127	-11921	1.857	1.128	1.136	6327.00	0.672	330	2.987
3500	0.4921	0.4155	-5.723	-12275	1.872	1.458	1.174	2191.00	0.749	359	2.862
4000	2.097	1.410	-5.249	-12934	1.921	1.755	1.218	530.800	0.767	403	2.815
4500	8.150	4.204	-4.855	-14057	2.045	2.029	1.257	140.200	0.737	462	2.982
5000	27.89	10.60	-4.453	-15749	2.259	2.256	1.288	41.680	0.668	540	3.387
5500	84.87	23.35	-4.003	-18086	2.560	2.422	1.312	13.840	0.582	637	4.065
6000	231.1	46.89	-3.512	-21003	2.915	2.501	1.336	5.114	0.501	746	5.013
6500	564.3	88.85	-3.015	-24157	3.253	2.567	1.355	2.104	0.440	851	6.093
7000	1234	162.5	-2.568	-26886	3.474	2.523	1.379	0.964	0.394	938	6.952
7500	2395	288.9	-2.220	-28430	3.495	2.401	1.408	0.492	0.363	996	7.130
8000	4098	493.4	-1.985	-28433	3.306	2.231	1.446	0.279	0.344	1025	6.490
8500	6179	793.4	-1.831	-27220	2.987	2.059	1.501	0.175	0.335	1039	5.383
9000	8240	1192	-1.719	-25345	2.625	1.915	1.591	0.124	0.341	1039	4.206
9500	9628	1712	-1.642	-23057	2.254	1.815	1.773	0.103	0.383	1004	3.104
9700	9757	1982	-1.629	-21990	2.099	1.798	1.915	0.102	0.424	972	2.677
9800	9664	2144	-1.629	-21397	2.017	1.800	2.023	0.104	0.456	951	2.460
9900	9412	2342	-1.636	-20733	1.929	1.813	2.178	0.109	0.503	924	2.233
9950	9192	2462	-1.644	-20355	1.880	1.827	2.288	0.113	0.536	907	2.112
10000	8970	2515	-1.611	-20124	1.851	1.838	2.370	0.118	0.562	892	2.030

Table A.28. Vapour Coexisting with the Boiling Liquid ($(O/U)^{(liq)} = 2.04$)

T K	P bar	ρ $kg\,m^{-3}$	H $10^{3} kJ\,kg^{-1}$	G $kJ\,kg^{-1}$	S $kJ\,kg^{-1}\,K^{-1}$	C_P $kJ\,kg^{-1}\,K^{-1}$	C_P/C_V	β_T $10^{-3} bar^{-1}$	α_T $10^{-3} K^{-1}$	V_S $m\,s^{-1}$	O/U
3000	0.1602	0.1654	-6.292	-11876	1.862	0.905	1.122	6444.00	0.592	324	3.087
3120	0.2227	0.2185	-6.193	-12021	1.868	1.045	1.132	4681.00	0.633	333	3.060
3500	0.6152	0.5108	-5.819	-12446	1.893	1.440	1.172	1749.00	0.734	362	2.964
4000	2.402	1.577	-5.333	-13145	1.953	1.776	1.219	463.300	0.764	408	2.925
4500	8.928	4.477	-4.912	-14285	2.083	2.032	1.257	127.800	0.727	469	3.094
5000	29.93	11.04	-4.491	-16007	2.303	2.247	1.287	38.740	0.655	548	3.513
5500	90.02	24.08	-4.031	-18383	2.610	2.413	1.310	13.010	0.571	647	4.216
6000	243.1	48.13	-3.533	-21332	2.966	2.494	1.335	4.849	0.493	756	5.190
6500	589.5	90.98	-3.033	-24488	3.301	2.564	1.354	2.008	0.434	861	6.285
7000	1280	166.2	-2.585	-27175	3.513	2.519	1.378	0.925	0.390	946	7.128
7500	2470	294.9	-2.238	-28638	3.520	2.395	1.407	0.475	0.360	1002	7.256
8000	4199	502.1	-2.003	-28555	3.319	2.224	1.444	0.271	0.341	1031	6.559
8500	6294	804.5	-1.847	-27281	2.992	2.052	1.499	0.171	0.332	1045	5.415
9000	8354	1205	-1.732	-25372	2.627	1.909	1.589	0.121	0.339	1044	4.219
9500	9715	1725	-1.654	-23059	2.253	1.812	1.772	0.101	0.381	1008	3.106
9700	9822	1996	-1.640	-21981	2.097	1.796	1.916	0.101	0.422	975	2.675
9800	9714	2161	-1.640	-21381	2.014	1.798	2.025	0.103	0.455	953	2.456
9900	9439	2361	-1.648	-20707	1.925	1.812	2.183	0.108	0.503	925	2.227
9950	9203	2481	-1.655	-20326	1.877	1.827	2.295	0.112	0.537	908	2.105
10000	9098	2529	-1.622	-20160	1.854	1.827	2.353	0.115	0.553	900	2.040

Table A.28. Vapour Coexisting with the Boiling Liquid ($(O/U)^{(liq)} = 2.05$)

T K	P bar	ρ $kg\,m^{-3}$	H $10^{3} kJ\,kg^{-1}$	G $kJ\,kg^{-1}$	S $kJ\,kg^{-1}$	C_P K^{-1}	C_P/C_V	β_T $10^{-4} bar^{-1}$	α_T $10^{-3} K^{-1}$	V_S $m\,s^{-1}$	O/U
3000	0.2096	0.2132	-6.330	-11947	1.872	0.832	1.120	4.908_{10}^{+4}	0.558	327	3.144
3120	0.2871	0.2777	-6.240	-12105	1.880	0.966	1.129	3.615_{10}^{+4}	0.597	335	3.123
3500	0.7544	0.6160	-5.894	-12589	1.913	1.390	1.168	1.421_{10}^{+4}	0.707	365	3.053
4000	2.757	1.766	-5.407	-13347	1.985	1.771	1.218	4030.00	0.751	414	3.034
4500	9.823	4.784	-4.966	-14517	2.122	2.028	1.256	1159.00	0.714	476	3.212
5000	32.22	11.53	-4.529	-16275	2.349	2.237	1.285	358.900	0.641	557	3.647
5500	95.67	24.88	-4.058	-18692	2.661	2.405	1.309	122.100	0.560	657	4.375
6000	256.0	49.45	-3.554	-21669	3.019	2.489	1.334	45.920	0.486	767	5.375
6500	616.3	93.24	-3.051	-24823	3.349	2.561	1.353	19.160	0.428	870	6.483
7000	1329	170.0	-2.603	-27463	3.551	2.515	1.377	8.886	0.386	955	7.308
7500	2547	301.1	-2.257	-28841	3.545	2.388	1.406	4.585	0.356	1009	7.382
8000	4301	511.0	-2.020	-28672	3.331	2.215	1.443	2.626	0.338	1037	6.627
8500	6411	815.6	-1.862	-27341	2.997	2.044	1.497	1.664	0.329	1050	5.446
9000	8469	1217	-1.745	-25399	2.628	1.904	1.587	1.185	0.336	1049	4.232
9500	9802	1739	-1.666	-23061	2.252	1.809	1.771	0.996	0.379	1011	3.108
9700	9886	2011	-1.652	-21971	2.095	1.793	1.916	0.997	0.421	978	2.673
9800	9762	2177	-1.652	-21364	2.011	1.796	2.027	1.022	0.454	955	2.452
9900	9465	2380	-1.660	-20680	1.921	1.811	2.189	1.073	0.504	926	2.220
9950	9211	2497	-1.663	-20301	1.873	1.827	2.301	1.117	0.538	908	2.099
10000	9207	2537	-1.629	-20197	1.857	1.818	2.336	1.121	0.544	906	2.050

Table A.28. Vapour Coexisting with the Boiling Liquid ($(O/U)^{(liq)} = 2.06$)

T K	P bar	ρ $kg\,m^{-3}$	H $10^{3} kJ\,kg^{-1}$	G $kJ\,kg^{-1}$	S $kJ\,kg^{-1}\,K^{-1}$	C_P	C_P/C_V	β_T $10^{-4} bar^{-1}$	α_T $10^{-3} K^{-1}$	V_S $m\,s^{-1}$	O/U
3000	0.2650	0.2651	-6.361	-12017	1.885	0.776	1.120	3.872_{10}^{+4}	0.533	330	3.200
3120	0.3589	0.3417	-6.276	-12182	1.893	0.901	1.128	2.882_{10}^{+4}	0.567	338	3.182
3500	0.9099	0.7301	-5.953	-12715	1.932	1.328	1.164	1.173_{10}^{+4}	0.677	369	3.133
4000	3.161	1.976	-5.474	-13540	2.016	1.748	1.216	3505.00	0.733	419	3.140
4500	10.84	5.126	-5.019	-14754	2.163	2.017	1.255	1047.00	0.699	483	3.334
5000	34.78	12.07	-4.567	-16553	2.397	2.227	1.284	331.600	0.628	566	3.790
5500	101.9	25.74	-4.085	-19010	2.714	2.398	1.308	114.300	0.550	667	4.543
6000	269.9	50.87	-3.575	-22013	3.073	2.484	1.333	43.440	0.478	777	5.570
6500	644.6	95.63	-3.070	-25161	3.399	2.558	1.352	18.270	0.423	880	6.689
7000	1380	174.1	-2.622	-27750	3.590	2.510	1.376	8.529	0.381	963	7.491
7500	2626	307.5	-2.275	-29042	3.569	2.381	1.404	4.428	0.353	1016	7.508
8000	4406	520.1	-2.039	-28787	3.344	2.207	1.441	2.548	0.335	1043	6.694
8500	6530	826.9	-1.878	-27399	3.002	2.037	1.496	1.622	0.326	1056	5.476
9000	8584	1229	-1.759	-25424	2.630	1.898	1.586	1.161	0.333	1054	4.244
9500	9888	1752	-1.677	-23062	2.251	1.805	1.770	0.981	0.377	1015	3.110
9700	9949	2026	-1.663	-21961	2.093	1.791	1.917	0.985	0.420	980	2.671
9800	9809	2193	-1.663	-21347	2.009	1.794	2.029	1.011	0.454	957	2.448
9900	9489	2399	-1.671	-20652	1.917	1.811	2.194	1.064	0.504	927	2.213
9950	9215	2503	-1.667	-20291	1.872	1.826	2.304	1.115	0.538	909	2.096
10000	9283	2536	-1.631	-20234	1.860	1.812	2.322	1.104	0.538	911	2.060

Table A.28. Vapour Coexisting with the Boiling Liquid ($(O/U)^{(liq)} = 2.08$)

T K	P bar	ρ $kg\,m^{-3}$	H $10^{3} kJ\,kg^{-1}$	G $kJ\,kg^{-1}$	S $kJ\,kg^{-1}\,K^{-1}$	C_P $kJ\,kg^{-1}\,K^{-1}$	C_P/C_V	β_T $10^{-4} bar^{-1}$	α_T $10^{-3} K^{-1}$	V_S $m\,s^{-1}$	O/U
3000	0.3916	0.3776	-6.412	-12163	1.917	0.707	1.120	2.611^{+4}_{10}	0.499	337	3.314
3120	0.5227	0.4807	-6.334	-12336	1.924	0.812	1.126	1.969^{+4}_{10}	0.526	345	3.298
3500	1.268	0.9809	-6.043	-12942	1.971	1.209	1.158	8352.00	0.622	376	3.281
4000	4.120	2.450	-5.588	-13907	2.080	1.675	1.210	2670.00	0.689	430	3.350
4500	13.29	5.911	-5.118	-15239	2.249	1.982	1.251	849.600	0.666	499	3.594
5000	40.80	13.30	-4.642	-17135	2.499	2.206	1.281	281.100	0.602	585	4.099
5500	116.0	27.67	-4.141	-19675	2.824	2.387	1.306	99.870	0.530	687	4.909
6000	300.8	53.99	-3.620	-22724	3.184	2.477	1.332	38.780	0.464	798	5.986
6500	706.3	100.8	-3.110	-25847	3.498	2.554	1.350	16.590	0.413	898	7.121
7000	1490	182.6	-2.661	-28320	3.666	2.501	1.374	7.853	0.373	979	7.866
7500	2793	320.9	-2.314	-29432	3.616	2.366	1.402	4.127	0.346	1029	7.760
8000	4622	538.9	-2.075	-29006	3.366	2.190	1.438	2.400	0.329	1055	6.824
8500	6772	849.7	-1.909	-27509	3.012	2.022	1.492	1.541	0.321	1068	5.535
9000	8817	1254	-1.785	-25474	2.632	1.887	1.582	1.113	0.328	1064	4.268
9500	10060	1779	-1.700	-23063	2.249	1.798	1.768	0.952	0.373	1022	3.113
9700	10070	2056	-1.685	-21939	2.088	1.786	1.918	0.961	0.417	985	2.667
9800	9898	2226	-1.685	-21310	2.002	1.791	2.034	0.991	0.453	960	2.439
9900	9529	2438	-1.694	-20592	1.909	1.809	2.207	1.049	0.506	929	2.199
9950	9194	2480	-1.654	-20323	1.876	1.828	2.296	1.125	0.538	907	2.104
10000	9329	2504	-1.615	-20314	1.870	1.808	2.298	1.101	0.531	913	2.080

Table A.28. Vapour Coexisting with the Boiling Liquid ($(O/U)^{(liq)} = 2.10$)

T K	P bar	ρ $kg\,m^{-3}$	H $kJ\,kg^{-1}$	G $kJ\,kg^{-1}$	S $kJ\,kg^{-1}\,K^{-1}$	C_P	C_P/C_V	β_T $10^{-4}bar^{-1}$	α_T K^{-1}	V_S $m\,s^{-1}$	O/U
3000	0.5322	0.4937	-6.456	-12317	1.954	0.673	1.122	1.917^{+4}_{10}	0.480	344	3.433
3120	0.7059	0.6255	-6.381	-12493	1.959	0.763	1.127	1.454^{+4}_{10}	0.500	352	3.416
3500	1.681	1.251	-6.112	-13155	2.012	1.122	1.155	6263.00	0.580	384	3.424
4000	5.273	2.982	-5.682	-14257	2.144	1.598	1.204	2071.00	0.647	442	3.559
4500	16.29	6.818	-5.209	-15734	2.339	1.943	1.247	688.500	0.633	515	3.873
5000	48.12	14.72	-4.716	-17750	2.607	2.191	1.279	237.000	0.577	605	4.441
5500	132.7	29.88	-4.199	-20376	2.941	2.382	1.304	86.840	0.512	709	5.313
6000	336.3	57.53	-3.667	-23460	3.299	2.473	1.331	34.510	0.451	819	6.440
6500	775.3	106.6	-3.153	-26541	3.598	2.550	1.349	15.030	0.403	917	7.581
7000	1609	192.0	-2.702	-28882	3.740	2.491	1.372	7.224	0.366	994	8.252
7500	2970	335.1	-2.354	-29805	3.660	2.350	1.399	3.846	0.340	1042	8.009
8000	4847	558.3	-2.111	-29213	3.388	2.172	1.436	2.260	0.322	1067	6.950
8500	7020	872.7	-1.940	-27613	3.020	2.006	1.489	1.465	0.315	1079	5.592
9000	9053	1279	-1.810	-25520	2.634	1.875	1.578	1.068	0.323	1075	4.292
9500	10230	1806	-1.722	-23062	2.246	1.791	1.766	0.924	0.369	1029	3.116
9700	10190	2085	-1.707	-21915	2.083	1.781	1.919	0.939	0.415	990	2.662
9800	9982	2259	-1.707	-21270	1.996	1.788	2.039	0.972	0.452	964	2.430
9900	9560	2477	-1.716	-20528	1.900	1.808	2.221	1.035	0.507	930	2.183
9950	9331	2527	-1.682	-20301	1.871	1.818	2.295	1.084	0.531	915	2.101
10000	9576	2527	-1.633	-20385	1.875	1.791	2.269	1.044	0.514	927	2.100

Table A.28. Vapour Coexisting with the Boiling Liquid ($(O/U)^{(vap)} = 2.125$)

T K	P bar	ρ $kg\,m^{-3}$	H $10^{3} kJ\,kg^{-1}$	G $kJ\,kg^{-1}$	S $kJ\,kg^{-1}\,K^{-1}$	C_P $kJ\,kg^{-1}\,K^{-1}$	C_P/C_V	β_T $10^{-4} bar^{-1}$	α_T $10^{-3} K^{-1}$	V_S $m\,s^{-1}$	O/U
3000	0.7126	0.6305	-6.505	-12509	2.001	0.655	1.124	1.430^{+4}_{10}	0.466	353	3.580
3120	0.9453	0.8000	-6.432	-12690	2.006	0.733	1.128	1.084^{+4}_{10}	0.481	361	3.562
3500	2.253	1.597	-6.181	-13414	2.067	1.054	1.153	4651.00	0.545	394	3.602
4000	6.966	3.700	-5.780	-14684	2.226	1.522	1.199	1557.00	0.603	456	3.824
4500	20.86	8.093	-5.314	-16363	2.455	1.902	1.242	534.000	0.597	536	4.245
5000	59.26	16.77	-4.808	-18552	2.749	2.179	1.276	191.200	0.550	631	4.915
5500	157.7	33.07	-4.275	-21292	3.094	2.383	1.303	72.660	0.492	736	5.875
6000	387.9	62.57	-3.732	-24407	3.446	2.473	1.330	29.740	0.436	845	7.061
6500	872.8	114.7	-3.211	-27414	3.723	2.547	1.347	13.260	0.392	941	8.194
7000	1773	204.8	-2.757	-29568	3.830	2.478	1.369	6.501	0.357	1014	8.747
7500	3207	354.2	-2.406	-30248	3.712	2.329	1.396	3.519	0.331	1058	8.316
8000	5141	583.5	-2.157	-29453	3.412	2.149	1.432	2.096	0.315	1082	7.101
8500	7339	902.0	-1.978	-27734	3.030	1.987	1.484	1.375	0.308	1094	5.660
9000	9353	1310	-1.841	-25575	2.637	1.861	1.574	1.015	0.317	1088	4.320
9500	10440	1839	-1.750	-23059	2.243	1.782	1.764	0.891	0.365	1038	3.118
9700	10340	2122	-1.734	-21881	2.077	1.775	1.921	0.913	0.412	996	2.654
9800	10080	2300	-1.735	-21216	1.988	1.783	2.046	0.950	0.451	968	2.417
9900	9581	2511	-1.735	-20472	1.893	1.807	2.233	1.025	0.509	932	2.169
9950	9610	2547	-1.700	-20391	1.878	1.797	2.259	1.023	0.512	931	2.125
10000	9861	2547	-1.650	-20475	1.882	1.772	2.234	0.987	0.496	943	2.125

Table A.28. Vapour Coexisting with the Boiling Liquid ($(O/U)^{(liq)} = 2.15$)

T K	P bar	ρ $kg\,m^{-3}$	H $10^{3} kJ\,kg^{-1}$	G $kJ\,kg^{-1}$	S $kJ\,kg^{-1}\,K^{-1}$	C_P $kJ\,kg^{-1}\,K^{-1}$	C_P/C_V	β_T $10^{-4} bar^{-1}$	α_T $10^{-3} K^{-1}$	V_S $m\,s^{-1}$	O/U
3000	0.8843	0.7502	-6.548	-12689	2.047	0.649	1.126	1.151^{+4}_{10}	0.457	361	3.718
3120	1.180	0.9572	-6.477	-12878	2.052	0.721	1.130	8670.00	0.470	369	3.703
3500	2.860	1.932	-6.241	-13668	2.122	1.017	1.153	3652.00	0.521	404	3.778
4000	8.899	4.446	-5.865	-15106	2.310	1.472	1.196	1212.00	0.571	471	4.096
4500	26.30	9.487	-5.412	-16999	2.575	1.874	1.239	420.900	0.567	557	4.642
5000	72.73	19.09	-4.901	-19381	2.896	2.177	1.274	154.900	0.526	656	5.437
5500	187.5	36.74	-4.356	-22241	3.252	2.392	1.301	60.740	0.474	764	6.500
6000	448.3	68.35	-3.802	-25375	3.596	2.476	1.329	25.580	0.423	872	7.744
6500	984.0	123.9	-3.274	-28284	3.848	2.543	1.345	11.680	0.381	964	8.847
7000	1954	219.0	-2.816	-30231	3.917	2.463	1.367	5.845	0.348	1033	9.254
7500	3461	374.6	-2.459	-30664	3.761	2.306	1.393	3.219	0.324	1075	8.615
8000	5449	609.7	-2.202	-29675	3.434	2.126	1.428	1.944	0.307	1098	7.245
8500	7668	931.6	-2.014	-27848	3.039	1.967	1.480	1.290	0.300	1110	5.726
9000	9659	1340	-1.870	-25627	2.640	1.847	1.570	0.965	0.310	1102	4.347
9500	10650	1872	-1.776	-23053	2.240	1.773	1.762	0.860	0.360	1046	3.120
9700	10480	2159	-1.761	-21844	2.070	1.769	1.924	0.889	0.410	1001	2.646
9800	10160	2342	-1.761	-21157	1.979	1.779	2.054	0.930	0.451	971	2.402
9900	9561	2491	-1.724	-20502	1.897	1.808	2.227	1.033	0.509	930	2.176
9950	9691	2515	-1.684	-20487	1.890	1.792	2.232	1.014	0.503	936	2.150

Table A.28. Vapour Coexisting with the Boiling Liquid ($(O/U)^{(vap)} = 2.20$)

T K	P bar	ρ $kg\,m^{-3}$	H $10^{3} kJ\,kg^{-1}$	G $kJ\,kg^{-1}$	S $kJ\,kg^{-1}\,K^{-1}$	C_P	C_P/C_V	β_T $10^{-4} bar^{-1}$	α_T $10^{-3} K^{-1}$	V_S $m\,s^{-1}$	O/U
3000	1.177	0.9334	-6.618	-12992	2.125	0.650	1.129	8638.00	0.447	374	3.952
3120	1.602	1.210	-6.552	-13209	2.134	0.716	1.133	6378.00	0.456	383	3.953
3500	4.105	2.538	-6.342	-14154	2.232	0.989	1.154	2535.00	0.492	424	4.123
4000	13.34	5.935	-6.015	-15937	2.481	1.430	1.194	802.200	0.526	501	4.662
4500	39.68	12.50	-5.596	-18278	2.818	1.856	1.235	276.200	0.521	598	5.508
5000	107.0	24.42	-5.092	-21075	3.197	2.193	1.270	104.200	0.487	706	6.618
5500	263.7	45.51	-4.533	-24183	3.573	2.419	1.298	42.730	0.443	817	7.934
6000	599.5	82.34	-3.961	-27325	3.894	2.486	1.327	18.910	0.399	923	9.288
6500	1253	145.8	-3.417	-29980	4.087	2.534	1.341	9.045	0.362	1009	10.265
7000	2372	251.7	-2.942	-31469	4.075	2.429	1.362	4.718	0.331	1071	10.282
7500	4023	419.4	-2.567	-31408	3.845	2.258	1.387	2.691	0.308	1109	9.187
8000	6110	664.8	-2.290	-30070	3.472	2.078	1.421	1.671	0.292	1131	7.514
8500	8359	991.5	-2.084	-28058	3.056	1.927	1.472	1.138	0.286	1142	5.850
9000	10290	1401	-1.927	-25726	2.644	1.818	1.561	0.873	0.298	1130	4.400
9500	11070	1937	-1.828	-23034	2.232	1.756	1.758	0.803	0.352	1063	3.121
9700	10730	2234	-1.812	-21760	2.056	1.758	1.930	0.845	0.406	1011	2.626
9800	10300	2427	-1.814	-21024	1.960	1.772	2.074	0.896	0.451	977	2.369
9900	9882	2523	-1.748	-20583	1.902	1.787	2.194	0.967	0.490	948	2.200

Table A.28. Vapour Coexisting with the Boiling Liquid ($(O/U)^{(vap)} = 2.25$)

T K	P bar	ρ $kg\,m^{-3}$	H $10^3 kJ\,kg^{-1}$	G $kJ\,kg^{-1}$	S $kJ\,kg^{-1}\,K^{-1}$	C_P $kJ\,kg^{-1}\,K^{-1}$	C_P/C_V	β_T $10^{-4} bar^{-1}$	α_T $10^{-3} K^{-1}$	V_S $m\,s^{-1}$	O/U
3000	1.402	1.058	-6.670	-13226	2.185	0.656	1.132	7251.00	0.442	384	4.136
3120	1.952	1.395	-6.611	-13484	2.203	0.721	1.135	5228.00	0.448	395	4.164
3500	5.343	3.056	-6.430	-14610	2.337	0.988	1.155	1942.00	0.475	441	4.459
4000	18.41	7.374	-6.150	-16760	2.652	1.427	1.193	577.900	0.496	529	5.261
4500	56.29	15.70	-5.773	-19557	3.063	1.866	1.232	193.100	0.487	638	6.471
5000	151.5	30.58	-5.289	-22772	3.496	2.219	1.266	72.860	0.456	754	7.978
5500	364.6	56.27	-4.729	-26116	3.889	2.446	1.294	30.580	0.417	867	9.609
6000	797.6	99.94	-4.142	-29223	4.180	2.489	1.324	14.030	0.377	972	11.059
6500	1592	172.9	-3.577	-31564	4.306	2.516	1.336	7.000	0.343	1051	11.795
7000	2874	290.4	-3.077	-32562	4.212	2.387	1.357	3.803	0.315	1108	11.301
7500	4663	469.6	-2.676	-32039	3.915	2.206	1.381	2.247	0.293	1144	9.716
8000	6833	722.9	-2.375	-30411	3.505	2.029	1.414	1.437	0.278	1167	7.762
8500	9098	1052	-2.148	-28253	3.071	1.889	1.464	1.005	0.273	1177	5.970
9000	10940	1460	-1.979	-25821	2.649	1.790	1.553	0.792	0.286	1159	4.452
9500	11480	2001	-1.877	-23007	2.224	1.739	1.755	0.752	0.344	1080	3.119
9700	10960	2308	-1.862	-21661	2.041	1.747	1.939	0.807	0.403	1020	2.601
9800	10390	2511	-1.863	-20877	1.940	1.767	2.098	0.870	0.454	980	2.331
9900	10510	2571	-1.786	-20754	1.916	1.749	2.134	0.861	0.456	982	2.250
9950	10770	2571	-1.734	-20840	1.920	1.730	2.117	0.834	0.443	994	2.250

Table A.29. Liquid Phase

T	P	ρ	H	G	S	C_P	C_P/C_V	β_T	α_T	V_S	O/U
K	bar	$kg\ m^{-3}$	$10^3 kJ\ kg^{-1}$	$kJ\ kg^{-1}$		$kJ\ kg^{-1}\ K^{-1}$		$10^{-5} bar^{-1}$	$10^{-4} K^{-1}$	$m\ s^{-1}$	
3500	3	8896	-5.648	-9827	1.194	0.426	1.234	0.428	0.937	1801	1.750
3500	3	8824	-5.767	-9966	1.200	0.429	1.233	0.430	0.938	1802	1.800
3500	3	8753	-5.884	-10101	1.205	0.432	1.233	0.433	0.940	1804	1.850
3500	3	8718	-5.943	-10168	1.207	0.434	1.233	0.434	0.942	1804	1.875
3500	3	8684	-6.001	-10234	1.209	0.436	1.233	0.436	0.944	1805	1.900
3500	3	8656	-6.047	-10286	1.211	0.439	1.233	0.437	0.947	1805	1.920
3500	3	8628	-6.093	-10337	1.212	0.442	1.234	0.439	0.952	1805	1.940
3500	3	8614	-6.116	-10362	1.213	0.444	1.234	0.440	0.955	1806	1.950
3500	3	8600	-6.139	-10387	1.214	0.447	1.235	0.441	0.960	1805	1.960
3500	3	8585	-6.161	-10412	1.215	0.450	1.236	0.442	0.966	1805	1.970
3500	3	8569	-6.182	-10437	1.215	0.456	1.238	0.444	0.976	1804	1.980
3500	3	8550	-6.202	-10461	1.217	0.461	1.240	0.446	0.987	1803	1.990
3500	3	8528	-6.220	-10484	1.218	0.464	1.241	0.449	0.993	1800	2.000
3500	3	8499	-6.233	-10507	1.221	0.459	1.239	0.452	0.986	1795	2.010
3500	3	8467	-6.243	-10529	1.225	0.450	1.236	0.456	0.974	1789	2.020
3500	3	8432	-6.252	-10551	1.228	0.443	1.234	0.461	0.965	1782	2.030
3500	3	8396	-6.261	-10573	1.232	0.437	1.232	0.466	0.958	1775	2.040
3500	3	8359	-6.268	-10594	1.236	0.431	1.231	0.471	0.954	1767	2.050
3500	3	8323	-6.276	-10615	1.240	0.427	1.230	0.477	0.951	1760	2.060
3500	3	8250	-6.290	-10656	1.247	0.419	1.228	0.489	0.947	1744	2.080
3500	3	8178	-6.305	-10697	1.255	0.411	1.228	0.502	0.946	1729	2.100
3500	3	8089	-6.323	-10746	1.264	0.403	1.228	0.519	0.947	1711	2.125
3500	3	8001	-6.341	-10795	1.273	0.396	1.228	0.536	0.949	1693	2.150
3500	3	7830	-6.379	-10890	1.289	0.383	1.230	0.572	0.957	1658	2.200
3500	3	7665	-6.418	-10984	1.304	0.372	1.233	0.610	0.968	1624	2.250

Table A.29. Liquid Phase (continued)

T K	P bar	ρ $kg\ m^{-3}$	H $10^3 kJ\ kg^{-1}$	G $kJ\ kg^{-1}$	S	C_P $kJ\ kg^{-1}\ K^{-1}$	C_P/C_V	β_T $10^{-5} bar^{-1}$	α_T $10^{-4} K^{-1}$	V_S $m\ s^{-1}$	O/U
3500	10	8896	-5.648	-9827	1.194	0.426	1.234	0.428	0.937	1801	1.750
3500	10	8824	-5.766	-9965	1.200	0.429	1.233	0.430	0.938	1802	1.800
3500	10	8753	-5.884	-10101	1.205	0.432	1.233	0.433	0.940	1804	1.850
3500	10	8718	-5.943	-10168	1.207	0.434	1.233	0.434	0.942	1804	1.875
3500	10	8684	-6.001	-10234	1.209	0.436	1.233	0.436	0.944	1805	1.900
3500	10	8656	-6.047	-10286	1.211	0.439	1.233	0.437	0.947	1805	1.920
3500	10	8628	-6.093	-10337	1.212	0.442	1.234	0.439	0.952	1806	1.940
3500	10	8614	-6.116	-10362	1.213	0.444	1.234	0.440	0.955	1806	1.950
3500	10	8600	-6.139	-10387	1.214	0.447	1.235	0.441	0.960	1806	1.960
3500	10	8585	-6.161	-10412	1.215	0.450	1.236	0.442	0.966	1805	1.970
3500	10	8569	-6.182	-10437	1.215	0.456	1.238	0.444	0.976	1805	1.980
3500	10	8551	-6.202	-10461	1.217	0.461	1.240	0.446	0.987	1803	1.990
3500	10	8528	-6.219	-10484	1.218	0.464	1.241	0.449	0.993	1800	2.000
3500	10	8500	-6.233	-10507	1.221	0.459	1.239	0.452	0.986	1795	2.010
3500	10	8467	-6.243	-10529	1.225	0.450	1.236	0.456	0.974	1789	2.020
3500	10	8432	-6.252	-10551	1.228	0.443	1.234	0.461	0.965	1782	2.030
3500	10	8396	-6.261	-10573	1.232	0.437	1.232	0.466	0.958	1775	2.040
3500	10	8360	-6.268	-10594	1.236	0.431	1.230	0.471	0.954	1767	2.050
3500	10	8323	-6.276	-10615	1.240	0.427	1.230	0.477	0.951	1760	2.060
3500	10	8250	-6.290	-10656	1.247	0.419	1.228	0.489	0.947	1745	2.080
3500	10	8178	-6.305	-10697	1.255	0.411	1.228	0.502	0.946	1730	2.100
3500	10	8089	-6.323	-10746	1.264	0.403	1.228	0.519	0.947	1711	2.125
3500	10	8001	-6.341	-10795	1.273	0.396	1.228	0.536	0.949	1693	2.150
3500	10	7830	-6.379	-10890	1.289	0.383	1.230	0.572	0.957	1658	2.200
3500	10	7665	-6.418	-10984	1.304	0.372	1.233	0.610	0.968	1624	2.250

Table A.29. Liquid Phase (continued)

T	P	ρ	H	G	S	C_P	C_P/C_V	β_T	α_T	V_S	O/U
K	bar	$kg\ m^{-3}$	$10^3 kJ\ kg^{-1}$	$kJ\ kg^{-1}$		$kJ\ kg^{-1}\ K^{-1}$		$10^{-5} bar^{-1}$	$10^{-4} K^{-1}$	$m\ s^{-1}$	
3500	30	8897	-5.648	-9827	1.194	0.426	1.234	0.427	0.936	1801	1.750
3500	30	8825	-5.766	-9965	1.200	0.429	1.233	0.430	0.938	1803	1.800
3500	30	8754	-5.884	-10101	1.205	0.432	1.233	0.433	0.940	1804	1.850
3500	30	8719	-5.943	-10167	1.207	0.434	1.233	0.434	0.941	1805	1.875
3500	30	8685	-6.001	-10233	1.209	0.436	1.233	0.436	0.944	1805	1.900
3500	30	8657	-6.047	-10285	1.211	0.439	1.233	0.437	0.947	1806	1.920
3500	30	8629	-6.093	-10337	1.212	0.442	1.234	0.438	0.951	1806	1.940
3500	30	8615	-6.116	-10362	1.213	0.444	1.234	0.439	0.955	1806	1.950
3500	30	8601	-6.139	-10387	1.214	0.447	1.235	0.440	0.959	1806	1.960
3500	30	8586	-6.161	-10412	1.215	0.450	1.236	0.442	0.966	1806	1.970
3500	30	8570	-6.182	-10436	1.215	0.456	1.238	0.443	0.975	1805	1.980
3500	30	8552	-6.202	-10460	1.217	0.461	1.240	0.446	0.987	1804	1.990
3500	30	8529	-6.219	-10484	1.218	0.464	1.241	0.449	0.993	1801	2.000
3500	30	8500	-6.233	-10507	1.221	0.459	1.239	0.452	0.986	1796	2.010
3500	30	8468	-6.243	-10529	1.225	0.450	1.236	0.456	0.974	1790	2.020
3500	30	8433	-6.252	-10551	1.228	0.443	1.234	0.460	0.964	1783	2.030
3500	30	8397	-6.260	-10573	1.232	0.437	1.232	0.466	0.958	1775	2.040
3500	30	8360	-6.268	-10594	1.236	0.431	1.230	0.471	0.953	1768	2.050
3500	30	8324	-6.276	-10615	1.240	0.427	1.230	0.477	0.950	1760	2.060
3500	30	8251	-6.290	-10656	1.247	0.419	1.228	0.489	0.947	1745	2.080
3500	30	8179	-6.305	-10696	1.255	0.411	1.228	0.502	0.946	1730	2.100
3500	30	8090	-6.323	-10746	1.264	0.403	1.228	0.518	0.947	1711	2.125
3500	30	8002	-6.341	-10795	1.273	0.396	1.228	0.535	0.949	1693	2.150
3500	30	7831	-6.378	-10890	1.289	0.383	1.230	0.571	0.957	1658	2.200
3500	30	7666	-6.418	-10984	1.304	0.372	1.233	0.609	0.967	1625	2.250

Table A.29. Liquid Phase (continued)

T K	P bar	ρ $kg\ m^{-3}$	H $10^3 kJ\ kg^{-1}$	G $kJ\ kg^{-1}$	S	C_P $kJ\ kg^{-1}\ K^{-1}$	C_P/C_V	β_T $10^{-5} bar^{-1}$	α_T $10^{-4} K^{-1}$	V_S $m\ s^{-1}$	O/U
3500	100	8899	-5.647	-9826	1.194	0.426	1.233	0.427	0.935	1803	1.750
3500	100	8827	-5.766	-9964	1.200	0.429	1.233	0.429	0.937	1804	1.800
3500	100	8757	-5.883	-10100	1.205	0.432	1.233	0.432	0.939	1805	1.850
3500	100	8722	-5.942	-10167	1.207	0.434	1.233	0.433	0.941	1806	1.875
3500	100	8687	-6.000	-10233	1.209	0.436	1.233	0.435	0.943	1807	1.900
3500	100	8659	-6.047	-10285	1.211	0.439	1.233	0.436	0.946	1807	1.920
3500	100	8632	-6.093	-10336	1.212	0.442	1.234	0.438	0.950	1807	1.940
3500	100	8618	-6.115	-10361	1.213	0.444	1.234	0.439	0.954	1807	1.950
3500	100	8603	-6.138	-10386	1.214	0.447	1.235	0.440	0.958	1807	1.960
3500	100	8588	-6.160	-10411	1.215	0.450	1.236	0.441	0.965	1807	1.970
3500	100	8572	-6.182	-10435	1.215	0.456	1.238	0.443	0.974	1806	1.980
3500	100	8554	-6.202	-10460	1.217	0.461	1.240	0.445	0.986	1805	1.990
3500	100	8532	-6.219	-10483	1.218	0.464	1.241	0.448	0.992	1802	2.000
3500	100	8503	-6.232	-10506	1.221	0.459	1.239	0.451	0.985	1797	2.010
3500	100	8470	-6.243	-10528	1.224	0.450	1.236	0.455	0.973	1791	2.020
3500	100	8435	-6.252	-10550	1.228	0.443	1.233	0.460	0.963	1784	2.030
3500	100	8399	-6.260	-10572	1.232	0.437	1.232	0.465	0.957	1777	2.040
3500	100	8363	-6.268	-10593	1.236	0.431	1.230	0.470	0.952	1769	2.050
3500	100	8327	-6.275	-10614	1.240	0.427	1.229	0.476	0.949	1762	2.060
3500	100	8254	-6.290	-10655	1.247	0.419	1.228	0.488	0.946	1746	2.080
3500	100	8182	-6.304	-10695	1.255	0.411	1.228	0.501	0.945	1731	2.100
3500	100	8093	-6.322	-10745	1.264	0.403	1.228	0.517	0.946	1713	2.125
3500	100	8005	-6.340	-10794	1.272	0.396	1.228	0.534	0.948	1695	2.150
3500	100	7834	-6.378	-10889	1.289	0.383	1.230	0.570	0.955	1660	2.200
3500	100	7669	-6.417	-10983	1.304	0.372	1.233	0.608	0.966	1626	2.250

Table A.29. Liquid Phase (continued)

T K	P bar	ρ $kg\ m^{-3}$	H $10^3 kJ\ kg^{-1}$	G $kJ\ kg^{-1}$	S $kJ\ kg^{-1}$	C_P $kJ\ kg^{-1}\ K^{-1}$	C_P/C_V	β_T $10^{-5} bar^{-1}$	α_T $10^{-4} K^{-1}$	V_S $m\ s^{-1}$	O/U
3500	300	8907	-5.646	-9824	1.194	0.426	1.233	0.424	0.933	1806	1.750
3500	300	8835	-5.764	-9962	1.199	0.429	1.233	0.427	0.934	1808	1.800
3500	300	8764	-5.882	-10098	1.204	0.432	1.232	0.430	0.936	1809	1.850
3500	300	8729	-5.940	-10164	1.207	0.434	1.232	0.431	0.938	1810	1.875
3500	300	8695	-5.999	-10230	1.209	0.437	1.232	0.432	0.940	1810	1.900
3500	300	8667	-6.045	-10282	1.211	0.439	1.233	0.434	0.943	1811	1.920
3500	300	8639	-6.091	-10333	1.212	0.442	1.233	0.435	0.948	1811	1.940
3500	300	8625	-6.114	-10359	1.213	0.444	1.234	0.436	0.951	1811	1.950
3500	300	8611	-6.136	-10384	1.214	0.447	1.235	0.437	0.956	1811	1.960
3500	300	8596	-6.159	-10409	1.214	0.451	1.236	0.439	0.962	1811	1.970
3500	300	8580	-6.180	-10433	1.215	0.456	1.238	0.440	0.971	1810	1.980
3500	300	8562	-6.200	-10457	1.216	0.461	1.240	0.443	0.983	1809	1.990
3500	300	8539	-6.217	-10481	1.218	0.464	1.241	0.446	0.989	1806	2.000
3500	300	8511	-6.231	-10504	1.221	0.459	1.239	0.449	0.982	1801	2.010
3500	300	8478	-6.241	-10526	1.224	0.450	1.235	0.452	0.970	1795	2.020
3500	300	8443	-6.250	-10548	1.228	0.443	1.233	0.457	0.960	1788	2.030
3500	300	8407	-6.258	-10569	1.232	0.437	1.231	0.462	0.954	1781	2.040
3500	300	8371	-6.266	-10591	1.236	0.431	1.230	0.467	0.949	1773	2.050
3500	300	8335	-6.274	-10612	1.239	0.427	1.229	0.473	0.946	1766	2.060
3500	300	8262	-6.288	-10653	1.247	0.419	1.228	0.485	0.943	1751	2.080
3500	300	8190	-6.302	-10693	1.254	0.411	1.227	0.498	0.942	1736	2.100
3500	300	8101	-6.320	-10743	1.263	0.403	1.227	0.514	0.942	1717	2.125
3500	300	8014	-6.338	-10791	1.272	0.396	1.228	0.531	0.945	1699	2.150
3500	300	7843	-6.376	-10887	1.289	0.383	1.229	0.566	0.952	1664	2.200
3500	300	7678	-6.416	-10980	1.304	0.372	1.232	0.603	0.962	1631	2.250

Table A.29. Liquid Phase (continued)

T K	P bar	ρ $kg\ m^{-3}$	H $10^3 kJ\ kg^{-1}$	G $kJ\ kg^{-1}$	S	C_P $kJ\ kg^{-1}\ K^{-1}$	C_P/C_V	β_T $10^{-5} bar^{-1}$	α_T $10^{-4} K^{-1}$	V_S $m\ s^{-1}$	O/U
3500	1000	8933	-5.641	-9816	1.193	0.427	1.232	0.417	0.924	1819	1.750
3500	1000	8861	-5.759	-9954	1.199	0.430	1.231	0.419	0.926	1821	1.800
3500	1000	8790	-5.877	-10090	1.204	0.433	1.231	0.422	0.928	1822	1.850
3500	1000	8756	-5.935	-10156	1.206	0.435	1.231	0.423	0.929	1823	1.875
3500	1000	8721	-5.993	-10222	1.208	0.437	1.231	0.425	0.931	1823	1.900
3500	1000	8693	-6.040	-10274	1.210	0.439	1.231	0.426	0.934	1824	1.920
3500	1000	8665	-6.086	-10325	1.211	0.442	1.232	0.427	0.939	1824	1.940
3500	1000	8651	-6.108	-10351	1.212	0.444	1.232	0.428	0.942	1824	1.950
3500	1000	8637	-6.131	-10376	1.213	0.447	1.233	0.429	0.946	1824	1.960
3500	1000	8622	-6.153	-10401	1.214	0.451	1.234	0.430	0.953	1824	1.970
3500	1000	8606	-6.175	-10425	1.214	0.456	1.236	0.432	0.962	1823	1.980
3500	1000	8588	-6.195	-10449	1.216	0.462	1.238	0.434	0.973	1822	1.990
3500	1000	8566	-6.212	-10472	1.217	0.464	1.239	0.437	0.979	1819	2.000
3500	1000	8537	-6.225	-10495	1.220	0.459	1.237	0.440	0.972	1814	2.010
3500	1000	8505	-6.236	-10518	1.223	0.451	1.234	0.444	0.960	1808	2.020
3500	1000	8470	-6.245	-10540	1.227	0.443	1.232	0.448	0.950	1801	2.030
3500	1000	8434	-6.253	-10561	1.231	0.437	1.230	0.453	0.944	1794	2.040
3500	1000	8398	-6.260	-10582	1.235	0.432	1.229	0.458	0.940	1787	2.050
3500	1000	8362	-6.268	-10603	1.239	0.427	1.228	0.464	0.937	1779	2.060
3500	1000	8290	-6.282	-10644	1.246	0.419	1.226	0.475	0.933	1765	2.080
3500	1000	8218	-6.297	-10684	1.254	0.412	1.226	0.487	0.932	1750	2.100
3500	1000	8130	-6.315	-10734	1.263	0.404	1.226	0.503	0.932	1732	2.125
3500	1000	8043	-6.333	-10782	1.271	0.396	1.226	0.519	0.934	1714	2.150
3500	1000	7874	-6.370	-10878	1.288	0.383	1.228	0.553	0.940	1680	2.200
3500	1000	7710	-6.410	-10971	1.303	0.372	1.231	0.588	0.950	1647	2.250

Table A.29. Liquid Phase (continued)

T K	P bar	ρ $kg\ m^{-3}$	H $10^3 kJ\ kg^{-1}$	G $kJ\ kg^{-1}$	S	C_P $kJ\ kg^{-1}\ K^{-1}$	C_P/C_V	β_T $10^{-5} bar^{-1}$	α_T $10^{-4} K^{-1}$	V_S $m\ s^{-1}$	O/U
3500	3000	9006	-5.625	-9794	1.191	0.428	1.229	0.397	0.902	1854	1.750
3500	3000	8934	-5.744	-9932	1.197	0.431	1.228	0.399	0.903	1856	1.800
3500	3000	8863	-5.861	-10067	1.202	0.434	1.228	0.401	0.904	1858	1.850
3500	3000	8828	-5.920	-10134	1.204	0.436	1.228	0.403	0.906	1859	1.875
3500	3000	8793	-5.978	-10199	1.206	0.438	1.228	0.404	0.908	1859	1.900
3500	3000	8766	-6.024	-10251	1.208	0.440	1.228	0.405	0.910	1860	1.920
3500	3000	8738	-6.070	-10302	1.209	0.443	1.228	0.406	0.914	1860	1.940
3500	3000	8724	-6.093	-10328	1.210	0.445	1.229	0.407	0.917	1861	1.950
3500	3000	8709	-6.115	-10353	1.211	0.448	1.230	0.408	0.921	1861	1.960
3500	3000	8695	-6.138	-10377	1.211	0.452	1.231	0.409	0.927	1860	1.970
3500	3000	8679	-6.159	-10402	1.212	0.456	1.232	0.410	0.936	1860	1.980
3500	3000	8661	-6.179	-10426	1.213	0.462	1.234	0.413	0.946	1859	1.990
3500	3000	8639	-6.196	-10449	1.215	0.465	1.235	0.415	0.952	1856	2.000
3500	3000	8611	-6.210	-10472	1.218	0.460	1.233	0.418	0.945	1851	2.010
3500	3000	8578	-6.220	-10494	1.221	0.451	1.230	0.421	0.933	1846	2.020
3500	3000	8544	-6.229	-10516	1.225	0.444	1.228	0.425	0.924	1839	2.030
3500	3000	8509	-6.237	-10537	1.229	0.438	1.226	0.429	0.918	1832	2.040
3500	3000	8473	-6.244	-10559	1.233	0.432	1.225	0.434	0.913	1825	2.050
3500	3000	8438	-6.252	-10579	1.236	0.428	1.224	0.439	0.910	1818	2.060
3500	3000	8367	-6.266	-10620	1.244	0.420	1.223	0.449	0.907	1804	2.080
3500	3000	8297	-6.280	-10660	1.251	0.413	1.222	0.460	0.905	1790	2.100
3500	3000	8210	-6.298	-10709	1.260	0.404	1.222	0.474	0.904	1772	2.125
3500	3000	8124	-6.316	-10758	1.269	0.397	1.223	0.488	0.905	1755	2.150
3500	3000	7959	-6.353	-10852	1.286	0.384	1.224	0.518	0.910	1723	2.200
3500	3000	7799	-6.392	-10945	1.301	0.372	1.227	0.549	0.917	1692	2.250

Table A.29. Liquid Phase (continued)

T K	P bar	ρ $kg\ m^{-3}$	H $10^3 kJ\ kg^{-1}$	G $kJ\ kg^{-1}$	S $kJ\ kg^{-1}\ K^{-1}$	C_P	C_P/C_V	β_T $10^{-5} bar^{-1}$	α_T $10^{-4} K^{-1}$	V_S $m\ s^{-1}$	O/U
3500	10000	9241	-5.572	-9717	1.184	0.433	1.219	0.341	0.836	1968	1.750
3500	10000	9168	-5.690	-9854	1.190	0.435	1.219	0.342	0.837	1971	1.800
3500	10000	9096	-5.807	-9989	1.195	0.438	1.218	0.344	0.838	1973	1.850
3500	10000	9061	-5.865	-10055	1.197	0.440	1.218	0.345	0.839	1974	1.875
3500	10000	9026	-5.923	-10121	1.199	0.442	1.218	0.346	0.840	1976	1.900
3500	10000	8998	-5.969	-10172	1.201	0.444	1.218	0.347	0.842	1976	1.920
3500	10000	8970	-6.015	-10223	1.202	0.447	1.218	0.347	0.845	1977	1.940
3500	10000	8956	-6.038	-10249	1.203	0.449	1.219	0.348	0.847	1978	1.950
3500	10000	8942	-6.061	-10273	1.204	0.451	1.219	0.349	0.850	1978	1.960
3500	10000	8927	-6.083	-10298	1.204	0.455	1.220	0.349	0.855	1978	1.970
3500	10000	8912	-6.105	-10322	1.205	0.459	1.221	0.350	0.862	1978	1.980
3500	10000	8895	-6.125	-10346	1.206	0.465	1.223	0.352	0.871	1977	1.990
3500	10000	8873	-6.142	-10369	1.208	0.468	1.224	0.354	0.877	1975	2.000
3500	10000	8846	-6.155	-10392	1.211	0.463	1.222	0.356	0.869	1971	2.010
3500	10000	8814	-6.165	-10414	1.214	0.454	1.219	0.358	0.858	1965	2.020
3500	10000	8781	-6.173	-10435	1.218	0.447	1.217	0.361	0.850	1959	2.030
3500	10000	8747	-6.181	-10456	1.222	0.441	1.216	0.364	0.844	1953	2.040
3500	10000	8713	-6.188	-10477	1.225	0.436	1.215	0.368	0.840	1947	2.050
3500	10000	8679	-6.195	-10498	1.229	0.431	1.214	0.371	0.837	1941	2.060
3500	10000	8612	-6.209	-10538	1.237	0.423	1.213	0.379	0.832	1928	2.080
3500	10000	8545	-6.222	-10577	1.244	0.416	1.212	0.387	0.830	1916	2.100
3500	10000	8462	-6.239	-10625	1.253	0.408	1.212	0.397	0.827	1901	2.125
3500	10000	8381	-6.257	-10673	1.262	0.400	1.213	0.407	0.826	1886	2.150
3500	10000	8224	-6.293	-10766	1.278	0.386	1.214	0.427	0.826	1859	2.200
3500	10000	8073	-6.331	-10857	1.293	0.374	1.216	0.448	0.829	1833	2.250

Table A.29. Liquid Phase (continued)

T K	P bar	ρ $kg\ m^{-3}$	H $10^3 kJ\ kg^{-1}$	G $kJ\ kg^{-1}$	S	C_P $kJ\ kg^{-1}\ K^{-1}$	C_P/C_V	β_T $10^{-5} bar^{-1}$	α_T $10^{-4} K^{-1}$	V_S $m\ s^{-1}$	O/U
4000	30	8507	-5.454	-10438	1.246	0.356	1.255	0.493	0.871	1729	1.750
4000	30	8438	-5.571	-10579	1.252	0.360	1.254	0.497	0.875	1730	1.800
4000	30	8368	-5.687	-10717	1.258	0.365	1.255	0.501	0.881	1730	1.850
4000	30	8334	-5.744	-10785	1.260	0.368	1.255	0.503	0.886	1730	1.875
4000	30	8299	-5.800	-10852	1.263	0.372	1.256	0.506	0.893	1730	1.900
4000	30	8270	-5.845	-10905	1.265	0.377	1.258	0.509	0.902	1729	1.920
4000	30	8240	-5.889	-10957	1.267	0.383	1.260	0.512	0.913	1728	1.940
4000	30	8224	-5.910	-10983	1.268	0.386	1.262	0.514	0.920	1727	1.950
4000	30	8208	-5.931	-11008	1.269	0.390	1.263	0.517	0.928	1726	1.960
4000	30	8190	-5.951	-11034	1.271	0.393	1.265	0.519	0.936	1725	1.970
4000	30	8171	-5.970	-11059	1.272	0.397	1.267	0.522	0.944	1723	1.980
4000	30	8151	-5.988	-11084	1.274	0.399	1.268	0.525	0.950	1721	1.990
4000	30	8127	-6.005	-11108	1.276	0.399	1.268	0.529	0.952	1718	2.000
4000	30	8102	-6.020	-11132	1.278	0.397	1.268	0.533	0.951	1714	2.010
4000	30	8074	-6.034	-11156	1.281	0.393	1.266	0.537	0.946	1710	2.020
4000	30	8044	-6.046	-11180	1.283	0.387	1.264	0.541	0.939	1705	2.030
4000	30	8013	-6.058	-11203	1.286	0.382	1.263	0.546	0.932	1699	2.040
4000	30	7981	-6.068	-11226	1.289	0.376	1.261	0.551	0.926	1693	2.050
4000	30	7947	-6.078	-11248	1.293	0.371	1.260	0.557	0.921	1687	2.060
4000	30	7880	-6.097	-11293	1.299	0.363	1.259	0.570	0.914	1675	2.080
4000	30	7811	-6.115	-11337	1.305	0.356	1.259	0.584	0.912	1662	2.100
4000	30	7726	-6.137	-11391	1.313	0.349	1.259	0.602	0.914	1645	2.125
4000	30	7640	-6.158	-11444	1.321	0.343	1.261	0.622	0.919	1628	2.150
4000	30	7472	-6.201	-11547	1.337	0.336	1.265	0.665	0.935	1595	2.200
4000	30	7308	-6.244	-11648	1.351	0.330	1.270	0.711	0.956	1563	2.250

Table A.29. Liquid Phase (continued)

T K	P bar	ρ $kg\ m^{-3}$	H $10^3 kJ\ kg^{-1}$	G $kJ\ kg^{-1}$	S	C_P $kJ\ kg^{-1}\ K^{-1}$	C_P/C_V	β_T $10^{-5} bar^{-1}$	α_T $10^{-4} K^{-1}$	V_S $m\ s^{-1}$	O/U
4000	100	8510	-5.453	-10437	1.246	0.356	1.255	0.492	0.870	1731	1.750
4000	100	8440	-5.570	-10578	1.252	0.360	1.254	0.496	0.874	1731	1.800
4000	100	8371	-5.686	-10716	1.258	0.365	1.254	0.500	0.880	1731	1.850
4000	100	8337	-5.743	-10784	1.260	0.368	1.255	0.502	0.885	1731	1.875
4000	100	8302	-5.800	-10851	1.263	0.372	1.256	0.505	0.892	1731	1.900
4000	100	8273	-5.845	-10904	1.265	0.377	1.258	0.508	0.901	1730	1.920
4000	100	8243	-5.888	-10956	1.267	0.383	1.260	0.511	0.912	1730	1.940
4000	100	8227	-5.910	-10982	1.268	0.386	1.262	0.513	0.919	1729	1.950
4000	100	8211	-5.930	-11008	1.269	0.390	1.263	0.515	0.927	1728	1.960
4000	100	8193	-5.951	-11033	1.271	0.393	1.265	0.518	0.935	1727	1.970
4000	100	8174	-5.970	-11058	1.272	0.397	1.267	0.521	0.943	1725	1.980
4000	100	8154	-5.988	-11083	1.274	0.399	1.268	0.524	0.949	1722	1.990
4000	100	8130	-6.004	-11107	1.276	0.399	1.268	0.528	0.951	1719	2.000
4000	100	8105	-6.020	-11131	1.278	0.397	1.268	0.531	0.949	1716	2.010
4000	100	8077	-6.033	-11155	1.280	0.393	1.266	0.535	0.944	1711	2.020
4000	100	8047	-6.046	-11179	1.283	0.387	1.264	0.540	0.938	1706	2.030
4000	100	8016	-6.057	-11202	1.286	0.382	1.262	0.545	0.930	1701	2.040
4000	100	7984	-6.068	-11225	1.289	0.376	1.261	0.550	0.924	1695	2.050
4000	100	7951	-6.078	-11248	1.292	0.371	1.260	0.556	0.919	1689	2.060
4000	100	7883	-6.097	-11292	1.299	0.362	1.259	0.568	0.913	1676	2.080
4000	100	7815	-6.114	-11336	1.305	0.356	1.258	0.582	0.911	1663	2.100
4000	100	7729	-6.136	-11390	1.313	0.349	1.259	0.601	0.913	1647	2.125
4000	100	7643	-6.157	-11443	1.321	0.343	1.261	0.621	0.918	1630	2.150
4000	100	7476	-6.200	-11546	1.336	0.336	1.265	0.663	0.934	1597	2.200
4000	100	7312	-6.244	-11647	1.351	0.330	1.270	0.709	0.954	1565	2.250

Table A.29. Liquid Phase (continued)

T	P	ρ	H	G	S	C_P	C_P/C_V	β_T	α_T	V_S	O/U
K	bar	$kg\ m^{-3}$	$10^3 kJ\ kg^{-1}$	$kJ\ kg^{-1}$	$kJ\ kg^{-1}\ K^{-1}$			$10^{-5} bar^{-1}$	$10^{-4} K^{-1}$	$m\ s^{-1}$	
4000	300	8519	-5.452	-10435	1.246	0.356	1.254	0.489	0.868	1735	1.750
4000	300	8449	-5.569	-10576	1.252	0.360	1.254	0.493	0.871	1735	1.800
4000	300	8380	-5.684	-10714	1.257	0.365	1.254	0.497	0.877	1735	1.850
4000	300	8345	-5.742	-10782	1.260	0.368	1.255	0.499	0.882	1735	1.875
4000	300	8310	-5.798	-10849	1.263	0.372	1.256	0.502	0.889	1735	1.900
4000	300	8281	-5.843	-10902	1.265	0.377	1.257	0.505	0.897	1735	1.920
4000	300	8251	-5.887	-10954	1.267	0.382	1.260	0.508	0.909	1734	1.940
4000	300	8236	-5.908	-10980	1.268	0.386	1.261	0.510	0.915	1733	1.950
4000	300	8219	-5.929	-11005	1.269	0.389	1.263	0.512	0.923	1732	1.960
4000	300	8202	-5.949	-11030	1.270	0.393	1.264	0.515	0.931	1731	1.970
4000	300	8183	-5.968	-11056	1.272	0.397	1.266	0.518	0.939	1729	1.980
4000	300	8162	-5.986	-11080	1.273	0.399	1.267	0.521	0.945	1727	1.990
4000	300	8139	-6.003	-11105	1.275	0.399	1.268	0.524	0.948	1724	2.000
4000	300	8113	-6.018	-11129	1.278	0.397	1.267	0.528	0.946	1720	2.010
4000	300	8086	-6.032	-11153	1.280	0.393	1.266	0.532	0.941	1716	2.020
4000	300	8056	-6.044	-11176	1.283	0.387	1.264	0.536	0.934	1711	2.030
4000	300	8025	-6.055	-11199	1.286	0.382	1.262	0.541	0.927	1705	2.040
4000	300	7992	-6.066	-11222	1.289	0.376	1.260	0.546	0.921	1699	2.050
4000	300	7959	-6.076	-11245	1.292	0.371	1.259	0.552	0.916	1693	2.060
4000	300	7892	-6.095	-11290	1.299	0.362	1.258	0.564	0.910	1681	2.080
4000	300	7824	-6.113	-11333	1.305	0.355	1.258	0.578	0.907	1668	2.100
4000	300	7738	-6.134	-11387	1.313	0.349	1.259	0.597	0.909	1651	2.125
4000	300	7653	-6.156	-11440	1.321	0.343	1.260	0.616	0.914	1635	2.150
4000	300	7485	-6.199	-11543	1.336	0.335	1.265	0.658	0.930	1602	2.200
4000	300	7322	-6.242	-11644	1.350	0.330	1.269	0.703	0.950	1570	2.250

Table A.29. Liquid Phase (continued)

T K	P bar	ρ $kg\ m^{-3}$	H $10^3 kJ\ kg^{-1}$	G $kJ\ kg^{-1}$	S	C_P $kJ\ kg^{-1}\ K^{-1}$	C_P/C_V	β_T $10^{-5} bar^{-1}$	α_T $10^{-4} K^{-1}$	V_S $m\ s^{-1}$	O/U
4000	1000	8547	-5.447	-10426	1.245	0.356	1.253	0.479	0.858	1749	1.750
4000	1000	8478	-5.563	-10567	1.251	0.360	1.252	0.483	0.861	1749	1.800
4000	1000	8409	-5.679	-10705	1.257	0.365	1.253	0.487	0.867	1750	1.850
4000	1000	8374	-5.736	-10773	1.259	0.368	1.253	0.489	0.872	1750	1.875
4000	1000	8339	-5.793	-10840	1.262	0.372	1.254	0.491	0.879	1750	1.900
4000	1000	8310	-5.838	-10893	1.264	0.376	1.256	0.494	0.887	1749	1.920
4000	1000	8281	-5.881	-10945	1.266	0.382	1.258	0.497	0.898	1748	1.940
4000	1000	8265	-5.903	-10971	1.267	0.385	1.259	0.499	0.904	1748	1.950
4000	1000	8249	-5.924	-10997	1.268	0.389	1.261	0.501	0.912	1747	1.960
4000	1000	8231	-5.944	-11022	1.270	0.393	1.263	0.504	0.920	1745	1.970
4000	1000	8212	-5.963	-11047	1.271	0.396	1.264	0.506	0.928	1744	1.980
4000	1000	8192	-5.981	-11072	1.273	0.398	1.265	0.509	0.933	1741	1.990
4000	1000	8169	-5.998	-11096	1.275	0.399	1.266	0.513	0.936	1739	2.000
4000	1000	8143	-6.013	-11120	1.277	0.396	1.265	0.516	0.934	1735	2.010
4000	1000	8115	-6.026	-11144	1.279	0.392	1.264	0.520	0.929	1731	2.020
4000	1000	8086	-6.039	-11168	1.282	0.387	1.262	0.524	0.922	1726	2.030
4000	1000	8055	-6.050	-11191	1.285	0.381	1.260	0.529	0.915	1720	2.040
4000	1000	8023	-6.061	-11214	1.288	0.376	1.259	0.534	0.909	1715	2.050
4000	1000	7990	-6.071	-11236	1.291	0.371	1.257	0.539	0.904	1709	2.060
4000	1000	7923	-6.089	-11281	1.298	0.362	1.256	0.551	0.898	1696	2.080
4000	1000	7855	-6.107	-11325	1.304	0.355	1.256	0.564	0.895	1684	2.100
4000	1000	7770	-6.129	-11378	1.312	0.348	1.257	0.582	0.897	1668	2.125
4000	1000	7686	-6.150	-11431	1.320	0.343	1.259	0.600	0.901	1652	2.150
4000	1000	7520	-6.193	-11534	1.335	0.335	1.263	0.640	0.916	1620	2.200
4000	1000	7358	-6.236	-11635	1.350	0.330	1.268	0.683	0.935	1588	2.250

Table A.29. Liquid Phase (continued)

T K	P bar	ρ $kg\ m^{-3}$	H $10^3 kJ\ kg^{-1}$	G $kJ\ kg^{-1}$	S	C_P $kJ\ kg^{-1}\ K^{-1}$	C_P/C_V	β_T $10^{-5} bar^{-1}$	α_T $10^{-4} K^{-1}$	V_S $m\ s^{-1}$	O/U
4000	3000	8627	-5.431	-10403	1.243	0.356	1.249	0.453	0.833	1788	1.750
4000	3000	8558	-5.548	-10544	1.249	0.359	1.248	0.456	0.835	1789	1.800
4000	3000	8488	-5.663	-10682	1.255	0.364	1.249	0.459	0.841	1789	1.850
4000	3000	8454	-5.721	-10750	1.257	0.367	1.249	0.461	0.845	1789	1.875
4000	3000	8419	-5.777	-10816	1.260	0.371	1.250	0.464	0.851	1789	1.900
4000	3000	8390	-5.822	-10869	1.262	0.376	1.251	0.466	0.859	1789	1.920
4000	3000	8361	-5.866	-10921	1.264	0.381	1.253	0.469	0.869	1788	1.940
4000	3000	8345	-5.887	-10947	1.265	0.384	1.255	0.470	0.875	1788	1.950
4000	3000	8329	-5.908	-10973	1.266	0.388	1.256	0.472	0.882	1787	1.960
4000	3000	8312	-5.928	-10998	1.267	0.392	1.258	0.474	0.890	1786	1.970
4000	3000	8293	-5.948	-11023	1.269	0.395	1.259	0.477	0.897	1784	1.980
4000	3000	8273	-5.966	-11047	1.270	0.398	1.260	0.480	0.903	1782	1.990
4000	3000	8250	-5.982	-11072	1.272	0.398	1.261	0.483	0.905	1780	2.000
4000	3000	8225	-5.997	-11096	1.275	0.395	1.260	0.486	0.903	1776	2.010
4000	3000	8198	-6.011	-11120	1.277	0.391	1.259	0.489	0.898	1772	2.020
4000	3000	8168	-6.023	-11143	1.280	0.386	1.257	0.493	0.891	1767	2.030
4000	3000	8138	-6.034	-11166	1.283	0.380	1.255	0.497	0.884	1762	2.040
4000	3000	8106	-6.045	-11189	1.286	0.375	1.254	0.501	0.877	1757	2.050
4000	3000	8074	-6.054	-11211	1.289	0.369	1.253	0.506	0.873	1751	2.060
4000	3000	8008	-6.073	-11256	1.296	0.361	1.252	0.517	0.866	1740	2.080
4000	3000	7941	-6.091	-11299	1.302	0.354	1.252	0.528	0.864	1728	2.100
4000	3000	7858	-6.112	-11353	1.310	0.347	1.253	0.544	0.865	1712	2.125
4000	3000	7775	-6.133	-11405	1.318	0.342	1.254	0.560	0.868	1697	2.150
4000	3000	7613	-6.176	-11508	1.333	0.334	1.258	0.594	0.880	1668	2.200
4000	3000	7455	-6.219	-11608	1.347	0.328	1.263	0.631	0.897	1638	2.250

Table A.29. Liquid Phase (continued)

T K	P bar	ρ $kg\ m^{-3}$	H $10^3 kJ\ kg^{-1}$	G $kJ\ kg^{-1}$	S $kJ\ kg^{-1}\ K^{-1}$	C_P	C_P/C_V	β_T $10^{-5} bar^{-1}$	α_T $10^{-4} K^{-1}$	V_S $m\ s^{-1}$	O/U
4000	10000	8882	-5.377	-10323	1.237	0.356	1.239	0.382	0.762	1912	1.750
4000	10000	8811	-5.493	-10463	1.243	0.359	1.238	0.384	0.764	1914	1.800
4000	10000	8742	-5.608	-10601	1.248	0.363	1.238	0.386	0.767	1915	1.850
4000	10000	8707	-5.665	-10668	1.251	0.366	1.238	0.387	0.770	1916	1.875
4000	10000	8672	-5.722	-10735	1.253	0.370	1.238	0.389	0.775	1916	1.900
4000	10000	8644	-5.767	-10787	1.255	0.374	1.239	0.390	0.781	1916	1.920
4000	10000	8615	-5.811	-10839	1.257	0.379	1.241	0.392	0.788	1916	1.940
4000	10000	8600	-5.832	-10864	1.258	0.382	1.242	0.393	0.794	1916	1.950
4000	10000	8584	-5.853	-10890	1.259	0.386	1.243	0.395	0.800	1916	1.960
4000	10000	8568	-5.873	-10915	1.260	0.390	1.245	0.396	0.806	1915	1.970
4000	10000	8550	-5.893	-10940	1.262	0.393	1.246	0.398	0.813	1914	1.980
4000	10000	8530	-5.911	-10964	1.263	0.396	1.247	0.400	0.818	1912	1.990
4000	10000	8508	-5.927	-10988	1.265	0.396	1.248	0.402	0.820	1910	2.000
4000	10000	8484	-5.942	-11012	1.267	0.394	1.247	0.404	0.817	1907	2.010
4000	10000	8457	-5.955	-11036	1.270	0.389	1.246	0.406	0.812	1904	2.020
4000	10000	8429	-5.967	-11059	1.273	0.383	1.244	0.409	0.805	1900	2.030
4000	10000	8399	-5.978	-11081	1.276	0.377	1.242	0.412	0.798	1895	2.040
4000	10000	8368	-5.988	-11104	1.279	0.372	1.241	0.415	0.792	1891	2.050
4000	10000	8337	-5.997	-11126	1.282	0.367	1.240	0.418	0.788	1886	2.060
4000	10000	8274	-6.015	-11170	1.289	0.359	1.240	0.426	0.782	1876	2.080
4000	10000	8211	-6.032	-11213	1.295	0.352	1.240	0.434	0.779	1866	2.100
4000	10000	8132	-6.053	-11265	1.303	0.345	1.241	0.444	0.778	1853	2.125
4000	10000	8054	-6.074	-11317	1.311	0.340	1.243	0.455	0.780	1841	2.150
4000	10000	7902	-6.116	-11417	1.325	0.331	1.247	0.478	0.788	1817	2.200
4000	10000	7754	-6.158	-11516	1.339	0.325	1.252	0.502	0.798	1793	2.250

Table A.29. Liquid Phase (continued)

T	P	ρ	H	G	S	C_P	C_P/C_V	β_T	α_T	V_S	O/U
K	bar	$kg\ m^{-3}$	$10^3 kJ\ kg^{-1}$	$kJ\ kg^{-1}$		$kJ\ kg^{-1}\ K^{-1}$		$10^{-5} bar^{-1}$	$10^{-3} K^{-1}$	$m\ s^{-1}$	
5000	300	7791	-5.120	-11719	1.320	0.328	1.365	0.666	0.095	1622	1.750
5000	300	7720	-5.231	-11867	1.327	0.336	1.369	0.676	0.097	1620	1.800
5000	300	7645	-5.339	-12011	1.334	0.345	1.374	0.688	0.099	1616	1.850
5000	300	7606	-5.392	-12083	1.338	0.351	1.377	0.695	0.101	1614	1.875
5000	300	7565	-5.443	-12153	1.342	0.356	1.380	0.703	0.102	1611	1.900
5000	300	7531	-5.483	-12208	1.345	0.360	1.383	0.711	0.103	1607	1.920
5000	300	7494	-5.521	-12263	1.348	0.364	1.385	0.719	0.105	1604	1.940
5000	300	7475	-5.540	-12290	1.350	0.366	1.386	0.723	0.105	1602	1.950
5000	300	7456	-5.558	-12318	1.352	0.368	1.387	0.727	0.106	1599	1.960
5000	300	7435	-5.576	-12344	1.354	0.369	1.388	0.732	0.106	1597	1.970
5000	300	7414	-5.594	-12371	1.355	0.370	1.388	0.737	0.106	1594	1.980
5000	300	7393	-5.611	-12398	1.357	0.371	1.388	0.742	0.107	1591	1.990
5000	300	7370	-5.627	-12424	1.359	0.371	1.388	0.747	0.107	1588	2.000
5000	300	7347	-5.643	-12450	1.361	0.371	1.388	0.753	0.107	1584	2.010
5000	300	7323	-5.659	-12477	1.363	0.370	1.387	0.759	0.107	1580	2.020
5000	300	7298	-5.674	-12502	1.366	0.370	1.386	0.765	0.107	1576	2.030
5000	300	7272	-5.688	-12528	1.368	0.368	1.385	0.771	0.107	1572	2.040
5000	300	7245	-5.702	-12554	1.370	0.367	1.384	0.778	0.107	1567	2.050
5000	300	7218	-5.716	-12579	1.373	0.365	1.383	0.785	0.107	1562	2.060
5000	300	7161	-5.741	-12629	1.378	0.361	1.379	0.800	0.107	1552	2.080
5000	300	7101	-5.765	-12679	1.383	0.357	1.376	0.816	0.106	1541	2.100
5000	300	7024	-5.793	-12740	1.389	0.352	1.372	0.840	0.106	1526	2.125
5000	300	6945	-5.819	-12800	1.396	0.348	1.369	0.865	0.106	1510	2.150
5000	300	6784	-5.868	-12918	1.410	0.342	1.365	0.924	0.107	1476	2.200
5000	300	6622	-5.914	-13032	1.424	0.340	1.362	0.991	0.109	1441	2.250

Table A.29. Liquid Phase (continued)

T K	P bar	ρ $kg\ m^{-3}$	H $10^3 kJ\ kg^{-1}$	G $kJ\ kg^{-1}$	S $kJ\ kg^{-1}\ K^{-1}$	C_P	C_P/C_V	β_T $10^{-5} bar^{-1}$	α_T $10^{-4} K^{-1}$	V_S $m\ s^{-1}$	O/U
5000	1000	7827	-5.115	-11710	1.319	0.327	1.362	0.647	0.939	1640	1.750
5000	1000	7756	-5.226	-11858	1.326	0.335	1.366	0.657	0.955	1638	1.800
5000	1000	7682	-5.335	-12002	1.334	0.344	1.371	0.668	0.977	1634	1.850
5000	1000	7643	-5.387	-12073	1.337	0.349	1.374	0.675	0.990	1632	1.875
5000	1000	7602	-5.438	-12144	1.341	0.355	1.377	0.683	1.004	1629	1.900
5000	1000	7568	-5.478	-12199	1.344	0.359	1.379	0.690	1.015	1626	1.920
5000	1000	7532	-5.517	-12254	1.347	0.363	1.382	0.697	1.026	1622	1.940
5000	1000	7513	-5.536	-12281	1.349	0.365	1.383	0.701	1.031	1620	1.950
5000	1000	7493	-5.554	-12308	1.351	0.366	1.383	0.705	1.036	1618	1.960
5000	1000	7473	-5.572	-12335	1.353	0.368	1.384	0.710	1.040	1615	1.970
5000	1000	7452	-5.589	-12362	1.354	0.369	1.384	0.714	1.044	1613	1.980
5000	1000	7430	-5.606	-12388	1.356	0.369	1.385	0.719	1.047	1610	1.990
5000	1000	7408	-5.623	-12415	1.358	0.370	1.385	0.724	1.049	1607	2.000
5000	1000	7385	-5.639	-12441	1.360	0.369	1.384	0.729	1.051	1603	2.010
5000	1000	7361	-5.655	-12467	1.362	0.369	1.384	0.735	1.052	1600	2.020
5000	1000	7336	-5.669	-12493	1.365	0.368	1.383	0.740	1.052	1596	2.030
5000	1000	7310	-5.684	-12519	1.367	0.367	1.382	0.746	1.051	1591	2.040
5000	1000	7284	-5.698	-12544	1.369	0.365	1.380	0.753	1.051	1587	2.050
5000	1000	7257	-5.711	-12569	1.372	0.364	1.379	0.759	1.049	1582	2.060
5000	1000	7200	-5.737	-12620	1.377	0.360	1.376	0.773	1.046	1572	2.080
5000	1000	7141	-5.760	-12669	1.382	0.356	1.372	0.789	1.042	1561	2.100
5000	1000	7065	-5.788	-12730	1.388	0.350	1.369	0.810	1.039	1546	2.125
5000	1000	6987	-5.814	-12790	1.395	0.346	1.365	0.834	1.039	1531	2.150
5000	1000	6827	-5.863	-12907	1.409	0.340	1.361	0.888	1.046	1498	2.200
5000	1000	6667	-5.909	-13022	1.422	0.338	1.358	0.951	1.064	1464	2.250

Table A.29. Liquid Phase (continued)

T K	P bar	ρ $kg\ m^{-3}$	H $10^3 kJ\ kg^{-1}$	G $kJ\ kg^{-1}$	S	C_P $kJ\ kg^{-1}\ K^{-1}$	C_P/C_V	β_T $10^{-5} bar^{-1}$	α_T $10^{-4} K^{-1}$	V_S $m\ s^{-1}$	O/U
5000	3000	7925	-5.101	-11685	1.317	0.324	1.355	0.600	0.898	1688	1.750
5000	3000	7854	-5.213	-11832	1.324	0.331	1.358	0.608	0.913	1686	1.800
5000	3000	7781	-5.321	-11976	1.331	0.341	1.362	0.618	0.933	1684	1.850
5000	3000	7743	-5.374	-12047	1.335	0.346	1.365	0.624	0.945	1682	1.875
5000	3000	7702	-5.425	-12117	1.338	0.351	1.368	0.630	0.958	1679	1.900
5000	3000	7669	-5.465	-12173	1.342	0.355	1.371	0.636	0.968	1676	1.920
5000	3000	7633	-5.504	-12228	1.345	0.359	1.373	0.643	0.978	1673	1.940
5000	3000	7614	-5.523	-12255	1.346	0.361	1.374	0.646	0.983	1671	1.950
5000	3000	7595	-5.541	-12282	1.348	0.363	1.374	0.650	0.987	1669	1.960
5000	3000	7575	-5.559	-12308	1.350	0.364	1.375	0.653	0.991	1667	1.970
5000	3000	7555	-5.576	-12335	1.352	0.365	1.375	0.657	0.994	1664	1.980
5000	3000	7534	-5.593	-12362	1.354	0.365	1.375	0.661	0.997	1662	1.990
5000	3000	7512	-5.610	-12388	1.356	0.366	1.375	0.665	0.999	1659	2.000
5000	3000	7489	-5.626	-12414	1.358	0.366	1.375	0.670	1.000	1656	2.010
5000	3000	7465	-5.641	-12440	1.360	0.365	1.374	0.674	1.001	1652	2.020
5000	3000	7441	-5.656	-12466	1.362	0.364	1.373	0.679	1.000	1648	2.030
5000	3000	7416	-5.671	-12491	1.364	0.363	1.372	0.684	0.999	1645	2.040
5000	3000	7390	-5.685	-12517	1.366	0.361	1.371	0.689	0.998	1640	2.050
5000	3000	7363	-5.698	-12542	1.369	0.360	1.369	0.695	0.996	1636	2.060
5000	3000	7307	-5.723	-12592	1.374	0.355	1.366	0.707	0.992	1626	2.080
5000	3000	7250	-5.747	-12641	1.379	0.351	1.363	0.720	0.988	1616	2.100
5000	3000	7175	-5.774	-12702	1.385	0.346	1.359	0.738	0.984	1602	2.125
5000	3000	7099	-5.800	-12762	1.392	0.342	1.356	0.758	0.982	1588	2.150
5000	3000	6943	-5.849	-12878	1.406	0.336	1.352	0.803	0.987	1558	2.200
5000	3000	6788	-5.895	-12992	1.419	0.334	1.349	0.853	1.000	1526	2.250

Table A.29. Liquid Phase (continued)

T K	P bar	ρ $kg\ m^{-3}$	H $10^3 kJ\ kg^{-1}$	G $kJ\ kg^{-1}$	S	C_P $kJ\ kg^{-1}\ K^{-1}$	C_P/C_V	β_T $10^{-5} bar^{-1}$	α_T $10^{-4} K^{-1}$	V_S $m\ s^{-1}$	O/U
5000	10000	8228	-5.051	-11598	1.309	0.316	1.338	0.481	0.794	1838	1.750
5000	10000	8158	-5.163	-11745	1.316	0.322	1.340	0.486	0.805	1838	1.800
5000	10000	8086	-5.271	-11888	1.323	0.331	1.343	0.492	0.820	1837	1.850
5000	10000	8049	-5.324	-11959	1.327	0.336	1.345	0.496	0.829	1836	1.875
5000	10000	8010	-5.376	-12028	1.330	0.341	1.347	0.500	0.839	1834	1.900
5000	10000	7978	-5.416	-12083	1.333	0.345	1.349	0.504	0.848	1832	1.920
5000	10000	7943	-5.455	-12138	1.337	0.349	1.351	0.508	0.856	1830	1.940
5000	10000	7926	-5.474	-12165	1.338	0.351	1.352	0.510	0.860	1829	1.950
5000	10000	7907	-5.492	-12191	1.340	0.353	1.353	0.512	0.863	1827	1.960
5000	10000	7888	-5.510	-12218	1.342	0.354	1.353	0.515	0.866	1826	1.970
5000	10000	7869	-5.528	-12244	1.343	0.355	1.353	0.517	0.869	1824	1.980
5000	10000	7848	-5.545	-12271	1.345	0.356	1.353	0.520	0.870	1822	1.990
5000	10000	7827	-5.561	-12297	1.347	0.356	1.353	0.522	0.872	1819	2.000
5000	10000	7805	-5.577	-12323	1.349	0.356	1.353	0.525	0.872	1817	2.010
5000	10000	7782	-5.592	-12348	1.351	0.355	1.352	0.528	0.872	1814	2.020
5000	10000	7759	-5.607	-12374	1.353	0.354	1.351	0.531	0.870	1811	2.030
5000	10000	7735	-5.621	-12399	1.356	0.352	1.350	0.534	0.869	1808	2.040
5000	10000	7710	-5.635	-12424	1.358	0.351	1.349	0.537	0.867	1804	2.050
5000	10000	7684	-5.648	-12449	1.360	0.349	1.347	0.541	0.864	1801	2.060
5000	10000	7632	-5.673	-12498	1.365	0.344	1.344	0.548	0.858	1793	2.080
5000	10000	7577	-5.696	-12547	1.370	0.340	1.341	0.556	0.853	1785	2.100
5000	10000	7506	-5.722	-12607	1.377	0.335	1.338	0.567	0.848	1774	2.125
5000	10000	7434	-5.748	-12665	1.384	0.330	1.336	0.579	0.845	1762	2.150
5000	10000	7289	-5.795	-12780	1.397	0.325	1.332	0.605	0.846	1737	2.200
5000	10000	7145	-5.841	-12892	1.410	0.323	1.330	0.635	0.853	1712	2.250

Table A.29. Liquid Phase (continued)

T K	P bar	ρ $kg\ m^{-3}$	H $10^3 kJ\ kg^{-1}$	G $kJ\ kg^{-1}$	S	C_P $kJ\ kg^{-1}\ K^{-1}$	C_P/C_V	β_T $10^{-4} bar^{-1}$	α_T $10^{-3} K^{-1}$	V_S $m\ s^{-1}$	O/U
6000	1000	7052	-4.770	-13061	1.382	0.372	1.484	0.095	0.116	1490	1.750
6000	1000	6972	-4.872	-13217	1.391	0.381	1.490	0.097	0.119	1483	1.800
6000	1000	6888	-4.971	-13369	1.400	0.390	1.496	0.100	0.122	1475	1.850
6000	1000	6844	-5.019	-13444	1.404	0.395	1.499	0.101	0.123	1470	1.875
6000	1000	6798	-5.065	-13519	1.409	0.399	1.501	0.103	0.125	1465	1.900
6000	1000	6760	-5.101	-13578	1.413	0.402	1.503	0.104	0.126	1460	1.920
6000	1000	6721	-5.137	-13636	1.417	0.405	1.504	0.106	0.127	1454	1.940
6000	1000	6701	-5.154	-13665	1.419	0.407	1.504	0.107	0.127	1451	1.950
6000	1000	6680	-5.171	-13694	1.421	0.408	1.505	0.107	0.128	1448	1.960
6000	1000	6659	-5.188	-13723	1.423	0.410	1.505	0.108	0.128	1445	1.970
6000	1000	6638	-5.204	-13752	1.425	0.411	1.505	0.109	0.129	1442	1.980
6000	1000	6617	-5.220	-13780	1.427	0.412	1.505	0.110	0.129	1438	1.990
6000	1000	6594	-5.236	-13809	1.429	0.413	1.505	0.111	0.130	1435	2.000
6000	1000	6572	-5.252	-13837	1.431	0.414	1.504	0.112	0.130	1431	2.010
6000	1000	6549	-5.267	-13865	1.433	0.414	1.504	0.113	0.131	1427	2.020
6000	1000	6526	-5.282	-13893	1.435	0.415	1.503	0.114	0.131	1423	2.030
6000	1000	6502	-5.297	-13921	1.437	0.416	1.503	0.115	0.132	1419	2.040
6000	1000	6477	-5.311	-13949	1.440	0.416	1.502	0.116	0.132	1414	2.050
6000	1000	6453	-5.325	-13976	1.442	0.416	1.501	0.117	0.132	1410	2.060
6000	1000	6402	-5.352	-14031	1.447	0.416	1.498	0.119	0.133	1400	2.080
6000	1000	6349	-5.378	-14086	1.451	0.416	1.495	0.122	0.133	1390	2.100
6000	1000	6281	-5.408	-14153	1.457	0.416	1.490	0.125	0.134	1376	2.125
6000	1000	6211	-5.437	-14219	1.464	0.415	1.484	0.129	0.135	1361	2.150
6000	1000	6065	-5.490	-14350	1.477	0.414	1.472	0.137	0.136	1329	2.200
6000	1000	5913	-5.537	-14478	1.490	0.416	1.460	0.147	0.138	1294	2.250

Table A.29. Liquid Phase (continued)

T K	P bar	ρ $kg\ m^{-3}$	H $10^3 kJ\ kg^{-1}$	G $kJ\ kg^{-1}$	S	C_P $kJ\ kg^{-1}\ K^{-1}$	C_P/C_V	β_T $10^{-4} bar^{-1}$	α_T $10^{-3} K^{-1}$	V_S $m\ s^{-1}$	O/U
6000	3000	7179	-4.760	-13033	1.379	0.365	1.469	0.085	0.109	1551	1.750
6000	3000	7101	-4.863	-13188	1.387	0.374	1.474	0.087	0.111	1546	1.800
6000	3000	7019	-4.962	-13341	1.396	0.383	1.479	0.089	0.114	1539	1.850
6000	3000	6976	-5.010	-13416	1.401	0.388	1.481	0.090	0.115	1534	1.875
6000	3000	6931	-5.057	-13490	1.405	0.392	1.483	0.092	0.116	1530	1.900
6000	3000	6894	-5.093	-13549	1.409	0.395	1.485	0.093	0.117	1525	1.920
6000	3000	6856	-5.129	-13607	1.413	0.398	1.486	0.094	0.118	1520	1.940
6000	3000	6836	-5.146	-13636	1.415	0.399	1.486	0.094	0.119	1518	1.950
6000	3000	6817	-5.163	-13665	1.417	0.401	1.486	0.095	0.119	1515	1.960
6000	3000	6796	-5.180	-13693	1.419	0.402	1.486	0.096	0.119	1512	1.970
6000	3000	6776	-5.197	-13722	1.421	0.403	1.486	0.096	0.120	1509	1.980
6000	3000	6754	-5.213	-13750	1.423	0.404	1.486	0.097	0.120	1506	1.990
6000	3000	6733	-5.229	-13779	1.425	0.405	1.486	0.098	0.120	1503	2.000
6000	3000	6711	-5.244	-13807	1.427	0.406	1.485	0.098	0.121	1499	2.010
6000	3000	6689	-5.260	-13835	1.429	0.406	1.484	0.099	0.121	1496	2.020
6000	3000	6666	-5.275	-13863	1.431	0.407	1.484	0.100	0.121	1492	2.030
6000	3000	6643	-5.289	-13891	1.434	0.407	1.483	0.101	0.122	1488	2.040
6000	3000	6619	-5.304	-13918	1.436	0.407	1.482	0.102	0.122	1484	2.050
6000	3000	6595	-5.318	-13946	1.438	0.408	1.480	0.102	0.122	1480	2.060
6000	3000	6546	-5.345	-14000	1.443	0.408	1.477	0.104	0.122	1471	2.080
6000	3000	6495	-5.371	-14054	1.447	0.407	1.474	0.106	0.123	1462	2.100
6000	3000	6429	-5.401	-14121	1.453	0.406	1.469	0.109	0.123	1449	2.125
6000	3000	6361	-5.430	-14187	1.460	0.405	1.463	0.112	0.123	1436	2.150
6000	3000	6221	-5.482	-14317	1.472	0.403	1.450	0.118	0.124	1407	2.200
6000	3000	6076	-5.530	-14445	1.486	0.404	1.437	0.125	0.125	1375	2.250

Table A.29. Liquid Phase (continued)

T K	P bar	ρ $kg\ m^{-3}$	H $10^3 kJ\ kg^{-1}$	G $kJ\ kg^{-1}$	S	C_P $kJ\ kg^{-1}\ K^{-1}$	C_P/C_V	β_T $10^{-4} bar^{-1}$	α_T $10^{-3} K^{-1}$	V_S $m\ s^{-1}$	O/U
6000	10000	7555	-4.722	-12938	1.369	0.350	1.435	0.631	0.918	1735	1.750
6000	10000	7480	-4.826	-13092	1.378	0.358	1.439	0.641	0.935	1732	1.800
6000	10000	7402	-4.926	-13244	1.386	0.367	1.443	0.653	0.952	1727	1.850
6000	10000	7361	-4.974	-13318	1.391	0.371	1.444	0.660	0.961	1725	1.875
6000	10000	7318	-5.021	-13392	1.395	0.375	1.446	0.667	0.969	1721	1.900
6000	10000	7283	-5.058	-13450	1.399	0.378	1.446	0.673	0.976	1718	1.920
6000	10000	7247	-5.093	-13508	1.402	0.381	1.447	0.679	0.982	1715	1.940
6000	10000	7229	-5.111	-13536	1.404	0.382	1.447	0.682	0.985	1713	1.950
6000	10000	7210	-5.128	-13565	1.406	0.383	1.447	0.685	0.987	1711	1.960
6000	10000	7191	-5.145	-13593	1.408	0.384	1.447	0.689	0.990	1709	1.970
6000	10000	7172	-5.161	-13622	1.410	0.385	1.446	0.692	0.992	1707	1.980
6000	10000	7152	-5.178	-13650	1.412	0.386	1.446	0.696	0.994	1704	1.990
6000	10000	7132	-5.194	-13678	1.414	0.387	1.445	0.700	0.995	1702	2.000
6000	10000	7111	-5.209	-13706	1.416	0.387	1.445	0.704	0.997	1699	2.010
6000	10000	7090	-5.225	-13733	1.418	0.388	1.444	0.707	0.998	1697	2.020
6000	10000	7069	-5.240	-13761	1.420	0.388	1.443	0.712	0.999	1694	2.030
6000	10000	7047	-5.254	-13788	1.422	0.388	1.442	0.716	1.000	1691	2.040
6000	10000	7025	-5.269	-13816	1.425	0.388	1.440	0.720	1.000	1688	2.050
6000	10000	7002	-5.283	-13843	1.427	0.388	1.439	0.724	1.000	1685	2.060
6000	10000	6956	-5.310	-13897	1.431	0.387	1.436	0.733	1.000	1678	2.080
6000	10000	6909	-5.335	-13950	1.436	0.386	1.432	0.743	0.998	1670	2.100
6000	10000	6847	-5.365	-14016	1.442	0.385	1.427	0.756	0.996	1660	2.125
6000	10000	6784	-5.394	-14081	1.448	0.383	1.421	0.769	0.993	1650	2.150
6000	10000	6654	-5.445	-14209	1.461	0.380	1.409	0.800	0.988	1627	2.200
6000	10000	6522	-5.492	-14334	1.474	0.379	1.396	0.834	0.988	1602	2.250

Table A.29. Liquid Phase (continued)

T K	P bar	ρ $kg\ m^{-3}$	H $10^3 kJ\ kg^{-1}$	G $kJ\ kg^{-1}$	S	C_P $kJ\ kg^{-1}\ K^{-1}$	C_P/C_V	β_T $10^{-4} bar^{-1}$	α_T $10^{-3} K^{-1}$	V_S $m\ s^{-1}$	O/U
7000	3000	6357	-4.370	-14442	1.439	0.424	1.584	0.132	0.137	1376	1.750
7000	3000	6270	-4.463	-14607	1.449	0.437	1.589	0.136	0.140	1365	1.800
7000	3000	6179	-4.552	-14768	1.460	0.451	1.594	0.141	0.145	1353	1.850
7000	3000	6132	-4.594	-14848	1.465	0.458	1.595	0.144	0.147	1346	1.875
7000	3000	6084	-4.635	-14928	1.470	0.466	1.597	0.147	0.149	1338	1.900
7000	3000	6044	-4.668	-14990	1.475	0.472	1.598	0.149	0.151	1332	1.920
7000	3000	6003	-4.699	-15053	1.479	0.479	1.598	0.152	0.153	1325	1.940
7000	3000	5982	-4.714	-15084	1.481	0.482	1.599	0.153	0.154	1322	1.950
7000	3000	5961	-4.729	-15115	1.484	0.485	1.599	0.154	0.155	1318	1.960
7000	3000	5940	-4.744	-15146	1.486	0.489	1.599	0.156	0.156	1314	1.970
7000	3000	5918	-4.758	-15176	1.488	0.492	1.599	0.157	0.157	1310	1.980
7000	3000	5896	-4.773	-15207	1.491	0.496	1.599	0.159	0.158	1306	1.990
7000	3000	5874	-4.787	-15237	1.493	0.499	1.598	0.161	0.159	1302	2.000
7000	3000	5851	-4.800	-15268	1.495	0.503	1.598	0.162	0.160	1298	2.010
7000	3000	5828	-4.814	-15298	1.498	0.507	1.598	0.164	0.161	1294	2.020
7000	3000	5805	-4.827	-15328	1.500	0.510	1.597	0.166	0.162	1289	2.030
7000	3000	5782	-4.840	-15358	1.503	0.514	1.597	0.167	0.163	1285	2.040
7000	3000	5758	-4.853	-15388	1.505	0.518	1.596	0.169	0.164	1280	2.050
7000	3000	5734	-4.865	-15418	1.508	0.522	1.595	0.171	0.165	1275	2.060
7000	3000	5685	-4.889	-15477	1.513	0.529	1.593	0.175	0.167	1265	2.080
7000	3000	5635	-4.911	-15536	1.518	0.538	1.591	0.179	0.170	1255	2.100
7000	3000	5571	-4.937	-15610	1.525	0.548	1.588	0.185	0.173	1241	2.125
7000	3000	5504	-4.962	-15682	1.531	0.559	1.585	0.191	0.176	1227	2.150
7000	3000	5367	-5.005	-15825	1.546	0.583	1.577	0.205	0.183	1197	2.200
7000	3000	5223	-5.040	-15966	1.561	0.608	1.571	0.221	0.191	1165	2.250

Table A.29. Liquid Phase (end)

T K	P bar	ρ $kg\,m^{-3}$	H $10^3 kJ\,kg^{-1}$	G $kJ\,kg^{-1}$	S $kJ\,kg^{-1}\,K^{-1}$	C_P	C_P/C_V	β_T $10^{-4} bar^{-1}$	α_T $10^{-3} K^{-1}$	V_S $m\,s^{-1}$	O/U
7000	10000	6844	-4.353	-14336	1.426	0.394	1.519	0.086	0.106	1606	1.750
7000	10000	6764	-4.448	-14499	1.436	0.405	1.521	0.088	0.109	1599	1.800
7000	10000	6680	-4.538	-14660	1.446	0.416	1.522	0.090	0.111	1591	1.850
7000	10000	6637	-4.582	-14739	1.451	0.422	1.523	0.091	0.112	1586	1.875
7000	10000	6593	-4.624	-14817	1.456	0.428	1.522	0.092	0.113	1581	1.900
7000	10000	6557	-4.657	-14879	1.460	0.433	1.522	0.093	0.114	1577	1.920
7000	10000	6520	-4.690	-14941	1.465	0.437	1.521	0.094	0.115	1572	1.940
7000	10000	6502	-4.705	-14972	1.467	0.440	1.521	0.095	0.115	1570	1.950
7000	10000	6483	-4.721	-15002	1.469	0.442	1.520	0.096	0.116	1567	1.960
7000	10000	6463	-4.736	-15033	1.471	0.445	1.520	0.096	0.116	1564	1.970
7000	10000	6444	-4.751	-15063	1.473	0.447	1.519	0.097	0.117	1562	1.980
7000	10000	6424	-4.766	-15093	1.475	0.450	1.518	0.097	0.117	1559	1.990
7000	10000	6405	-4.780	-15124	1.478	0.452	1.517	0.098	0.118	1556	2.000
7000	10000	6384	-4.795	-15154	1.480	0.455	1.516	0.099	0.118	1553	2.010
7000	10000	6364	-4.809	-15183	1.482	0.458	1.515	0.099	0.118	1550	2.020
7000	10000	6344	-4.822	-15213	1.484	0.460	1.514	0.100	0.119	1546	2.030
7000	10000	6323	-4.836	-15243	1.487	0.463	1.512	0.100	0.119	1543	2.040
7000	10000	6302	-4.849	-15272	1.489	0.465	1.511	0.101	0.120	1540	2.050
7000	10000	6280	-4.862	-15302	1.491	0.468	1.509	0.102	0.120	1536	2.060
7000	10000	6237	-4.887	-15360	1.496	0.473	1.506	0.103	0.121	1529	2.080
7000	10000	6193	-4.910	-15418	1.501	0.478	1.502	0.105	0.122	1522	2.100
7000	10000	6136	-4.938	-15490	1.507	0.485	1.497	0.107	0.123	1512	2.125
7000	10000	6078	-4.964	-15562	1.514	0.492	1.491	0.109	0.124	1501	2.150
7000	10000	5959	-5.010	-15702	1.527	0.507	1.478	0.113	0.126	1479	2.200
7000	10000	5836	-5.049	-15840	1.542	0.523	1.465	0.118	0.128	1456	2.250

Table A.30. Gas Phase

T K	P bar	ρ $kg\ m^{-3}$	H $10^3 kJ\ kg^{-1}$	G $kJ\ kg^{-1}$	S $kJ\ kg^{-1}\ K^{-1}$	C_P	C_P/C_V	β_T bar^{-1}	α_T $10^{-3} K^{-1}$	V_S $m\ s^{-1}$	O/U
4000	1	0.7790	-3.877	-10451	1.644	0.664	1.124	1.026	0.382	375	1.750
4000	1	0.7783	-3.975	-10589	1.653	0.717	1.125	1.029	0.400	375	1.800
4000	1	0.7769	-4.070	-10725	1.664	0.776	1.128	1.033	0.420	375	1.850
4000	1	0.7760	-4.116	-10792	1.669	0.806	1.130	1.035	0.431	375	1.875
4000	1	0.7749	-4.161	-10859	1.675	0.839	1.132	1.038	0.443	375	1.900
4000	1	0.7739	-4.195	-10912	1.679	0.865	1.133	1.040	0.453	375	1.920
4000	1	0.7728	-4.230	-10965	1.684	0.892	1.135	1.042	0.462	376	1.940
4000	1	0.7722	-4.246	-10991	1.686	0.906	1.136	1.043	0.468	376	1.950
4000	1	0.7716	-4.263	-11018	1.689	0.920	1.137	1.044	0.473	376	1.960
4000	1	0.7709	-4.279	-11044	1.691	0.934	1.138	1.045	0.478	376	1.970
4000	1	0.7702	-4.295	-11070	1.694	0.949	1.139	1.046	0.483	376	1.980
4000	1	0.7694	-4.311	-11096	1.696	0.963	1.140	1.047	0.489	376	1.990
4000	1	0.7687	-4.327	-11122	1.699	0.978	1.141	1.048	0.494	376	2.000
4000	1	0.7679	-4.343	-11148	1.701	0.992	1.143	1.049	0.500	377	2.010
4000	1	0.7670	-4.358	-11174	1.704	1.007	1.144	1.051	0.505	377	2.020
4000	1	0.7661	-4.373	-11200	1.707	1.022	1.145	1.052	0.511	377	2.030
4000	1	0.7652	-4.388	-11225	1.709	1.037	1.146	1.053	0.516	377	2.040
4000	1	0.7643	-4.403	-11251	1.712	1.052	1.148	1.054	0.522	377	2.050
4000	1	0.7633	-4.417	-11277	1.715	1.068	1.149	1.055	0.528	378	2.060
4000	1	0.7612	-4.445	-11328	1.720	1.099	1.152	1.058	0.540	378	2.080
4000	1	0.7589	-4.473	-11378	1.726	1.130	1.155	1.061	0.552	379	2.100
4000	1	0.7559	-4.506	-11441	1.734	1.170	1.158	1.064	0.567	379	2.125
4000	1	0.7527	-4.538	-11504	1.741	1.210	1.162	1.067	0.582	380	2.150
4000	1	0.7456	-4.598	-11628	1.758	1.293	1.170	1.074	0.613	382	2.200
4000	1	0.7377	-4.652	-11750	1.775	1.377	1.178	1.081	0.643	384	2.250

Table A.30. Gas Phase (continued)

T K	P bar	ρ $kg\ m^{-3}$	H $10^3 kJ\ kg^{-1}$	G $kJ\ kg^{-1}$	S	C_P $kJ\ kg^{-1}\ K^{-1}$	C_P/C_V	β_T bar^{-1}	α_T $10^{-3} K^{-1}$	V_S $m\ s^{-1}$	O/U
5000	1	0.4036	-2.003	-12266	2.053	2.429	1.251	1.128	0.666	524	1.750
5000	1	0.3949	-2.006	-12424	2.084	2.493	1.253	1.129	0.670	530	1.800
5000	1	0.3864	-2.007	-12581	2.115	2.554	1.254	1.130	0.672	536	1.850
5000	1	0.3823	-2.007	-12659	2.130	2.583	1.255	1.130	0.673	539	1.875
5000	1	0.3782	-2.007	-12737	2.146	2.611	1.255	1.130	0.674	542	1.900
5000	1	0.3751	-2.006	-12799	2.158	2.633	1.256	1.130	0.674	544	1.920
5000	1	0.3719	-2.006	-12860	2.171	2.654	1.256	1.130	0.674	547	1.940
5000	1	0.3703	-2.005	-12891	2.177	2.665	1.256	1.130	0.674	548	1.950
5000	1	0.3688	-2.005	-12922	2.183	2.675	1.256	1.130	0.674	549	1.960
5000	1	0.3673	-2.004	-12953	2.190	2.685	1.256	1.130	0.674	550	1.970
5000	1	0.3657	-2.004	-12983	2.196	2.695	1.256	1.130	0.674	551	1.980
5000	1	0.3642	-2.003	-13014	2.202	2.705	1.256	1.130	0.674	552	1.990
5000	1	0.3627	-2.002	-13044	2.208	2.715	1.257	1.130	0.674	554	2.000
5000	1	0.3612	-2.001	-13075	2.215	2.725	1.257	1.130	0.674	555	2.010
5000	1	0.3598	-2.001	-13105	2.221	2.734	1.257	1.130	0.674	556	2.020
5000	1	0.3583	-2.000	-13136	2.227	2.744	1.257	1.130	0.674	557	2.030
5000	1	0.3568	-1.999	-13166	2.233	2.753	1.257	1.130	0.674	558	2.040
5000	1	0.3554	-1.998	-13197	2.240	2.762	1.257	1.130	0.674	559	2.050
5000	1	0.3539	-1.997	-13227	2.246	2.771	1.257	1.130	0.673	561	2.060
5000	1	0.3511	-1.995	-13287	2.258	2.789	1.257	1.130	0.673	563	2.080
5000	1	0.3483	-1.993	-13348	2.271	2.807	1.257	1.130	0.672	565	2.100
5000	1	0.3448	-1.990	-13423	2.287	2.828	1.257	1.130	0.672	568	2.125
5000	1	0.3414	-1.986	-13498	2.302	2.848	1.257	1.130	0.671	571	2.150
5000	1	0.3348	-1.979	-13647	2.333	2.888	1.257	1.129	0.669	577	2.200
5000	1	0.3285	-1.971	-13794	2.365	2.924	1.257	1.129	0.666	582	2.250

Table A.30. Gas Phase (continued)

T K	P bar	ρ $kg\ m^{-3}$	H $10^3 kJ\ kg^{-1}$	G $kJ\ kg^{-1}$	S	C_P $kJ\ kg^{-1}\ K^{-1}$	C_P/C_V	β_T bar^{-1}	α_T $10^{-3} K^{-1}$	V_S $m\ s^{-1}$	O/U
5000	3	1.412	-2.643	-12013	1.874	2.330	1.264	0.382	0.725	484	1.750
5000	3	1.386	-2.671	-12166	1.899	2.427	1.269	0.384	0.740	489	1.800
5000	3	1.359	-2.696	-12317	1.924	2.518	1.274	0.385	0.753	494	1.850
5000	3	1.346	-2.707	-12393	1.937	2.562	1.276	0.385	0.758	496	1.875
5000	3	1.333	-2.717	-12468	1.950	2.605	1.278	0.386	0.763	498	1.900
5000	3	1.322	-2.725	-12527	1.960	2.638	1.279	0.386	0.767	500	1.920
5000	3	1.312	-2.733	-12587	1.971	2.670	1.281	0.387	0.771	502	1.940
5000	3	1.307	-2.736	-12616	1.976	2.686	1.281	0.387	0.772	503	1.950
5000	3	1.302	-2.740	-12646	1.981	2.701	1.282	0.387	0.774	504	1.960
5000	3	1.297	-2.743	-12676	1.987	2.716	1.283	0.387	0.775	505	1.970
5000	3	1.292	-2.746	-12705	1.992	2.732	1.283	0.387	0.777	506	1.980
5000	3	1.287	-2.749	-12735	1.997	2.747	1.284	0.387	0.778	507	1.990
5000	3	1.282	-2.752	-12764	2.002	2.761	1.284	0.388	0.779	509	2.000
5000	3	1.277	-2.755	-12794	2.008	2.776	1.285	0.388	0.780	510	2.010
5000	3	1.271	-2.758	-12823	2.013	2.790	1.285	0.388	0.782	511	2.020
5000	3	1.266	-2.761	-12852	2.018	2.804	1.286	0.388	0.783	512	2.030
5000	3	1.262	-2.764	-12881	2.024	2.818	1.286	0.388	0.784	513	2.040
5000	3	1.257	-2.766	-12911	2.029	2.832	1.287	0.388	0.785	514	2.050
5000	3	1.252	-2.769	-12940	2.034	2.845	1.287	0.388	0.785	515	2.060
5000	3	1.242	-2.773	-12998	2.045	2.872	1.288	0.388	0.787	517	2.080
5000	3	1.232	-2.778	-13056	2.056	2.898	1.289	0.389	0.788	519	2.100
5000	3	1.220	-2.783	-13128	2.069	2.929	1.290	0.389	0.790	521	2.125
5000	3	1.208	-2.787	-13200	2.083	2.959	1.290	0.389	0.791	524	2.150
5000	3	1.185	-2.795	-13343	2.110	3.015	1.292	0.389	0.792	529	2.200
5000	3	1.162	-2.800	-13485	2.137	3.068	1.293	0.389	0.793	535	2.250

Table A.30. Gas Phase (continued)

T K	P bar	ρ $kg\ m^{-3}$	H $10^3 kJ\ kg^{-1}$	G $kJ\ kg^{-1}$	S	C_P $kJ\ kg^{-1}\ K^{-1}$	C_P/C_V	β_T bar^{-1}	α_T $10^{-3} K^{-1}$	V_S $m\ s^{-1}$	O/U
5000	10	5.538	-3.202	-11778	1.715	1.544	1.222	0.112	0.589	445	1.750
5000	10	5.474	-3.263	-11927	1.733	1.638	1.229	0.112	0.613	447	1.800
5000	10	5.405	-3.320	-12074	1.751	1.729	1.237	0.113	0.636	450	1.850
5000	10	5.370	-3.347	-12148	1.760	1.774	1.241	0.113	0.647	451	1.875
5000	10	5.334	-3.373	-12220	1.770	1.819	1.244	0.114	0.658	453	1.900
5000	10	5.304	-3.393	-12278	1.777	1.854	1.247	0.114	0.666	454	1.920
5000	10	5.274	-3.412	-12336	1.785	1.888	1.250	0.114	0.674	456	1.940
5000	10	5.259	-3.421	-12365	1.789	1.905	1.251	0.114	0.678	456	1.950
5000	10	5.244	-3.430	-12394	1.793	1.922	1.253	0.114	0.682	457	1.960
5000	10	5.229	-3.440	-12422	1.797	1.939	1.254	0.115	0.686	458	1.970
5000	10	5.213	-3.449	-12451	1.801	1.955	1.255	0.115	0.690	458	1.980
5000	10	5.198	-3.457	-12480	1.804	1.972	1.257	0.115	0.693	459	1.990
5000	10	5.182	-3.466	-12508	1.808	1.988	1.258	0.115	0.697	460	2.000
5000	10	5.166	-3.474	-12537	1.812	2.004	1.259	0.115	0.700	460	2.010
5000	10	5.150	-3.483	-12565	1.816	2.020	1.260	0.115	0.704	461	2.020
5000	10	5.134	-3.491	-12593	1.821	2.036	1.262	0.115	0.707	462	2.030
5000	10	5.118	-3.499	-12622	1.825	2.052	1.263	0.115	0.710	463	2.040
5000	10	5.102	-3.507	-12650	1.829	2.067	1.264	0.115	0.713	463	2.050
5000	10	5.086	-3.514	-12678	1.833	2.083	1.265	0.116	0.716	464	2.060
5000	10	5.054	-3.529	-12734	1.841	2.113	1.268	0.116	0.722	466	2.080
5000	10	5.021	-3.544	-12790	1.849	2.142	1.270	0.116	0.728	467	2.100
5000	10	4.980	-3.561	-12860	1.860	2.178	1.272	0.116	0.734	469	2.125
5000	10	4.939	-3.577	-12930	1.871	2.213	1.275	0.116	0.740	471	2.150
5000	10	4.855	-3.607	-13068	1.892	2.278	1.279	0.117	0.751	475	2.200
5000	10	4.771	-3.633	-13204	1.914	2.340	1.284	0.117	0.760	480	2.250

Table A.30. Gas Phase (continued)

T K	P bar	ρ $kg\ m^{-3}$	H $10^2 kJ\ kg^{-1}$	G $kJ\ kg^{-1}$	S	C_P $kJ\ kg^{-1}\ K^{-1}$	C_P/C_V	β_T bar^{-1}	α_T $10^{-3}K^{-1}$	V_S $m\ s^{-1}$	O/U
6000	1	0.2361	2.769	-14526	2.467	2.373	1.289	1.136	0.488	693	1.750
6000	1	0.2320	2.888	-14717	2.501	2.378	1.287	1.134	0.482	699	1.800
6000	1	0.2281	3.014	-14908	2.535	2.381	1.286	1.132	0.477	706	1.850
6000	1	0.2261	3.079	-15003	2.552	2.383	1.285	1.131	0.474	709	1.875
6000	1	0.2243	3.146	-15097	2.569	2.384	1.284	1.130	0.472	712	1.900
6000	1	0.2228	3.201	-15172	2.582	2.385	1.283	1.129	0.470	714	1.920
6000	1	0.2213	3.257	-15247	2.595	2.386	1.282	1.128	0.468	717	1.940
6000	1	0.2206	3.285	-15285	2.602	2.387	1.282	1.127	0.467	718	1.950
6000	1	0.2199	3.314	-15322	2.609	2.387	1.282	1.127	0.466	719	1.960
6000	1	0.2191	3.342	-15359	2.616	2.388	1.281	1.126	0.465	720	1.970
6000	1	0.2184	3.371	-15397	2.622	2.388	1.281	1.126	0.464	722	1.980
6000	1	0.2177	3.400	-15434	2.629	2.388	1.281	1.126	0.463	723	1.990
6000	1	0.2170	3.429	-15471	2.636	2.389	1.280	1.125	0.462	724	2.000
6000	1	0.2163	3.459	-15508	2.642	2.389	1.280	1.125	0.461	725	2.010
6000	1	0.2156	3.488	-15546	2.649	2.389	1.280	1.124	0.460	727	2.020
6000	1	0.2149	3.518	-15583	2.656	2.390	1.280	1.124	0.459	728	2.030
6000	1	0.2143	3.548	-15620	2.662	2.390	1.279	1.123	0.458	729	2.040
6000	1	0.2136	3.578	-15657	2.669	2.390	1.279	1.123	0.457	730	2.050
6000	1	0.2129	3.609	-15693	2.676	2.391	1.279	1.123	0.456	731	2.060
6000	1	0.2116	3.670	-15767	2.689	2.391	1.278	1.122	0.454	734	2.080
6000	1	0.2103	3.732	-15841	2.702	2.392	1.277	1.121	0.452	736	2.100
6000	1	0.2086	3.810	-15932	2.719	2.392	1.277	1.120	0.449	739	2.125
6000	1	0.2070	3.889	-16023	2.735	2.392	1.276	1.119	0.447	742	2.150
6000	1	0.2039	4.051	-16205	2.768	2.393	1.275	1.117	0.442	748	2.200
6000	1	0.2009	4.215	-16386	2.801	2.393	1.273	1.115	0.438	754	2.250

Table A.30. Gas Phase (continued)

T K	P bar	ρ $kg\ m^{-3}$	H $10^2 kJ\ kg^{-1}$	G $kJ\ kg^{-1}$	S	C_P $kJ\ kg^{-1}\ K^{-1}$	C_P/C_V	β_T bar^{-1}	α_T $10^{-3} K^{-1}$	V_S $m\ s^{-1}$	O/U
6000	3	0.8121	-5.653	-14092	2.254	1.789	1.258	0.371	0.429	646	1.750
6000	3	0.7960	-5.525	-14275	2.287	1.793	1.256	0.370	0.424	653	1.800
6000	3	0.7805	-5.387	-14458	2.320	1.797	1.255	0.369	0.419	660	1.850
6000	3	0.7730	-5.315	-14548	2.336	1.799	1.255	0.369	0.417	663	1.875
6000	3	0.7656	-5.241	-14638	2.352	1.801	1.254	0.369	0.414	667	1.900
6000	3	0.7598	-5.180	-14710	2.365	1.803	1.254	0.368	0.412	669	1.920
6000	3	0.7541	-5.118	-14782	2.378	1.804	1.253	0.368	0.411	672	1.940
6000	3	0.7512	-5.087	-14818	2.385	1.805	1.253	0.368	0.410	673	1.950
6000	3	0.7484	-5.056	-14854	2.391	1.805	1.253	0.368	0.409	675	1.960
6000	3	0.7456	-5.024	-14889	2.398	1.806	1.253	0.368	0.408	676	1.970
6000	3	0.7429	-4.992	-14925	2.404	1.807	1.252	0.368	0.407	677	1.980
6000	3	0.7401	-4.960	-14961	2.411	1.807	1.252	0.367	0.406	679	1.990
6000	3	0.7374	-4.927	-14996	2.417	1.808	1.252	0.367	0.405	680	2.000
6000	3	0.7347	-4.895	-15032	2.424	1.809	1.252	0.367	0.404	681	2.010
6000	3	0.7320	-4.862	-15067	2.430	1.809	1.251	0.367	0.404	683	2.020
6000	3	0.7293	-4.829	-15103	2.437	1.810	1.251	0.367	0.403	684	2.030
6000	3	0.7267	-4.795	-15138	2.443	1.811	1.251	0.367	0.402	685	2.040
6000	3	0.7240	-4.762	-15173	2.450	1.811	1.251	0.367	0.401	686	2.050
6000	3	0.7214	-4.728	-15209	2.456	1.812	1.251	0.367	0.400	688	2.060
6000	3	0.7163	-4.660	-15279	2.469	1.813	1.250	0.366	0.398	690	2.080
6000	3	0.7112	-4.591	-15349	2.482	1.814	1.250	0.366	0.397	693	2.100
6000	3	0.7049	-4.504	-15437	2.498	1.816	1.250	0.366	0.395	696	2.125
6000	3	0.6988	-4.416	-15524	2.514	1.817	1.249	0.365	0.393	699	2.150
6000	3	0.6869	-4.237	-15698	2.546	1.820	1.248	0.365	0.389	706	2.200
6000	3	0.6753	-4.055	-15870	2.577	1.823	1.248	0.364	0.385	712	2.250

Table A.30. Gas Phase (continued)

T K	P bar	ρ $kg\ m^{-3}$	H $10^2 kJ\ kg^{-1}$	G $kJ\ kg^{-1}$	S	C_P $kJ\ kg^{-1}\ K^{-1}$	C_P/C_V	β_T bar^{-1}	α_T $10^{-3} K^{-1}$	V_S $m\ s^{-1}$	O/U
6000	10	3.080	-1.254	-13675	2.070	1.647	1.252	0.111	0.435	605	1.750
6000	10	3.014	-1.245	-13849	2.101	1.674	1.252	0.111	0.434	612	1.800
6000	10	2.950	-1.235	-14023	2.131	1.700	1.252	0.111	0.432	619	1.850
6000	10	2.919	-1.230	-14109	2.147	1.713	1.253	0.111	0.432	622	1.875
6000	10	2.888	-1.224	-14195	2.162	1.725	1.253	0.111	0.431	625	1.900
6000	10	2.864	-1.219	-14263	2.174	1.735	1.253	0.111	0.431	628	1.920
6000	10	2.841	-1.215	-14331	2.186	1.744	1.253	0.111	0.430	631	1.940
6000	10	2.829	-1.212	-14365	2.192	1.749	1.253	0.111	0.430	632	1.950
6000	10	2.818	-1.210	-14399	2.198	1.754	1.253	0.111	0.429	633	1.960
6000	10	2.806	-1.207	-14433	2.204	1.758	1.253	0.111	0.429	635	1.970
6000	10	2.795	-1.205	-14467	2.210	1.763	1.253	0.111	0.429	636	1.980
6000	10	2.784	-1.202	-14501	2.216	1.767	1.253	0.111	0.428	637	1.990
6000	10	2.772	-1.199	-14535	2.223	1.772	1.253	0.111	0.428	639	2.000
6000	10	2.761	-1.197	-14568	2.229	1.776	1.253	0.111	0.428	640	2.010
6000	10	2.750	-1.194	-14602	2.235	1.781	1.253	0.111	0.427	641	2.020
6000	10	2.739	-1.192	-14636	2.241	1.785	1.253	0.111	0.427	643	2.030
6000	10	2.729	-1.189	-14669	2.247	1.789	1.253	0.111	0.427	644	2.040
6000	10	2.718	-1.186	-14703	2.253	1.793	1.253	0.111	0.426	645	2.050
6000	10	2.707	-1.183	-14736	2.259	1.798	1.253	0.111	0.426	647	2.060
6000	10	2.686	-1.178	-14803	2.271	1.806	1.253	0.111	0.425	649	2.080
6000	10	2.665	-1.172	-14870	2.283	1.814	1.253	0.111	0.425	652	2.100
6000	10	2.640	-1.165	-14953	2.298	1.824	1.253	0.111	0.424	655	2.125
6000	10	2.615	-1.157	-15035	2.313	1.834	1.253	0.111	0.423	658	2.150
6000	10	2.566	-1.142	-15200	2.343	1.853	1.253	0.111	0.421	665	2.200
6000	10	2.520	-1.127	-15364	2.373	1.871	1.253	0.110	0.419	671	2.250

Table A.30. Gas Phase (continued)

T	P	ρ	H	G	S	C_P	C_P/C_V	β_T	α_T	V_S	O/U
K	bar	$kg\ m^{-3}$	$10^3 kJ\ kg^{-1}$	$kJ\ kg^{-1}$		$kJ\ kg^{-1}\ K^{-1}$		$10^{-1} bar^{-1}$	$10^{-3} K^{-1}$	$m\ s^{-1}$	
6000	30	10.660	-1.898	-13341	1.907	1.863	1.286	0.385	0.532	560	1.750
6000	30	10.440	-1.907	-13508	1.934	1.924	1.289	0.385	0.538	566	1.800
6000	30	10.220	-1.913	-13674	1.960	1.981	1.292	0.386	0.543	572	1.850
6000	30	10.120	-1.916	-13757	1.973	2.009	1.294	0.386	0.545	575	1.875
6000	30	10.010	-1.918	-13839	1.987	2.036	1.295	0.387	0.547	578	1.900
6000	30	9.932	-1.920	-13904	1.997	2.057	1.296	0.387	0.549	581	1.920
6000	30	9.851	-1.921	-13969	2.008	2.078	1.297	0.387	0.550	583	1.940
6000	30	9.811	-1.921	-14002	2.013	2.088	1.298	0.387	0.550	585	1.950
6000	30	9.772	-1.922	-14034	2.019	2.098	1.298	0.387	0.551	586	1.960
6000	30	9.732	-1.922	-14067	2.024	2.108	1.298	0.387	0.552	587	1.970
6000	30	9.693	-1.923	-14099	2.029	2.117	1.299	0.387	0.552	588	1.980
6000	30	9.654	-1.923	-14132	2.035	2.127	1.299	0.387	0.553	589	1.990
6000	30	9.615	-1.923	-14164	2.040	2.137	1.300	0.387	0.553	591	2.000
6000	30	9.577	-1.923	-14196	2.045	2.146	1.300	0.388	0.553	592	2.010
6000	30	9.539	-1.923	-14228	2.051	2.156	1.300	0.388	0.554	593	2.020
6000	30	9.501	-1.923	-14260	2.056	2.165	1.301	0.388	0.554	594	2.030
6000	30	9.463	-1.923	-14292	2.062	2.174	1.301	0.388	0.555	596	2.040
6000	30	9.425	-1.923	-14325	2.067	2.183	1.301	0.388	0.555	597	2.050
6000	30	9.388	-1.923	-14356	2.072	2.192	1.302	0.388	0.555	598	2.060
6000	30	9.315	-1.923	-14420	2.083	2.210	1.302	0.388	0.556	600	2.080
6000	30	9.242	-1.922	-14484	2.094	2.227	1.303	0.388	0.556	603	2.100
6000	30	9.152	-1.921	-14563	2.107	2.248	1.303	0.388	0.557	606	2.125
6000	30	9.064	-1.920	-14642	2.120	2.269	1.304	0.388	0.557	609	2.150
6000	30	8.892	-1.917	-14799	2.147	2.308	1.305	0.388	0.557	615	2.200
6000	30	8.726	-1.912	-14955	2.174	2.345	1.306	0.388	0.557	621	2.250

Table A.30. Gas Phase (continued)

T K	P bar	ρ $kg\ m^{-3}$	H $10^3 kJ\ kg^{-1}$	G $kJ\ kg^{-1}$	S	C_P $kJ\ kg^{-1}\ K^{-1}$	C_P/C_V	β_T bar^{-1}	α_T $10^{-3} K^{-1}$	V_S $m\ s^{-1}$	O/U
7000	1	0.1525	2.807	-17204	2.859	1.972	1.294	1.099	0.327	879	1.750
7000	1	0.1504	2.828	-17430	2.894	1.980	1.293	1.098	0.326	885	1.800
7000	1	0.1484	2.848	-17655	2.929	1.988	1.293	1.097	0.324	891	1.850
7000	1	0.1474	2.858	-17767	2.946	1.992	1.292	1.096	0.323	894	1.875
7000	1	0.1464	2.868	-17879	2.964	1.995	1.292	1.096	0.322	897	1.900
7000	1	0.1457	2.877	-17967	2.978	1.998	1.292	1.095	0.321	900	1.920
7000	1	0.1449	2.885	-18056	2.992	2.001	1.292	1.095	0.320	902	1.940
7000	1	0.1445	2.889	-18100	2.998	2.003	1.292	1.095	0.320	903	1.950
7000	1	0.1442	2.893	-18145	3.005	2.004	1.291	1.095	0.319	905	1.960
7000	1	0.1438	2.897	-18189	3.012	2.005	1.291	1.094	0.319	906	1.970
7000	1	0.1434	2.901	-18233	3.019	2.007	1.291	1.094	0.319	907	1.980
7000	1	0.1431	2.905	-18277	3.026	2.008	1.291	1.094	0.318	908	1.990
7000	1	0.1427	2.909	-18321	3.033	2.009	1.291	1.094	0.318	909	2.000
7000	1	0.1423	2.913	-18365	3.040	2.011	1.291	1.094	0.317	911	2.010
7000	1	0.1420	2.917	-18409	3.047	2.012	1.291	1.093	0.317	912	2.020
7000	1	0.1416	2.921	-18452	3.053	2.013	1.291	1.093	0.317	913	2.030
7000	1	0.1413	2.925	-18496	3.060	2.014	1.291	1.093	0.316	914	2.040
7000	1	0.1409	2.929	-18540	3.067	2.016	1.290	1.093	0.316	915	2.050
7000	1	0.1406	2.933	-18583	3.074	2.017	1.290	1.093	0.316	917	2.060
7000	1	0.1399	2.941	-18670	3.087	2.019	1.290	1.092	0.315	919	2.080
7000	1	0.1392	2.950	-18757	3.101	2.022	1.290	1.092	0.314	921	2.100
7000	1	0.1384	2.960	-18866	3.118	2.025	1.290	1.091	0.313	924	2.125
7000	1	0.1375	2.970	-18974	3.135	2.027	1.289	1.091	0.312	927	2.150
7000	1	0.1359	2.990	-19188	3.168	2.033	1.289	1.090	0.311	933	2.200
7000	1	0.1343	3.010	-19402	3.202	2.038	1.288	1.089	0.309	939	2.250

Table A.30. Gas Phase (continued)

T K	P bar	ρ $kg\ m^{-3}$	H $10^3 kJ\ kg^{-1}$	G $kJ\ kg^{-1}$	S	C_P $kJ\ kg^{-1}\ K^{-1}$	C_P/C_V	β_T bar^{-1}	α_T $10^{-3} K^{-1}$	V_S $m\ s^{-1}$	O/U
7000	3	0.5192	1.772	-16525	2.614	2.326	1.312	0.377	0.393	819	1.750
7000	3	0.5114	1.788	-16742	2.647	2.335	1.311	0.376	0.390	825	1.800
7000	3	0.5038	1.806	-16957	2.680	2.344	1.310	0.376	0.387	832	1.850
7000	3	0.5001	1.814	-17064	2.697	2.349	1.309	0.376	0.386	835	1.875
7000	3	0.4965	1.823	-17170	2.713	2.353	1.309	0.375	0.385	838	1.900
7000	3	0.4937	1.830	-17255	2.726	2.356	1.309	0.375	0.383	841	1.920
7000	3	0.4908	1.837	-17340	2.740	2.359	1.308	0.375	0.382	843	1.940
7000	3	0.4894	1.841	-17383	2.746	2.361	1.308	0.375	0.382	844	1.950
7000	3	0.4880	1.844	-17425	2.753	2.362	1.308	0.375	0.381	846	1.960
7000	3	0.4867	1.848	-17467	2.759	2.364	1.307	0.375	0.380	847	1.970
7000	3	0.4853	1.851	-17509	2.766	2.365	1.307	0.375	0.380	848	1.980
7000	3	0.4839	1.855	-17551	2.772	2.366	1.307	0.374	0.379	849	1.990
7000	3	0.4826	1.858	-17593	2.779	2.368	1.307	0.374	0.379	851	2.000
7000	3	0.4812	1.862	-17635	2.785	2.369	1.307	0.374	0.378	852	2.010
7000	3	0.4799	1.866	-17677	2.792	2.371	1.306	0.374	0.378	853	2.020
7000	3	0.4786	1.869	-17719	2.798	2.372	1.306	0.374	0.377	854	2.030
7000	3	0.4773	1.873	-17761	2.805	2.373	1.306	0.374	0.377	856	2.040
7000	3	0.4759	1.876	-17803	2.811	2.374	1.306	0.374	0.376	857	2.050
7000	3	0.4746	1.880	-17844	2.818	2.376	1.306	0.374	0.375	858	2.060
7000	3	0.4721	1.887	-17928	2.831	2.378	1.305	0.374	0.374	860	2.080
7000	3	0.4695	1.895	-18011	2.844	2.381	1.305	0.373	0.373	863	2.100
7000	3	0.4664	1.904	-18114	2.860	2.384	1.304	0.373	0.372	866	2.125
7000	3	0.4633	1.913	-18217	2.876	2.386	1.304	0.373	0.370	869	2.150
7000	3	0.4572	1.932	-18423	2.908	2.392	1.303	0.372	0.368	875	2.200
7000	3	0.4513	1.950	-18627	2.940	2.397	1.302	0.372	0.365	881	2.250

Table A.30. Gas Phase (continued)

T K	P bar	ρ $kg\ m^{-3}$	H $10^3 kJ\ kg^{-1}$	G $kJ\ kg^{-1}$	S	C_P $kJ\ kg^{-1}\ K^{-1}$	C_P/C_V	β_T bar^{-1}	α_T $10^{-3} K^{-1}$	V_S $m\ s^{-1}$	O/U
7000	10	2.040	5.847	-15884	2.353	2.109	1.308	0.114	0.406	751	1.750
7000	10	2.005	5.990	-16090	2.384	2.112	1.307	0.113	0.401	758	1.800
7000	10	1.971	6.141	-16294	2.415	2.114	1.305	0.113	0.397	765	1.850
7000	10	1.954	6.219	-16396	2.431	2.115	1.304	0.113	0.395	768	1.875
7000	10	1.938	6.299	-16497	2.447	2.116	1.303	0.113	0.392	772	1.900
7000	10	1.925	6.363	-16578	2.459	2.117	1.303	0.113	0.391	774	1.920
7000	10	1.913	6.429	-16659	2.472	2.118	1.302	0.113	0.389	777	1.940
7000	10	1.906	6.462	-16699	2.478	2.118	1.302	0.113	0.388	778	1.950
7000	10	1.900	6.495	-16739	2.484	2.118	1.301	0.113	0.387	780	1.960
7000	10	1.894	6.529	-16779	2.490	2.118	1.301	0.113	0.386	781	1.970
7000	10	1.888	6.563	-16819	2.497	2.118	1.301	0.113	0.386	782	1.980
7000	10	1.882	6.597	-16859	2.503	2.119	1.300	0.113	0.385	784	1.990
7000	10	1.875	6.631	-16899	2.509	2.119	1.300	0.113	0.384	785	2.000
7000	10	1.869	6.665	-16939	2.515	2.119	1.300	0.112	0.383	786	2.010
7000	10	1.863	6.700	-16979	2.521	2.119	1.300	0.112	0.382	788	2.020
7000	10	1.857	6.734	-17019	2.527	2.119	1.299	0.112	0.382	789	2.030
7000	10	1.852	6.769	-17059	2.534	2.119	1.299	0.112	0.381	790	2.040
7000	10	1.846	6.804	-17098	2.540	2.120	1.299	0.112	0.380	792	2.050
7000	10	1.840	6.839	-17138	2.546	2.120	1.298	0.112	0.379	793	2.060
7000	10	1.828	6.910	-17217	2.558	2.120	1.298	0.112	0.378	796	2.080
7000	10	1.817	6.982	-17296	2.571	2.120	1.297	0.112	0.376	798	2.100
7000	10	1.803	7.072	-17394	2.586	2.120	1.297	0.112	0.374	801	2.125
7000	10	1.789	7.164	-17492	2.601	2.120	1.296	0.112	0.372	805	2.150
7000	10	1.762	7.349	-17687	2.632	2.120	1.295	0.112	0.368	811	2.200
7000	10	1.736	7.537	-17881	2.662	2.120	1.293	0.112	0.365	817	2.250

Table A.30. Gas Phase (continued)

T K	P bar	ρ $kg\ m^{-3}$	H $10^2 kJ\ kg^{-1}$	G $kJ\ kg^{-1}$	S	C_P $kJ\ kg^{-1}\ K^{-1}$	C_P/C_V	β_T $10^{-1} bar^{-1}$	α_T $10^{-3} K^{-1}$	V_S $m\ s^{-1}$	O/U
7000	30	7.028	-2.658	-15382	2.159	1.605	1.280	0.372	0.362	700	1.750
7000	30	6.889	-2.508	-15578	2.190	1.607	1.278	0.371	0.358	707	1.800
7000	30	6.755	-2.350	-15773	2.220	1.609	1.277	0.371	0.353	714	1.850
7000	30	6.690	-2.268	-15870	2.235	1.610	1.277	0.370	0.351	718	1.875
7000	30	6.626	-2.184	-15967	2.250	1.611	1.276	0.370	0.349	722	1.900
7000	30	6.576	-2.115	-16044	2.262	1.612	1.276	0.370	0.348	724	1.920
7000	30	6.526	-2.045	-16121	2.274	1.613	1.275	0.369	0.346	727	1.940
7000	30	6.502	-2.010	-16159	2.280	1.613	1.275	0.369	0.345	729	1.950
7000	30	6.478	-1.974	-16197	2.286	1.613	1.275	0.369	0.345	730	1.960
7000	30	6.454	-1.939	-16236	2.292	1.614	1.275	0.369	0.344	732	1.970
7000	30	6.430	-1.903	-16274	2.298	1.614	1.274	0.369	0.343	733	1.980
7000	30	6.406	-1.866	-16312	2.304	1.614	1.274	0.369	0.342	735	1.990
7000	30	6.382	-1.830	-16350	2.310	1.615	1.274	0.369	0.342	736	2.000
7000	30	6.359	-1.793	-16388	2.316	1.615	1.274	0.368	0.341	737	2.010
7000	30	6.336	-1.756	-16426	2.321	1.615	1.274	0.368	0.340	739	2.020
7000	30	6.313	-1.719	-16464	2.327	1.615	1.273	0.368	0.339	740	2.030
7000	30	6.290	-1.682	-16502	2.333	1.616	1.273	0.368	0.339	742	2.040
7000	30	6.267	-1.644	-16539	2.339	1.616	1.273	0.368	0.338	743	2.050
7000	30	6.244	-1.607	-16577	2.345	1.616	1.273	0.368	0.337	744	2.060
7000	30	6.200	-1.530	-16652	2.357	1.617	1.273	0.368	0.336	747	2.080
7000	30	6.156	-1.454	-16728	2.369	1.618	1.272	0.367	0.334	750	2.100
7000	30	6.102	-1.357	-16821	2.384	1.618	1.272	0.367	0.333	754	2.125
7000	30	6.049	-1.259	-16914	2.398	1.619	1.271	0.367	0.331	757	2.150
7000	30	5.946	-1.060	-17100	2.428	1.621	1.271	0.366	0.328	764	2.200
7000	30	5.846	-0.857	-17284	2.457	1.622	1.270	0.366	0.325	771	2.250

Table A.30. Gas Phase (continued)

T K	P bar	ρ $kg\ m^{-3}$	H $10^2 kJ\ kg^{-1}$	G $kJ\ kg^{-1}$	S	C_P $kJ\ kg^{-1}\ K^{-1}$	C_P/C_V	β_T $10^{-1} bar^{-1}$	α_T $10^{-3} K^{-1}$	V_S $m\ s^{-1}$	O/U
7000	100	27.01	-9.856	-14902	1.988	1.441	1.284	0.114	0.374	647	1.750
7000	100	26.43	-9.760	-15088	2.016	1.463	1.285	0.113	0.373	655	1.800
7000	100	25.87	-9.652	-15273	2.044	1.485	1.285	0.113	0.372	662	1.850
7000	100	25.60	-9.594	-15365	2.058	1.495	1.286	0.113	0.371	666	1.875
7000	100	25.33	-9.534	-15456	2.072	1.505	1.286	0.113	0.370	669	1.900
7000	100	25.12	-9.484	-15529	2.083	1.513	1.286	0.113	0.370	672	1.920
7000	100	24.91	-9.433	-15602	2.094	1.521	1.286	0.113	0.369	675	1.940
7000	100	24.81	-9.407	-15638	2.100	1.525	1.287	0.113	0.369	677	1.950
7000	100	24.71	-9.380	-15675	2.105	1.529	1.287	0.113	0.369	678	1.960
7000	100	24.61	-9.354	-15711	2.111	1.533	1.287	0.113	0.369	680	1.970
7000	100	24.51	-9.327	-15747	2.116	1.537	1.287	0.113	0.368	681	1.980
7000	100	24.41	-9.299	-15783	2.122	1.540	1.287	0.113	0.368	683	1.990
7000	100	24.31	-9.272	-15819	2.127	1.544	1.287	0.113	0.368	684	2.000
7000	100	24.22	-9.244	-15855	2.133	1.548	1.287	0.113	0.368	686	2.010
7000	100	24.12	-9.216	-15891	2.138	1.552	1.287	0.113	0.367	687	2.020
7000	100	24.03	-9.187	-15927	2.144	1.555	1.287	0.113	0.367	688	2.030
7000	100	23.93	-9.159	-15962	2.149	1.559	1.287	0.113	0.367	690	2.040
7000	100	23.84	-9.130	-15998	2.155	1.562	1.287	0.113	0.366	691	2.050
7000	100	23.74	-9.101	-16034	2.161	1.566	1.287	0.113	0.366	693	2.060
7000	100	23.56	-9.041	-16105	2.172	1.573	1.288	0.113	0.366	696	2.080
7000	100	23.38	-8.981	-16176	2.183	1.580	1.288	0.113	0.365	698	2.100
7000	100	23.15	-8.905	-16265	2.196	1.589	1.288	0.113	0.364	702	2.125
7000	100	22.93	-8.827	-16353	2.210	1.597	1.288	0.113	0.363	706	2.150
7000	100	22.51	-8.668	-16528	2.237	1.613	1.288	0.113	0.362	713	2.200
7000	100	22.10	-8.504	-16702	2.265	1.629	1.289	0.113	0.360	719	2.250

Table A.30. Gas Phase (continued)

T K	P bar	ρ $kg\ m^{-3}$	H $10^2 kJ\ kg^{-1}$	G $kJ\ kg^{-1}$	S	C_P $kJ\ kg^{-1}\ K^{-1}$	C_P/C_V	β_T $10^{-2} bar^{-1}$	α_T $10^{-3} K^{-1}$	V_S $m\ s^{-1}$	O/U
7000	300	99.90	-1.736	-14530	1.828	1.612	1.374	0.426	0.517	568	1.750
7000	300	97.77	-1.747	-14708	1.852	1.664	1.379	0.427	0.522	575	1.800
7000	300	95.70	-1.757	-14885	1.875	1.713	1.382	0.427	0.526	582	1.850
7000	300	94.68	-1.761	-14973	1.887	1.737	1.384	0.427	0.527	585	1.875
7000	300	93.68	-1.764	-15060	1.899	1.760	1.385	0.427	0.529	589	1.900
7000	300	92.89	-1.766	-15129	1.909	1.778	1.386	0.427	0.530	592	1.920
7000	300	92.11	-1.769	-15199	1.919	1.796	1.387	0.426	0.530	594	1.940
7000	300	91.72	-1.769	-15233	1.923	1.804	1.388	0.426	0.531	596	1.950
7000	300	91.33	-1.770	-15268	1.928	1.813	1.388	0.426	0.531	597	1.960
7000	300	90.95	-1.771	-15302	1.933	1.821	1.388	0.426	0.531	598	1.970
7000	300	90.57	-1.772	-15337	1.938	1.830	1.389	0.426	0.532	600	1.980
7000	300	90.19	-1.773	-15371	1.943	1.838	1.389	0.426	0.532	601	1.990
7000	300	89.82	-1.773	-15406	1.947	1.846	1.390	0.426	0.532	603	2.000
7000	300	89.44	-1.774	-15440	1.952	1.854	1.390	0.426	0.532	604	2.010
7000	300	89.07	-1.774	-15474	1.957	1.862	1.390	0.426	0.532	605	2.020
7000	300	88.70	-1.775	-15508	1.962	1.870	1.390	0.426	0.532	607	2.030
7000	300	88.34	-1.775	-15542	1.967	1.878	1.391	0.426	0.532	608	2.040
7000	300	87.97	-1.775	-15576	1.972	1.886	1.391	0.426	0.532	610	2.050
7000	300	87.61	-1.776	-15610	1.976	1.894	1.391	0.425	0.533	611	2.060
7000	300	86.89	-1.776	-15678	1.986	1.909	1.392	0.425	0.533	614	2.080
7000	300	86.19	-1.776	-15746	1.996	1.923	1.392	0.425	0.532	617	2.100
7000	300	85.32	-1.776	-15830	2.008	1.941	1.392	0.425	0.532	620	2.125
7000	300	84.46	-1.775	-15914	2.020	1.959	1.393	0.424	0.532	623	2.150
7000	300	82.79	-1.773	-16081	2.044	1.992	1.393	0.424	0.531	630	2.200
7000	300	81.18	-1.770	-16246	2.068	2.024	1.393	0.423	0.529	637	2.250

Table A.30. Gas Phase (continued)

T K	P bar	ρ $kg\ m^{-3}$	H $10^3 kJ\ kg^{-1}$	G $kJ\ kg^{-1}$	S	C_P $kJ\ kg^{-1}\ K^{-1}$	C_P/C_V	β_T bar^{-1}	α_T $10^{-3} K^{-1}$	V_S $m\ s^{-1}$	O/U
8000	1	0.1182	4.270	-20173	3.055	1.084	1.317	1.057	0.202	1026	1.750
8000	1	0.1167	4.295	-20435	3.091	1.085	1.317	1.057	0.201	1033	1.800
8000	1	0.1154	4.319	-20695	3.127	1.086	1.317	1.056	0.200	1040	1.850
8000	1	0.1147	4.331	-20825	3.144	1.086	1.317	1.055	0.199	1043	1.875
8000	1	0.1140	4.343	-20954	3.162	1.087	1.317	1.055	0.198	1047	1.900
8000	1	0.1135	4.353	-21057	3.176	1.087	1.317	1.054	0.198	1049	1.920
8000	1	0.1129	4.362	-21159	3.190	1.088	1.317	1.054	0.198	1052	1.940
8000	1	0.1127	4.367	-21210	3.197	1.088	1.317	1.054	0.197	1053	1.950
8000	1	0.1124	4.372	-21262	3.204	1.088	1.317	1.054	0.197	1054	1.960
8000	1	0.1122	4.376	-21313	3.211	1.088	1.317	1.054	0.197	1056	1.970
8000	1	0.1119	4.381	-21364	3.218	1.088	1.317	1.054	0.197	1057	1.980
8000	1	0.1117	4.386	-21415	3.225	1.088	1.317	1.053	0.196	1058	1.990
8000	1	0.1114	4.390	-21466	3.232	1.089	1.318	1.053	0.196	1060	2.000
8000	1	0.1112	4.395	-21516	3.239	1.089	1.318	1.053	0.196	1061	2.010
8000	1	0.1109	4.400	-21567	3.246	1.089	1.318	1.053	0.196	1062	2.020
8000	1	0.1107	4.405	-21618	3.253	1.089	1.318	1.053	0.196	1063	2.030
8000	1	0.1104	4.409	-21668	3.260	1.089	1.318	1.053	0.195	1065	2.040
8000	1	0.1102	4.414	-21719	3.267	1.089	1.318	1.052	0.195	1066	2.050
8000	1	0.1099	4.419	-21769	3.273	1.090	1.318	1.052	0.195	1067	2.060
8000	1	0.1095	4.428	-21870	3.287	1.090	1.318	1.052	0.195	1070	2.080
8000	1	0.1090	4.437	-21971	3.301	1.090	1.318	1.052	0.194	1072	2.100
8000	1	0.1084	4.449	-22096	3.318	1.091	1.318	1.051	0.194	1076	2.125
8000	1	0.1078	4.460	-22221	3.335	1.091	1.318	1.051	0.193	1079	2.150
8000	1	0.1067	4.483	-22470	3.369	1.092	1.319	1.050	0.192	1085	2.200
8000	1	0.1056	4.506	-22717	3.403	1.092	1.319	1.049	0.191	1091	2.250

Table A.30. Gas Phase (continued)

T K	P bar	ρ $kg\ m^{-3}$	H $10^3 kJ\ kg^{-1}$	G $kJ\ kg^{-1}$	S $kJ\ kg^{-1}\ K^{-1}$	C_P	C_P/C_V	β_T bar^{-1}	α_T $10^{-3} K^{-1}$	V_S $m\ s^{-1}$	O/U
8000	3	0.3821	3.600	-19276	2.859	1.372	1.311	0.360	0.237	977	1.750
8000	3	0.3771	3.625	-19526	2.894	1.377	1.311	0.360	0.235	983	1.800
8000	3	0.3722	3.650	-19775	2.928	1.382	1.311	0.359	0.234	990	1.850
8000	3	0.3699	3.662	-19899	2.945	1.384	1.311	0.359	0.233	993	1.875
8000	3	0.3675	3.674	-20022	2.962	1.386	1.311	0.359	0.233	997	1.900
8000	3	0.3657	3.684	-20120	2.976	1.388	1.311	0.359	0.232	999	1.920
8000	3	0.3639	3.694	-20219	2.989	1.390	1.311	0.359	0.232	1002	1.940
8000	3	0.3629	3.699	-20267	2.996	1.391	1.310	0.359	0.232	1003	1.950
8000	3	0.3621	3.704	-20316	3.003	1.392	1.310	0.359	0.231	1005	1.960
8000	3	0.3612	3.709	-20365	3.009	1.392	1.310	0.359	0.231	1006	1.970
8000	3	0.3603	3.714	-20414	3.016	1.393	1.310	0.358	0.231	1007	1.980
8000	3	0.3594	3.718	-20463	3.023	1.394	1.310	0.358	0.231	1009	1.990
8000	3	0.3585	3.723	-20511	3.029	1.395	1.310	0.358	0.230	1010	2.000
8000	3	0.3576	3.728	-20560	3.036	1.396	1.310	0.358	0.230	1011	2.010
8000	3	0.3568	3.733	-20608	3.043	1.397	1.310	0.358	0.230	1013	2.020
8000	3	0.3559	3.738	-20657	3.049	1.397	1.310	0.358	0.230	1014	2.030
8000	3	0.3551	3.743	-20705	3.056	1.398	1.310	0.358	0.229	1015	2.040
8000	3	0.3542	3.748	-20754	3.063	1.399	1.310	0.358	0.229	1016	2.050
8000	3	0.3534	3.752	-20802	3.069	1.400	1.310	0.358	0.229	1018	2.060
8000	3	0.3517	3.762	-20898	3.083	1.401	1.310	0.358	0.228	1020	2.080
8000	3	0.3500	3.772	-20994	3.096	1.403	1.310	0.358	0.228	1023	2.100
8000	3	0.3480	3.784	-21114	3.112	1.405	1.310	0.358	0.227	1026	2.125
8000	3	0.3460	3.796	-21234	3.129	1.407	1.310	0.358	0.227	1029	2.150
8000	3	0.3421	3.819	-21471	3.161	1.410	1.310	0.357	0.226	1035	2.200
8000	3	0.3382	3.843	-21708	3.194	1.414	1.310	0.357	0.225	1042	2.250

Table A.30. Gas Phase (continued)

T K	P bar	ρ $kg\ m^{-3}$	H $10^3 kJ\ kg^{-1}$	G $kJ\ kg^{-1}$	S	C_P $kJ\ kg^{-1}\ K^{-1}$	C_P/C_V	β_T bar^{-1}	α_T $10^{-3} K^{-1}$	V_S $m\ s^{-1}$	O/U
8000	10	1.428	2.620	-18380	2.625	1.748	1.322	0.111	0.291	912	1.750
8000	10	1.408	2.642	-18618	2.657	1.756	1.321	0.111	0.289	919	1.800
8000	10	1.388	2.664	-18854	2.690	1.765	1.321	0.111	0.287	926	1.850
8000	10	1.378	2.675	-18972	2.706	1.769	1.321	0.111	0.287	929	1.875
8000	10	1.369	2.686	-19089	2.722	1.773	1.320	0.111	0.286	933	1.900
8000	10	1.361	2.695	-19182	2.735	1.776	1.320	0.111	0.285	935	1.920
8000	10	1.354	2.704	-19275	2.747	1.779	1.320	0.111	0.284	938	1.940
8000	10	1.350	2.708	-19322	2.754	1.780	1.320	0.111	0.284	939	1.950
8000	10	1.346	2.713	-19368	2.760	1.782	1.319	0.111	0.284	941	1.960
8000	10	1.343	2.717	-19415	2.766	1.783	1.319	0.111	0.283	942	1.970
8000	10	1.339	2.722	-19461	2.773	1.784	1.319	0.111	0.283	943	1.980
8000	10	1.336	2.726	-19507	2.779	1.786	1.319	0.111	0.283	945	1.990
8000	10	1.332	2.730	-19554	2.785	1.787	1.319	0.111	0.282	946	2.000
8000	10	1.328	2.735	-19600	2.792	1.789	1.319	0.111	0.282	947	2.010
8000	10	1.325	2.739	-19646	2.798	1.790	1.319	0.111	0.282	949	2.020
8000	10	1.321	2.744	-19692	2.804	1.791	1.319	0.111	0.281	950	2.030
8000	10	1.318	2.748	-19738	2.811	1.793	1.319	0.111	0.281	951	2.040
8000	10	1.314	2.753	-19784	2.817	1.794	1.318	0.111	0.281	952	2.050
8000	10	1.311	2.757	-19830	2.823	1.795	1.318	0.111	0.280	954	2.060
8000	10	1.304	2.766	-19921	2.836	1.798	1.318	0.111	0.280	956	2.080
8000	10	1.297	2.775	-20012	2.848	1.800	1.318	0.111	0.279	959	2.100
8000	10	1.289	2.786	-20126	2.864	1.804	1.318	0.110	0.278	962	2.125
8000	10	1.281	2.797	-20240	2.880	1.807	1.317	0.110	0.277	965	2.150
8000	10	1.265	2.820	-20465	2.911	1.812	1.317	0.110	0.276	972	2.200
8000	10	1.250	2.842	-20690	2.941	1.818	1.316	0.110	0.274	978	2.250

Table A.30. Gas Phase (continued)

T K	P bar	ρ $kg\ m^{-3}$	H $10^3 kJ\ kg^{-1}$	G $kJ\ kg^{-1}$	S $kJ\ kg^{-1}\ K^{-1}$	C_P	C_P/C_V	β_T $10^{-1} bar^{-1}$	α_T $10^{-3} K^{-1}$	V_S $m\ s^{-1}$	O/U
8000	30	4.929	1.533	-17660	2.399	1.873	1.334	0.379	0.331	845	1.750
8000	30	4.850	1.551	-17887	2.430	1.879	1.333	0.379	0.328	852	1.800
8000	30	4.774	1.570	-18112	2.460	1.886	1.332	0.378	0.326	859	1.850
8000	30	4.736	1.579	-18224	2.475	1.889	1.331	0.378	0.324	862	1.875
8000	30	4.700	1.589	-18336	2.491	1.892	1.331	0.378	0.323	866	1.900
8000	30	4.671	1.597	-18425	2.503	1.894	1.330	0.377	0.322	869	1.920
8000	30	4.643	1.604	-18514	2.515	1.896	1.330	0.377	0.321	871	1.940
8000	30	4.628	1.608	-18558	2.521	1.897	1.329	0.377	0.320	873	1.950
8000	30	4.614	1.612	-18603	2.527	1.898	1.329	0.377	0.320	874	1.960
8000	30	4.601	1.616	-18647	2.533	1.899	1.329	0.377	0.319	876	1.970
8000	30	4.587	1.620	-18691	2.539	1.900	1.329	0.377	0.319	877	1.980
8000	30	4.573	1.624	-18735	2.545	1.901	1.328	0.377	0.318	878	1.990
8000	30	4.559	1.628	-18779	2.551	1.902	1.328	0.376	0.318	880	2.000
8000	30	4.546	1.632	-18823	2.557	1.903	1.328	0.376	0.317	881	2.010
8000	30	4.532	1.636	-18867	2.563	1.903	1.328	0.376	0.317	882	2.020
8000	30	4.519	1.640	-18911	2.569	1.904	1.328	0.376	0.316	884	2.030
8000	30	4.506	1.644	-18955	2.575	1.905	1.327	0.376	0.315	885	2.040
8000	30	4.492	1.648	-18998	2.581	1.906	1.327	0.376	0.315	887	2.050
8000	30	4.479	1.652	-19042	2.587	1.907	1.327	0.376	0.314	888	2.060
8000	30	4.453	1.660	-19129	2.599	1.909	1.326	0.376	0.313	891	2.080
8000	30	4.428	1.668	-19216	2.611	1.910	1.326	0.375	0.312	893	2.100
8000	30	4.396	1.679	-19325	2.625	1.912	1.325	0.375	0.311	897	2.125
8000	30	4.365	1.689	-19433	2.640	1.914	1.325	0.375	0.310	900	2.150
8000	30	4.304	1.710	-19648	2.670	1.917	1.324	0.374	0.307	907	2.200
8000	30	4.245	1.731	-19861	2.699	1.920	1.323	0.374	0.305	913	2.250

Table A.30. Gas Phase (continued)

T K	P bar	ρ $kg\ m^{-3}$	H $10^3 kJ\ kg^{-1}$	G $kJ\ kg^{-1}$	S	C_P $kJ\ kg^{-1}\ K^{-1}$	C_P/C_V	β_T $10^{-1} bar^{-1}$	α_T $10^{-3} K^{-1}$	V_S $m\ s^{-1}$	O/U
8000	100	19.40	4.352	-16986	2.178	1.526	1.313	0.113	0.316	773	1.750
8000	100	19.04	4.522	-17201	2.207	1.528	1.312	0.113	0.313	781	1.800
8000	100	18.69	4.701	-17414	2.236	1.529	1.311	0.113	0.309	788	1.850
8000	100	18.52	4.793	-17520	2.250	1.530	1.310	0.113	0.307	792	1.875
8000	100	18.35	4.887	-17626	2.264	1.531	1.309	0.113	0.306	796	1.900
8000	100	18.22	4.964	-17710	2.276	1.531	1.309	0.113	0.304	799	1.920
8000	100	18.09	5.041	-17794	2.287	1.532	1.308	0.112	0.303	802	1.940
8000	100	18.03	5.080	-17836	2.293	1.532	1.308	0.112	0.302	803	1.950
8000	100	17.97	5.119	-17878	2.299	1.532	1.308	0.112	0.302	805	1.960
8000	100	17.90	5.159	-17920	2.305	1.532	1.307	0.112	0.301	806	1.970
8000	100	17.84	5.199	-17962	2.310	1.533	1.307	0.112	0.300	808	1.980
8000	100	17.78	5.239	-18004	2.316	1.533	1.307	0.112	0.300	809	1.990
8000	100	17.72	5.279	-18045	2.322	1.533	1.307	0.112	0.299	811	2.000
8000	100	17.65	5.319	-18087	2.327	1.533	1.306	0.112	0.298	812	2.010
8000	100	17.59	5.360	-18129	2.333	1.533	1.306	0.112	0.298	814	2.020
8000	100	17.53	5.401	-18170	2.339	1.533	1.306	0.112	0.297	815	2.030
8000	100	17.47	5.442	-18211	2.344	1.534	1.306	0.112	0.296	817	2.040
8000	100	17.41	5.483	-18253	2.350	1.534	1.306	0.112	0.296	818	2.050
8000	100	17.35	5.524	-18294	2.356	1.534	1.305	0.112	0.295	820	2.060
8000	100	17.24	5.607	-18377	2.367	1.534	1.305	0.112	0.294	823	2.080
8000	100	17.12	5.691	-18459	2.378	1.535	1.304	0.112	0.293	826	2.100
8000	100	16.98	5.797	-18561	2.393	1.535	1.304	0.112	0.291	829	2.125
8000	100	16.84	5.904	-18664	2.407	1.535	1.303	0.112	0.290	833	2.150
8000	100	16.57	6.120	-18867	2.435	1.536	1.302	0.111	0.287	840	2.200
8000	100	16.31	6.340	-19069	2.463	1.536	1.302	0.111	0.284	847	2.250

Table A.30. Gas Phase (continued)

T K	P bar	ρ $kg\ m^{-3}$	H $10^2 kJ\ kg^{-1}$	G $kJ\ kg^{-1}$	S $kJ\ kg^{-1}\ K^{-1}$	C_P	C_P/C_V	β_T $10^{-2} bar^{-1}$	α_T $10^{-3} K^{-1}$	V_S $m\ s^{-1}$	O/U
8000	300	67.54	-3.308	-16459	2.016	1.252	1.306	0.384	0.308	710	1.750
8000	300	66.10	-3.148	-16663	2.044	1.261	1.305	0.383	0.305	718	1.800
8000	300	64.72	-2.978	-16866	2.071	1.270	1.305	0.382	0.303	727	1.850
8000	300	64.04	-2.890	-16966	2.085	1.274	1.305	0.381	0.301	731	1.875
8000	300	63.39	-2.800	-17067	2.098	1.278	1.305	0.381	0.300	735	1.900
8000	300	62.87	-2.727	-17147	2.109	1.281	1.304	0.381	0.299	738	1.920
8000	300	62.36	-2.653	-17227	2.120	1.285	1.304	0.380	0.298	742	1.940
8000	300	62.11	-2.615	-17267	2.126	1.286	1.304	0.380	0.298	743	1.950
8000	300	61.86	-2.577	-17306	2.131	1.288	1.304	0.380	0.297	745	1.960
8000	300	61.61	-2.539	-17346	2.136	1.289	1.304	0.380	0.297	746	1.970
8000	300	61.37	-2.501	-17386	2.142	1.291	1.304	0.380	0.296	748	1.980
8000	300	61.12	-2.462	-17425	2.147	1.293	1.304	0.380	0.296	750	1.990
8000	300	60.88	-2.424	-17465	2.153	1.294	1.304	0.380	0.295	751	2.000
8000	300	60.64	-2.385	-17504	2.158	1.296	1.304	0.379	0.295	753	2.010
8000	300	60.41	-2.346	-17544	2.164	1.297	1.304	0.379	0.294	754	2.020
8000	300	60.17	-2.306	-17583	2.169	1.299	1.304	0.379	0.294	756	2.030
8000	300	59.94	-2.267	-17622	2.174	1.300	1.304	0.379	0.293	758	2.040
8000	300	59.71	-2.227	-17661	2.180	1.302	1.304	0.379	0.293	759	2.050
8000	300	59.48	-2.187	-17701	2.185	1.303	1.304	0.379	0.292	761	2.060
8000	300	59.02	-2.106	-17779	2.196	1.306	1.303	0.378	0.291	764	2.080
8000	300	58.58	-2.025	-17857	2.207	1.309	1.303	0.378	0.291	767	2.100
8000	300	58.03	-1.922	-17954	2.220	1.313	1.303	0.378	0.289	771	2.125
8000	300	57.49	-1.819	-18051	2.234	1.317	1.303	0.378	0.288	775	2.150
8000	300	56.44	-1.609	-18243	2.260	1.324	1.303	0.377	0.286	783	2.200
8000	300	55.44	-1.396	-18435	2.287	1.331	1.303	0.376	0.284	790	2.250

Table A.30. Gas Phase (continued)

T K	P bar	ρ $kg\ m^{-3}$	H $10^2 kJ\ kg^{-1}$	G $kJ\ kg^{-1}$	S $kJ\ kg^{-1}\ K^{-1}$	C_P	C_P/C_V	β_T $10^{-2} bar^{-1}$	α_T $10^{-3} K^{-1}$	V_S $m\ s^{-1}$	O/U
8000	500	123.1	-6.772	-16241	1.945	1.284	1.342	0.242	0.349	671	1.750
8000	500	120.3	-6.648	-16440	1.972	1.302	1.342	0.241	0.347	680	1.800
8000	500	117.6	-6.513	-16638	1.998	1.319	1.342	0.241	0.345	688	1.850
8000	500	116.3	-6.442	-16737	2.012	1.327	1.342	0.240	0.344	693	1.875
8000	500	115.0	-6.369	-16835	2.025	1.336	1.342	0.240	0.343	697	1.900
8000	500	114.0	-6.309	-16913	2.035	1.342	1.342	0.240	0.342	700	1.920
8000	500	113.1	-6.247	-16990	2.046	1.348	1.342	0.240	0.341	704	1.940
8000	500	112.6	-6.216	-17029	2.051	1.352	1.342	0.240	0.341	705	1.950
8000	500	112.1	-6.184	-17068	2.056	1.355	1.342	0.239	0.340	707	1.960
8000	500	111.6	-6.152	-17107	2.061	1.358	1.342	0.239	0.340	709	1.970
8000	500	111.2	-6.120	-17145	2.067	1.361	1.342	0.239	0.339	710	1.980
8000	500	110.7	-6.088	-17184	2.072	1.364	1.341	0.239	0.339	712	1.990
8000	500	110.2	-6.055	-17223	2.077	1.367	1.341	0.239	0.338	714	2.000
8000	500	109.8	-6.022	-17261	2.082	1.370	1.341	0.239	0.338	715	2.010
8000	500	109.3	-5.989	-17299	2.088	1.373	1.341	0.239	0.338	717	2.020
8000	500	108.9	-5.956	-17338	2.093	1.376	1.341	0.239	0.337	719	2.030
8000	500	108.4	-5.922	-17376	2.098	1.379	1.341	0.238	0.337	720	2.040
8000	500	108.0	-5.888	-17414	2.103	1.382	1.341	0.238	0.336	722	2.050
8000	500	107.5	-5.854	-17453	2.108	1.385	1.341	0.238	0.336	724	2.060
8000	500	106.6	-5.785	-17529	2.119	1.390	1.341	0.238	0.335	727	2.080
8000	500	105.8	-5.715	-17605	2.129	1.396	1.341	0.238	0.334	730	2.100
8000	500	104.7	-5.627	-17700	2.142	1.403	1.341	0.238	0.333	734	2.125
8000	500	103.7	-5.537	-17794	2.155	1.410	1.341	0.237	0.332	738	2.150
8000	500	101.7	-5.354	-17982	2.181	1.423	1.340	0.237	0.330	746	2.200
8000	500	99.8	-5.167	-18168	2.206	1.435	1.340	0.236	0.328	754	2.250

Table A.30. Gas Phase (continued)

T K	P bar	ρ $kg\ m^{-3}$	H $10^2 kJ\ kg^{-1}$	G $kJ\ kg^{-1}$	S $kJ\ kg^{-1}\ K^{-1}$	C_P $kJ\ kg^{-1}\ K^{-1}$	C_P/C_V	β_T $10^{-2} bar^{-1}$	α_T $10^{-3} K^{-1}$	V_S $m\ s^{-1}$	O/U
8000	700	187.5	-9.428	-16110	1.896	1.377	1.403	0.186	0.415	635	1.750
8000	700	182.9	-9.347	-16306	1.921	1.403	1.403	0.185	0.413	644	1.800
8000	700	178.6	-9.252	-16501	1.947	1.427	1.402	0.184	0.410	653	1.850
8000	700	176.5	-9.200	-16597	1.960	1.439	1.402	0.184	0.409	657	1.875
8000	700	174.5	-9.145	-16694	1.972	1.450	1.402	0.184	0.408	662	1.900
8000	700	172.9	-9.100	-16770	1.983	1.459	1.401	0.183	0.407	665	1.920
8000	700	171.3	-9.052	-16847	1.993	1.468	1.401	0.183	0.406	669	1.940
8000	700	170.5	-9.028	-16885	1.998	1.472	1.401	0.183	0.405	670	1.950
8000	700	169.8	-9.003	-16923	2.003	1.476	1.400	0.183	0.405	672	1.960
8000	700	169.0	-8.978	-16961	2.008	1.480	1.400	0.183	0.404	674	1.970
8000	700	168.2	-8.953	-17000	2.013	1.485	1.400	0.182	0.403	676	1.980
8000	700	167.5	-8.927	-17037	2.018	1.489	1.400	0.182	0.403	677	1.990
8000	700	166.7	-8.901	-17075	2.023	1.493	1.400	0.182	0.402	679	2.000
8000	700	166.0	-8.875	-17113	2.028	1.497	1.399	0.182	0.402	681	2.010
8000	700	165.3	-8.848	-17151	2.033	1.501	1.399	0.182	0.401	682	2.020
8000	700	164.5	-8.821	-17189	2.038	1.505	1.399	0.182	0.400	684	2.030
8000	700	163.8	-8.794	-17226	2.043	1.509	1.399	0.182	0.400	686	2.040
8000	700	163.1	-8.766	-17264	2.048	1.512	1.399	0.181	0.399	688	2.050
8000	700	162.4	-8.738	-17302	2.053	1.516	1.398	0.181	0.399	689	2.060
8000	700	161.0	-8.681	-17377	2.064	1.524	1.398	0.181	0.397	693	2.080
8000	700	159.6	-8.623	-17451	2.074	1.531	1.397	0.181	0.396	696	2.100
8000	700	158.0	-8.549	-17544	2.086	1.540	1.397	0.180	0.395	700	2.125
8000	700	156.3	-8.474	-17637	2.099	1.549	1.396	0.180	0.393	704	2.150
8000	700	153.1	-8.318	-17822	2.124	1.566	1.395	0.179	0.390	713	2.200
8000	700	150.1	-8.157	-18005	2.149	1.582	1.394	0.179	0.387	721	2.250

Table A.30. Gas Phase (continued)

T K	P bar	ρ $kg\ m^{-3}$	H $10^3 kJ\ kg^{-1}$	G $kJ\ kg^{-1}$	S $kJ\ kg^{-1}\ K^{-1}$	C_P	C_P/C_V	β_T bar^{-1}	α_T $10^{-3} K^{-1}$	V_S $m\ s^{-1}$	O/U
9000	1	0.09970	5.129	-23284	3.157	0.682	1.366	1.033	0.145	1151	1.750
9000	1	0.09858	5.154	-23582	3.193	0.683	1.367	1.033	0.144	1159	1.800
9000	1	0.09749	5.178	-23878	3.228	0.684	1.367	1.032	0.143	1166	1.850
9000	1	0.09695	5.191	-24025	3.246	0.684	1.368	1.032	0.143	1169	1.875
9000	1	0.09643	5.203	-24172	3.264	0.685	1.368	1.031	0.143	1173	1.900
9000	1	0.09601	5.213	-24289	3.278	0.685	1.369	1.031	0.143	1176	1.920
9000	1	0.09560	5.222	-24405	3.292	0.685	1.369	1.031	0.142	1179	1.940
9000	1	0.09540	5.227	-24463	3.299	0.685	1.369	1.031	0.142	1180	1.950
9000	1	0.09520	5.232	-24522	3.306	0.686	1.369	1.031	0.142	1181	1.960
9000	1	0.09499	5.237	-24580	3.313	0.686	1.370	1.031	0.142	1183	1.970
9000	1	0.09479	5.242	-24638	3.320	0.686	1.370	1.031	0.142	1184	1.980
9000	1	0.09460	5.247	-24696	3.327	0.686	1.370	1.030	0.142	1186	1.990
9000	1	0.09440	5.251	-24753	3.334	0.686	1.370	1.030	0.142	1187	2.000
9000	1	0.09420	5.256	-24811	3.341	0.686	1.370	1.030	0.141	1188	2.010
9000	1	0.09400	5.261	-24869	3.348	0.687	1.371	1.030	0.141	1190	2.020
9000	1	0.09381	5.266	-24926	3.355	0.687	1.371	1.030	0.141	1191	2.030
9000	1	0.09362	5.271	-24984	3.362	0.687	1.371	1.030	0.141	1192	2.040
9000	1	0.09342	5.275	-25041	3.369	0.687	1.371	1.030	0.141	1194	2.050
9000	1	0.09323	5.280	-25099	3.375	0.687	1.371	1.030	0.141	1195	2.060
9000	1	0.09285	5.290	-25213	3.389	0.688	1.372	1.029	0.141	1198	2.080
9000	1	0.09247	5.299	-25328	3.403	0.688	1.372	1.029	0.141	1201	2.100
9000	1	0.09201	5.311	-25470	3.420	0.688	1.373	1.029	0.140	1204	2.125
9000	1	0.09155	5.323	-25612	3.437	0.689	1.373	1.029	0.140	1207	2.150
9000	1	0.09065	5.346	-25895	3.471	0.690	1.374	1.028	0.139	1214	2.200
9000	1	0.08978	5.369	-26175	3.505	0.691	1.375	1.028	0.139	1221	2.250

Table A.30. Gas Phase (continued)

T K	P bar	ρ $kg\ m^{-3}$	H $10^3 kJ\ kg^{-1}$	G $kJ\ kg^{-1}$	S	C_P $kJ\ kg^{-1}\ K^{-1}$	C_P/C_V	β_T bar^{-1}	α_T $10^{-3} K^{-1}$	V_S $m\ s^{-1}$	O/U
9000	3	0.3136	4.698	-22206	2.989	0.884	1.347	0.351	0.167	1106	1.750
9000	3	0.3099	4.725	-22491	3.024	0.886	1.347	0.351	0.166	1113	1.800
9000	3	0.3062	4.753	-22774	3.059	0.887	1.348	0.351	0.165	1121	1.850
9000	3	0.3044	4.766	-22915	3.076	0.887	1.348	0.350	0.165	1124	1.875
9000	3	0.3026	4.780	-23056	3.093	0.888	1.348	0.350	0.164	1128	1.900
9000	3	0.3012	4.791	-23168	3.106	0.888	1.348	0.350	0.164	1131	1.920
9000	3	0.2998	4.801	-23279	3.120	0.889	1.348	0.350	0.164	1134	1.940
9000	3	0.2991	4.807	-23335	3.127	0.889	1.349	0.350	0.164	1135	1.950
9000	3	0.2985	4.812	-23391	3.134	0.889	1.349	0.350	0.163	1136	1.960
9000	3	0.2978	4.817	-23446	3.140	0.890	1.349	0.350	0.163	1138	1.970
9000	3	0.2971	4.823	-23502	3.147	0.890	1.349	0.350	0.163	1139	1.980
9000	3	0.2964	4.828	-23557	3.154	0.890	1.349	0.350	0.163	1141	1.990
9000	3	0.2958	4.833	-23613	3.161	0.890	1.349	0.350	0.163	1142	2.000
9000	3	0.2951	4.839	-23668	3.167	0.891	1.349	0.350	0.163	1143	2.010
9000	3	0.2945	4.844	-23723	3.174	0.891	1.349	0.350	0.162	1145	2.020
9000	3	0.2938	4.849	-23778	3.181	0.891	1.349	0.350	0.162	1146	2.030
9000	3	0.2932	4.855	-23833	3.188	0.891	1.349	0.350	0.162	1148	2.040
9000	3	0.2925	4.860	-23888	3.194	0.891	1.349	0.349	0.162	1149	2.050
9000	3	0.2919	4.865	-23943	3.201	0.892	1.349	0.349	0.162	1150	2.060
9000	3	0.2906	4.876	-24053	3.214	0.892	1.350	0.349	0.161	1153	2.080
9000	3	0.2893	4.886	-24162	3.228	0.893	1.350	0.349	0.161	1156	2.100
9000	3	0.2878	4.899	-24299	3.244	0.893	1.350	0.349	0.161	1159	2.125
9000	3	0.2863	4.912	-24435	3.261	0.894	1.350	0.349	0.160	1163	2.150
9000	3	0.2833	4.938	-24705	3.294	0.895	1.351	0.349	0.160	1169	2.200
9000	3	0.2803	4.964	-24974	3.326	0.896	1.351	0.349	0.159	1176	2.250

Table A.30. Gas Phase (continued)

T	P	ρ	H	G	S	C_P	C_P/C_V	β_T	α_T	V_S	O/U
K	bar	$kg\ m^{-3}$	$10^3 kJ\ kg^{-1}$	$kJ\ kg^{-1}$		$kJ\ kg^{-1}\ K^{-1}$		bar^{-1}	$10^{-3} K^{-1}$	$m\ s^{-1}$	
9000	10	1.128	4.014	-21095	2.790	1.107	1.335	0.107	0.193	1050	1.750
9000	10	1.113	4.042	-21366	2.823	1.111	1.335	0.107	0.192	1057	1.800
9000	10	1.099	4.070	-21635	2.856	1.115	1.335	0.107	0.191	1064	1.850
9000	10	1.092	4.084	-21769	2.873	1.117	1.335	0.107	0.191	1068	1.875
9000	10	1.085	4.098	-21902	2.889	1.119	1.335	0.107	0.191	1072	1.900
9000	10	1.080	4.110	-22009	2.902	1.121	1.335	0.107	0.190	1074	1.920
9000	10	1.074	4.121	-22115	2.915	1.122	1.335	0.107	0.190	1077	1.940
9000	10	1.072	4.126	-22168	2.922	1.123	1.335	0.107	0.190	1079	1.950
9000	10	1.069	4.132	-22221	2.928	1.124	1.335	0.107	0.189	1080	1.960
9000	10	1.066	4.137	-22273	2.935	1.125	1.335	0.107	0.189	1082	1.970
9000	10	1.064	4.143	-22326	2.941	1.125	1.335	0.107	0.189	1083	1.980
9000	10	1.061	4.149	-22379	2.947	1.126	1.335	0.107	0.189	1084	1.990
9000	10	1.059	4.154	-22431	2.954	1.127	1.335	0.107	0.189	1086	2.000
9000	10	1.056	4.160	-22484	2.960	1.128	1.335	0.107	0.189	1087	2.010
9000	10	1.054	4.165	-22536	2.967	1.128	1.335	0.107	0.188	1089	2.020
9000	10	1.051	4.171	-22589	2.973	1.129	1.335	0.107	0.188	1090	2.030
9000	10	1.049	4.176	-22641	2.980	1.130	1.335	0.107	0.188	1091	2.040
9000	10	1.046	4.182	-22693	2.986	1.131	1.335	0.107	0.188	1093	2.050
9000	10	1.044	4.187	-22746	2.993	1.131	1.335	0.107	0.188	1094	2.060
9000	10	1.039	4.198	-22850	3.005	1.133	1.335	0.107	0.187	1097	2.080
9000	10	1.034	4.209	-22954	3.018	1.134	1.335	0.107	0.187	1100	2.100
9000	10	1.028	4.223	-23083	3.034	1.136	1.335	0.107	0.187	1103	2.125
9000	10	1.022	4.236	-23213	3.050	1.138	1.335	0.107	0.186	1106	2.150
9000	10	1.010	4.263	-23470	3.081	1.141	1.335	0.107	0.185	1113	2.200
9000	10	0.999	4.290	-23725	3.113	1.144	1.335	0.107	0.184	1120	2.250

Table A.30. Gas Phase (continued)

T K	P bar	ρ $kg\ m^{-3}$	H $10^3 kJ\ kg^{-1}$	G $kJ\ kg^{-1}$	S $kJ\ kg^{-1}\ K^{-1}$	C_P	C_P/C_V	β_T $10^{-1} bar^{-1}$	α_T $10^{-3} K^{-1}$	V_S $m\ s^{-1}$	O/U
9000	30	3.724	3.186	-20164	2.594	1.374	1.338	0.367	0.230	989	1.750
9000	30	3.671	3.212	-20422	2.626	1.382	1.338	0.367	0.229	996	1.800
9000	30	3.620	3.238	-20678	2.657	1.389	1.338	0.367	0.227	1004	1.850
9000	30	3.595	3.251	-20806	2.673	1.392	1.337	0.367	0.227	1007	1.875
9000	30	3.571	3.264	-20933	2.689	1.395	1.337	0.366	0.226	1011	1.900
9000	30	3.552	3.275	-21034	2.701	1.398	1.337	0.366	0.226	1014	1.920
9000	30	3.533	3.285	-21135	2.713	1.401	1.337	0.366	0.225	1017	1.940
9000	30	3.523	3.290	-21186	2.720	1.402	1.337	0.366	0.225	1018	1.950
9000	30	3.514	3.295	-21236	2.726	1.403	1.337	0.366	0.225	1020	1.960
9000	30	3.504	3.301	-21286	2.732	1.405	1.337	0.366	0.225	1021	1.970
9000	30	3.495	3.306	-21337	2.738	1.406	1.337	0.366	0.224	1022	1.980
9000	30	3.486	3.311	-21387	2.744	1.407	1.337	0.366	0.224	1024	1.990
9000	30	3.477	3.316	-21437	2.750	1.408	1.337	0.366	0.224	1025	2.000
9000	30	3.468	3.321	-21487	2.756	1.409	1.336	0.366	0.224	1027	2.010
9000	30	3.459	3.327	-21537	2.763	1.411	1.336	0.366	0.223	1028	2.020
9000	30	3.450	3.332	-21587	2.769	1.412	1.336	0.366	0.223	1030	2.030
9000	30	3.441	3.337	-21637	2.775	1.413	1.336	0.365	0.223	1031	2.040
9000	30	3.432	3.342	-21686	2.781	1.414	1.336	0.365	0.223	1032	2.050
9000	30	3.423	3.347	-21736	2.787	1.415	1.336	0.365	0.222	1034	2.060
9000	30	3.406	3.358	-21835	2.799	1.418	1.336	0.365	0.222	1037	2.080
9000	30	3.388	3.368	-21934	2.811	1.420	1.336	0.365	0.222	1039	2.100
9000	30	3.367	3.381	-22058	2.827	1.423	1.336	0.365	0.221	1043	2.125
9000	30	3.346	3.394	-22181	2.842	1.426	1.336	0.365	0.220	1046	2.150
9000	30	3.305	3.420	-22425	2.872	1.431	1.335	0.364	0.219	1053	2.200
9000	30	3.265	3.446	-22669	2.902	1.436	1.335	0.364	0.218	1060	2.250

Table A.30. Gas Phase (continued)

T K	P bar	ρ $kg\ m^{-3}$	H $10^3 kJ\ kg^{-1}$	G $kJ\ kg^{-1}$	S $kJ\ kg^{-1}\ K^{-1}$	C_P	C_P/C_V	β_T $10^{-1}bar^{-1}$	α_T $10^{-3}K^{-1}$	V_S $m\ s^{-1}$	O/U
9000	100	14.37	2.032	-19258	2.366	1.568	1.354	0.114	0.273	910	1.750
9000	100	14.14	2.053	-19503	2.395	1.575	1.353	0.114	0.271	918	1.800
9000	100	13.92	2.075	-19745	2.425	1.582	1.352	0.114	0.269	925	1.850
9000	100	13.81	2.086	-19866	2.439	1.585	1.352	0.113	0.268	929	1.875
9000	100	13.71	2.098	-19986	2.454	1.588	1.351	0.113	0.267	933	1.900
9000	100	13.63	2.107	-20082	2.465	1.590	1.351	0.113	0.266	936	1.920
9000	100	13.54	2.116	-20178	2.477	1.593	1.351	0.113	0.265	939	1.940
9000	100	13.50	2.120	-20225	2.483	1.594	1.350	0.113	0.265	940	1.950
9000	100	13.46	2.125	-20273	2.489	1.595	1.350	0.113	0.265	942	1.960
9000	100	13.42	2.129	-20321	2.494	1.596	1.350	0.113	0.264	943	1.970
9000	100	13.38	2.134	-20368	2.500	1.597	1.350	0.113	0.264	944	1.980
9000	100	13.34	2.138	-20416	2.506	1.598	1.350	0.113	0.263	946	1.990
9000	100	13.31	2.143	-20463	2.512	1.599	1.349	0.113	0.263	947	2.000
9000	100	13.27	2.148	-20511	2.518	1.600	1.349	0.113	0.263	949	2.010
9000	100	13.23	2.152	-20558	2.523	1.601	1.349	0.113	0.262	950	2.020
9000	100	13.19	2.157	-20605	2.529	1.602	1.349	0.113	0.262	952	2.030
9000	100	13.15	2.162	-20652	2.535	1.603	1.349	0.113	0.261	953	2.040
9000	100	13.11	2.166	-20699	2.541	1.604	1.348	0.113	0.261	955	2.050
9000	100	13.08	2.171	-20746	2.546	1.605	1.348	0.113	0.261	956	2.060
9000	100	13.00	2.180	-20840	2.558	1.607	1.348	0.113	0.260	959	2.080
9000	100	12.93	2.190	-20934	2.569	1.609	1.348	0.113	0.259	962	2.100
9000	100	12.84	2.201	-21051	2.584	1.611	1.347	0.113	0.258	965	2.125
9000	100	12.75	2.213	-21167	2.598	1.613	1.347	0.113	0.257	969	2.150
9000	100	12.57	2.237	-21398	2.626	1.618	1.346	0.112	0.255	976	2.200
9000	100	12.40	2.261	-21628	2.654	1.621	1.345	0.112	0.254	983	2.250

Table A.30. Gas Phase (continued)

T K	P bar	ρ $kg\ m^{-3}$	H $10^3 kJ\ kg^{-1}$	G $kJ\ kg^{-1}$	S $kJ\ kg^{-1}\ K^{-1}$	C_P	C_P/C_V	β_T $10^{-2} bar^{-1}$	α_T $10^{-3} K^{-1}$	V_S $m\ s^{-1}$	O/U
9000	300	50.62	0.970	-18551	2.169	1.378	1.346	0.384	0.277	832	1.750
9000	300	49.70	0.990	-18783	2.197	1.381	1.344	0.383	0.274	840	1.800
9000	300	48.81	1.011	-19014	2.225	1.384	1.343	0.382	0.271	849	1.850
9000	300	48.37	1.021	-19128	2.239	1.385	1.342	0.382	0.269	853	1.875
9000	300	47.95	1.032	-19243	2.253	1.386	1.341	0.381	0.268	857	1.900
9000	300	47.61	1.041	-19334	2.264	1.387	1.341	0.381	0.267	860	1.920
9000	300	47.28	1.049	-19425	2.275	1.388	1.340	0.380	0.265	863	1.940
9000	300	47.12	1.054	-19470	2.280	1.389	1.340	0.380	0.265	865	1.950
9000	300	46.96	1.058	-19515	2.286	1.389	1.340	0.380	0.264	866	1.960
9000	300	46.80	1.063	-19560	2.291	1.389	1.339	0.380	0.264	868	1.970
9000	300	46.64	1.067	-19605	2.297	1.390	1.339	0.380	0.263	870	1.980
9000	300	46.48	1.072	-19650	2.302	1.390	1.339	0.380	0.263	871	1.990
9000	300	46.32	1.076	-19695	2.308	1.391	1.339	0.379	0.262	873	2.000
9000	300	46.16	1.081	-19740	2.313	1.391	1.338	0.379	0.262	874	2.010
9000	300	46.01	1.085	-19785	2.319	1.391	1.338	0.379	0.261	876	2.020
9000	300	45.85	1.090	-19830	2.324	1.392	1.338	0.379	0.261	877	2.030
9000	300	45.70	1.094	-19875	2.330	1.392	1.338	0.379	0.260	879	2.040
9000	300	45.55	1.099	-19920	2.335	1.392	1.337	0.379	0.259	881	2.050
9000	300	45.40	1.103	-19964	2.341	1.393	1.337	0.379	0.259	882	2.060
9000	300	45.10	1.113	-20053	2.352	1.393	1.337	0.378	0.258	885	2.080
9000	300	44.80	1.122	-20142	2.363	1.394	1.336	0.378	0.257	888	2.100
9000	300	44.44	1.134	-20253	2.376	1.395	1.336	0.377	0.256	892	2.125
9000	300	44.08	1.146	-20363	2.390	1.396	1.335	0.377	0.254	896	2.150
9000	300	43.39	1.169	-20583	2.417	1.397	1.334	0.376	0.252	904	2.200
9000	300	42.71	1.194	-20801	2.444	1.399	1.333	0.376	0.250	912	2.250

Table A.30. Gas Phase (continued)

T K	P bar	ρ $kg\ m^{-3}$	H $10^1 kJ\ kg^{-1}$	G $kJ\ kg^{-1}$	S $kJ\ kg^{-1}\ K^{-1}$	C_P $kJ\ kg^{-1}\ K^{-1}$	C_P/C_V	β_T $10^{-2} bar^{-1}$	α_T $10^{-3} K^{-1}$	V_S $m\ s^{-1}$	O/U
9000	1000	207.8	0.14	-17903	1.989	1.145	1.374	0.123	0.297	735	1.750
9000	1000	203.0	2.13	-18122	2.016	1.151	1.372	0.122	0.293	745	1.800
9000	1000	198.5	4.21	-18339	2.042	1.157	1.370	0.121	0.289	754	1.850
9000	1000	196.3	5.27	-18447	2.055	1.161	1.369	0.121	0.287	759	1.875
9000	1000	194.2	6.35	-18554	2.069	1.164	1.368	0.121	0.286	764	1.900
9000	1000	192.5	7.23	-18640	2.079	1.166	1.367	0.121	0.284	768	1.920
9000	1000	190.8	8.12	-18725	2.090	1.168	1.366	0.120	0.283	771	1.940
9000	1000	190.0	8.56	-18768	2.095	1.170	1.366	0.120	0.282	773	1.950
9000	1000	189.2	9.01	-18811	2.100	1.171	1.366	0.120	0.281	775	1.960
9000	1000	188.4	9.46	-18853	2.105	1.172	1.365	0.120	0.281	777	1.970
9000	1000	187.6	9.92	-18896	2.111	1.173	1.365	0.120	0.280	779	1.980
9000	1000	186.8	10.37	-18938	2.116	1.174	1.365	0.120	0.279	781	1.990
9000	1000	186.1	10.83	-18980	2.121	1.175	1.364	0.120	0.279	782	2.000
9000	1000	185.3	11.29	-19023	2.126	1.176	1.364	0.120	0.278	784	2.010
9000	1000	184.5	11.75	-19065	2.131	1.178	1.364	0.120	0.278	786	2.020
9000	1000	183.8	12.21	-19107	2.137	1.179	1.363	0.120	0.277	788	2.030
9000	1000	183.0	12.68	-19149	2.142	1.180	1.363	0.119	0.276	790	2.040
9000	1000	182.3	13.14	-19191	2.147	1.181	1.363	0.119	0.276	792	2.050
9000	1000	181.5	13.61	-19233	2.152	1.182	1.362	0.119	0.275	793	2.060
9000	1000	180.1	14.55	-19317	2.162	1.184	1.362	0.119	0.274	797	2.080
9000	1000	178.6	15.50	-19400	2.173	1.187	1.361	0.119	0.273	801	2.100
9000	1000	176.9	16.69	-19504	2.186	1.189	1.360	0.119	0.271	805	2.125
9000	1000	175.2	17.89	-19608	2.199	1.192	1.360	0.119	0.270	809	2.150
9000	1000	171.8	20.31	-19814	2.224	1.198	1.358	0.118	0.267	818	2.200
9000	1000	168.6	22.75	-20019	2.250	1.203	1.357	0.118	0.264	827	2.250

Table A.30. Gas Phase (continued)

T K	P bar	ρ $kg\ m^{-3}$	H $10^3 kJ\ kg^{-1}$	G $kJ\ kg^{-1}$	S	C_P $kJ\ kg^{-1}\ K^{-1}$	C_P/C_V	β_T bar^{-1}	α_T $10^{-3} K^{-1}$	V_S $m\ s^{-1}$	O/U
10000	1	0.08767	5.715	-26474	3.219	0.524	1.403	1.020	0.116	1253	1.750
10000	1	0.08673	5.742	-26808	3.255	0.527	1.403	1.019	0.116	1260	1.800
10000	1	0.08580	5.768	-27139	3.291	0.530	1.404	1.019	0.115	1267	1.850
10000	1	0.08535	5.782	-27304	3.309	0.531	1.404	1.019	0.115	1271	1.875
10000	1	0.08491	5.795	-27469	3.326	0.533	1.404	1.019	0.115	1274	1.900
10000	1	0.08456	5.805	-27600	3.340	0.534	1.404	1.018	0.115	1277	1.920
10000	1	0.08421	5.816	-27730	3.355	0.535	1.404	1.018	0.115	1280	1.940
10000	1	0.08403	5.821	-27796	3.362	0.536	1.404	1.018	0.115	1281	1.950
10000	1	0.08386	5.826	-27861	3.369	0.536	1.404	1.018	0.115	1282	1.960
10000	1	0.08369	5.831	-27926	3.376	0.537	1.404	1.018	0.115	1284	1.970
10000	1	0.08352	5.836	-27991	3.383	0.537	1.405	1.018	0.115	1285	1.980
10000	1	0.08335	5.842	-28056	3.390	0.538	1.405	1.018	0.115	1287	1.990
10000	1	0.08319	5.847	-28121	3.397	0.539	1.405	1.018	0.115	1288	2.000
10000	1	0.08302	5.852	-28185	3.404	0.539	1.405	1.018	0.115	1289	2.010
10000	1	0.08285	5.857	-28250	3.411	0.540	1.405	1.018	0.115	1291	2.020
10000	1	0.08269	5.862	-28314	3.418	0.541	1.405	1.018	0.115	1292	2.030
10000	1	0.08252	5.867	-28379	3.425	0.541	1.405	1.018	0.114	1293	2.040
10000	1	0.08236	5.873	-28443	3.432	0.542	1.405	1.018	0.114	1295	2.050
10000	1	0.08219	5.878	-28508	3.439	0.542	1.405	1.018	0.114	1296	2.060
10000	1	0.08187	5.888	-28636	3.452	0.544	1.405	1.017	0.114	1299	2.080
10000	1	0.08155	5.898	-28764	3.466	0.545	1.405	1.017	0.114	1302	2.100
10000	1	0.08116	5.911	-28924	3.483	0.546	1.405	1.017	0.114	1305	2.125
10000	1	0.08077	5.924	-29083	3.501	0.548	1.406	1.017	0.114	1308	2.150
10000	1	0.08000	5.949	-29399	3.535	0.551	1.406	1.017	0.114	1315	2.200
10000	1	0.07925	5.974	-29714	3.569	0.554	1.406	1.016	0.114	1321	2.250

Table A.30. Gas Phase (continued)

T K	P bar	ρ $kg\ m^{-3}$	H $10^3 kJ\ kg^{-1}$	G $kJ\ kg^{-1}$	S $kJ\ kg^{-1}\ K^{-1}$	C_P $kJ\ kg^{-1}\ K^{-1}$	C_P/C_V	β_T bar^{-1}	α_T $10^{-3}K^{-1}$	V_S $m\ s^{-1}$	O/U
10000	3	0.2710	5.444	-25238	3.068	0.636	1.389	0.345	0.129	1219	1.750
10000	3	0.2679	5.473	-25558	3.103	0.638	1.390	0.345	0.129	1227	1.800
10000	3	0.2649	5.501	-25875	3.138	0.639	1.391	0.345	0.128	1234	1.850
10000	3	0.2634	5.515	-26033	3.155	0.640	1.391	0.345	0.128	1238	1.875
10000	3	0.2620	5.530	-26191	3.172	0.641	1.392	0.345	0.128	1242	1.900
10000	3	0.2608	5.541	-26316	3.186	0.641	1.392	0.344	0.127	1245	1.920
10000	3	0.2597	5.552	-26442	3.199	0.642	1.392	0.344	0.127	1248	1.940
10000	3	0.2591	5.558	-26504	3.206	0.642	1.392	0.344	0.127	1249	1.950
10000	3	0.2586	5.563	-26567	3.213	0.642	1.392	0.344	0.127	1251	1.960
10000	3	0.2580	5.569	-26629	3.220	0.643	1.393	0.344	0.127	1252	1.970
10000	3	0.2575	5.574	-26691	3.227	0.643	1.393	0.344	0.127	1254	1.980
10000	3	0.2569	5.580	-26754	3.233	0.643	1.393	0.344	0.127	1255	1.990
10000	3	0.2564	5.586	-26816	3.240	0.644	1.393	0.344	0.127	1257	2.000
10000	3	0.2558	5.591	-26878	3.247	0.644	1.393	0.344	0.127	1258	2.010
10000	3	0.2553	5.597	-26940	3.254	0.644	1.393	0.344	0.126	1259	2.020
10000	3	0.2548	5.602	-27002	3.260	0.645	1.393	0.344	0.126	1261	2.030
10000	3	0.2542	5.608	-27063	3.267	0.645	1.393	0.344	0.126	1262	2.040
10000	3	0.2537	5.613	-27125	3.274	0.645	1.394	0.344	0.126	1264	2.050
10000	3	0.2532	5.619	-27187	3.281	0.646	1.394	0.344	0.126	1265	2.060
10000	3	0.2521	5.630	-27310	3.294	0.646	1.394	0.344	0.126	1268	2.080
10000	3	0.2511	5.641	-27433	3.307	0.647	1.394	0.344	0.126	1271	2.100
10000	3	0.2498	5.654	-27586	3.324	0.648	1.395	0.344	0.125	1275	2.125
10000	3	0.2486	5.668	-27738	3.341	0.649	1.395	0.344	0.125	1278	2.150
10000	3	0.2461	5.695	-28042	3.374	0.650	1.396	0.343	0.125	1285	2.200
10000	3	0.2437	5.722	-28343	3.406	0.652	1.396	0.343	0.124	1292	2.250

Table A.30. Gas Phase (continued)

T K	P bar	ρ $kg\ m^{-3}$	H $10^3 kJ\ kg^{-1}$	G $kJ\ kg^{-1}$	S	C_P $kJ\ kg^{-1}\ K^{-1}$	C_P/C_V	β_T bar^{-1}	α_T $10^{-3} K^{-1}$	V_S $m\ s^{-1}$	O/U
10000	10	0.9538	4.948	-23938	2.889	0.797	1.372	0.106	0.147	1168	1.750
10000	10	0.9420	4.979	-24242	2.922	0.799	1.372	0.105	0.147	1176	1.800
10000	10	0.9307	5.011	-24544	2.955	0.801	1.373	0.105	0.146	1183	1.850
10000	10	0.9251	5.026	-24695	2.972	0.802	1.373	0.105	0.146	1187	1.875
10000	10	0.9196	5.041	-24845	2.989	0.803	1.373	0.105	0.145	1191	1.900
10000	10	0.9153	5.054	-24964	3.002	0.804	1.373	0.105	0.145	1194	1.920
10000	10	0.9110	5.066	-25083	3.015	0.805	1.373	0.105	0.145	1197	1.940
10000	10	0.9089	5.072	-25143	3.021	0.805	1.373	0.105	0.145	1199	1.950
10000	10	0.9068	5.078	-25202	3.028	0.806	1.373	0.105	0.145	1200	1.960
10000	10	0.9047	5.084	-25262	3.035	0.806	1.373	0.105	0.144	1202	1.970
10000	10	0.9027	5.090	-25321	3.041	0.806	1.374	0.105	0.144	1203	1.980
10000	10	0.9006	5.096	-25380	3.048	0.807	1.374	0.105	0.144	1205	1.990
10000	10	0.8985	5.103	-25439	3.054	0.807	1.374	0.105	0.144	1206	2.000
10000	10	0.8965	5.109	-25498	3.061	0.808	1.374	0.105	0.144	1208	2.010
10000	10	0.8945	5.115	-25557	3.067	0.808	1.374	0.105	0.144	1209	2.020
10000	10	0.8925	5.121	-25616	3.074	0.808	1.374	0.105	0.144	1211	2.030
10000	10	0.8905	5.127	-25675	3.080	0.809	1.374	0.105	0.144	1212	2.040
10000	10	0.8885	5.133	-25734	3.087	0.809	1.374	0.105	0.143	1214	2.050
10000	10	0.8865	5.139	-25792	3.093	0.809	1.374	0.105	0.143	1215	2.060
10000	10	0.8825	5.151	-25909	3.106	0.810	1.374	0.105	0.143	1218	2.080
10000	10	0.8787	5.163	-26026	3.119	0.811	1.375	0.105	0.143	1221	2.100
10000	10	0.8739	5.177	-26172	3.135	0.812	1.375	0.105	0.142	1225	2.125
10000	10	0.8691	5.192	-26317	3.151	0.813	1.375	0.105	0.142	1228	2.150
10000	10	0.8599	5.222	-26605	3.183	0.815	1.375	0.105	0.142	1235	2.200
10000	10	0.8509	5.251	-26892	3.214	0.816	1.376	0.105	0.141	1242	2.250

Table A.30. Gas Phase (continued)

T K	P bar	ρ $kg\ m^{-3}$	H $10^3 kJ\ kg^{-1}$	G $kJ\ kg^{-1}$	S $kJ\ kg^{-1}\ K^{-1}$	C_P $kJ\ kg^{-1}\ K^{-1}$	C_P/C_V	β_T $10^{-1} bar^{-1}$	α_T $10^{-3} K^{-1}$	V_S $m\ s^{-1}$	O/U
10000	30	3.068	4.323	-22824	2.715	0.948	1.359	0.358	0.166	1113	1.750
10000	30	3.027	4.355	-23113	2.747	0.952	1.359	0.357	0.165	1121	1.800
10000	30	2.988	4.387	-23401	2.779	0.956	1.359	0.357	0.164	1129	1.850
10000	30	2.969	4.402	-23544	2.795	0.958	1.360	0.357	0.164	1133	1.875
10000	30	2.950	4.418	-23687	2.811	0.960	1.360	0.357	0.164	1137	1.900
10000	30	2.935	4.431	-23801	2.823	0.962	1.360	0.357	0.163	1140	1.920
10000	30	2.920	4.443	-23915	2.836	0.964	1.360	0.356	0.163	1143	1.940
10000	30	2.913	4.449	-23971	2.842	0.964	1.360	0.356	0.163	1144	1.950
10000	30	2.906	4.456	-24028	2.848	0.965	1.360	0.356	0.163	1146	1.960
10000	30	2.898	4.462	-24084	2.855	0.966	1.360	0.356	0.162	1147	1.970
10000	30	2.891	4.468	-24141	2.861	0.967	1.360	0.356	0.162	1149	1.980
10000	30	2.884	4.474	-24197	2.867	0.967	1.360	0.356	0.162	1151	1.990
10000	30	2.877	4.480	-24254	2.873	0.968	1.360	0.356	0.162	1152	2.000
10000	30	2.870	4.487	-24310	2.880	0.969	1.360	0.356	0.162	1154	2.010
10000	30	2.863	4.493	-24366	2.886	0.970	1.360	0.356	0.162	1155	2.020
10000	30	2.856	4.499	-24422	2.892	0.970	1.360	0.356	0.162	1157	2.030
10000	30	2.849	4.505	-24478	2.898	0.971	1.360	0.356	0.162	1158	2.040
10000	30	2.842	4.511	-24534	2.905	0.972	1.360	0.356	0.161	1160	2.050
10000	30	2.836	4.517	-24590	2.911	0.973	1.360	0.356	0.161	1161	2.060
10000	30	2.822	4.530	-24701	2.923	0.974	1.360	0.356	0.161	1164	2.080
10000	30	2.822	4.530	-24701	2.923	0.974	1.360	0.356	0.161	1164	2.080
10000	30	2.809	4.542	-24813	2.935	0.976	1.360	0.356	0.161	1167	2.100
10000	30	2.792	4.557	-24951	2.951	0.977	1.360	0.356	0.160	1171	2.125
10000	30	2.776	4.572	-25089	2.966	0.979	1.360	0.355	0.160	1174	2.150
10000	30	2.744	4.603	-25365	2.997	0.983	1.360	0.355	0.159	1181	2.200
10000	30	2.714	4.633	-25638	3.027	0.986	1.361	0.355	0.159	1189	2.250

Table A.30. Gas Phase (continued)

T K	P bar	ρ $kg\ m^{-3}$	H $10^3 kJ\ kg^{-1}$	G $kJ\ kg^{-1}$	S $kJ\ kg^{-1}\ K^{-1}$	C_P	C_P/C_V	β_T $10^{-1} bar^{-1}$	α_T $10^{-3} K^{-1}$	V_S $m\ s^{-1}$	O/U
10000	100	11.36	3.420	-21702	2.512	1.186	1.362	0.111	0.199	1042	1.750
10000	100	11.19	3.449	-21976	2.543	1.193	1.362	0.110	0.198	1050	1.800
10000	100	11.04	3.478	-22249	2.573	1.200	1.362	0.110	0.197	1058	1.850
10000	100	10.96	3.493	-22384	2.588	1.203	1.362	0.110	0.197	1062	1.875
10000	100	10.88	3.507	-22519	2.603	1.206	1.361	0.110	0.196	1065	1.900
10000	100	10.82	3.519	-22627	2.615	1.209	1.361	0.110	0.196	1069	1.920
10000	100	10.77	3.530	-22734	2.626	1.211	1.361	0.110	0.195	1072	1.940
10000	100	10.74	3.536	-22788	2.632	1.212	1.361	0.110	0.195	1073	1.950
10000	100	10.71	3.542	-22841	2.638	1.213	1.361	0.110	0.195	1075	1.960
10000	100	10.68	3.548	-22895	2.644	1.215	1.361	0.110	0.195	1076	1.970
10000	100	10.65	3.553	-22948	2.650	1.216	1.361	0.110	0.194	1078	1.980
10000	100	10.62	3.559	-23002	2.656	1.217	1.361	0.110	0.194	1079	1.990
10000	100	10.59	3.565	-23055	2.662	1.218	1.361	0.110	0.194	1081	2.000
10000	100	10.57	3.571	-23108	2.668	1.219	1.361	0.110	0.194	1082	2.010
10000	100	10.54	3.576	-23161	2.674	1.221	1.361	0.110	0.194	1084	2.020
10000	100	10.51	3.582	-23214	2.680	1.222	1.361	0.110	0.193	1085	2.030
10000	100	10.48	3.588	-23267	2.686	1.223	1.361	0.110	0.193	1087	2.040
10000	100	10.45	3.594	-23320	2.691	1.224	1.361	0.110	0.193	1088	2.050
10000	100	10.43	3.600	-23373	2.697	1.225	1.361	0.110	0.193	1090	2.060
10000	100	10.37	3.611	-23478	2.709	1.227	1.360	0.110	0.193	1093	2.080
10000	100	10.32	3.623	-23584	2.721	1.229	1.360	0.110	0.192	1096	2.100
10000	100	10.25	3.637	-23715	2.735	1.232	1.360	0.110	0.192	1099	2.125
10000	100	10.19	3.652	-23845	2.750	1.235	1.360	0.110	0.191	1103	2.150
10000	100	10.06	3.680	-24105	2.779	1.240	1.360	0.110	0.190	1110	2.200
10000	100	9.94	3.709	-24364	2.807	1.245	1.360	0.110	0.189	1118	2.250

Table A.30. Gas Phase (continued)

T K	P bar	ρ $kg\ m^{-3}$	H $10^3 kJ\ kg^{-1}$	G $kJ\ kg^{-1}$	S $kJ\ kg^{-1}\ K^{-1}$	C_P	C_P/C_V	β_T $10^{-2} bar^{-1}$	α_T $10^{-3} K^{-1}$	V_S $m\ s^{-1}$	O/U
10000	300	39.10	2.357	-20795	2.315	1.335	1.380	0.382	0.234	962	1.750
10000	300	38.48	2.382	-21055	2.344	1.342	1.379	0.381	0.233	970	1.800
10000	300	37.88	2.407	-21314	2.372	1.348	1.378	0.380	0.231	978	1.850
10000	300	37.59	2.419	-21443	2.386	1.351	1.378	0.380	0.230	982	1.875
10000	300	37.30	2.432	-21571	2.400	1.354	1.377	0.380	0.229	986	1.900
10000	300	37.07	2.442	-21673	2.412	1.356	1.377	0.380	0.229	989	1.920
10000	300	36.85	2.453	-21775	2.423	1.358	1.376	0.379	0.228	992	1.940
10000	300	36.74	2.458	-21826	2.428	1.360	1.376	0.379	0.228	994	1.950
10000	300	36.63	2.463	-21877	2.434	1.361	1.376	0.379	0.227	996	1.960
10000	300	36.52	2.468	-21927	2.440	1.362	1.376	0.379	0.227	997	1.970
10000	300	36.41	2.473	-21978	2.445	1.363	1.376	0.379	0.227	999	1.980
10000	300	36.30	2.478	-22029	2.451	1.364	1.376	0.379	0.226	1000	1.990
10000	300	36.19	2.484	-22079	2.456	1.365	1.375	0.379	0.226	1002	2.000
10000	300	36.08	2.489	-22130	2.462	1.366	1.375	0.378	0.226	1004	2.010
10000	300	35.98	2.494	-22180	2.467	1.367	1.375	0.378	0.225	1005	2.020
10000	300	35.87	2.499	-22230	2.473	1.368	1.375	0.378	0.225	1007	2.030
10000	300	35.77	2.505	-22281	2.479	1.369	1.375	0.378	0.225	1008	2.040
10000	300	35.66	2.510	-22331	2.484	1.370	1.374	0.378	0.224	1010	2.050
10000	300	35.56	2.515	-22381	2.490	1.371	1.374	0.378	0.224	1011	2.060
10000	300	35.36	2.526	-22481	2.501	1.373	1.374	0.378	0.223	1014	2.080
10000	300	35.16	2.536	-22581	2.512	1.374	1.374	0.377	0.223	1018	2.100
10000	300	34.91	2.550	-22705	2.525	1.377	1.373	0.377	0.222	1021	2.125
10000	300	34.66	2.563	-22829	2.539	1.379	1.373	0.377	0.221	1025	2.150
10000	300	34.18	2.590	-23076	2.567	1.383	1.372	0.376	0.220	1033	2.200
10000	300	33.72	2.616	-23321	2.594	1.387	1.371	0.376	0.218	1040	2.250

Table A.30. Gas Phase (continued)

T K	P bar	ρ $kg\ m^{-3}$	H $10^1 kJ\ kg^{-1}$	G $kJ\ kg^{-1}$	S $kJ\ kg^{-1}\ K^{-1}$	C_P	C_P/C_V	β_T $10^{-2} bar^{-1}$	α_T $10^{-3} K^{-1}$	V_S $m\ s^{-1}$	O/U
10000	1000	158.8	1.151	-19953	2.110	1.178	1.393	0.119	0.250	860	1.750
10000	1000	155.7	1.174	-20199	2.137	1.180	1.391	0.118	0.247	869	1.800
10000	1000	152.8	1.198	-20443	2.164	1.183	1.388	0.118	0.244	878	1.850
10000	1000	151.3	1.210	-20564	2.177	1.184	1.387	0.118	0.243	883	1.875
10000	1000	149.9	1.223	-20685	2.191	1.186	1.386	0.118	0.241	887	1.900
10000	1000	148.8	1.233	-20781	2.201	1.186	1.385	0.117	0.240	891	1.920
10000	1000	147.7	1.243	-20877	2.212	1.187	1.385	0.117	0.239	894	1.940
10000	1000	147.2	1.248	-20925	2.217	1.188	1.384	0.117	0.238	896	1.950
10000	1000	146.7	1.253	-20973	2.223	1.188	1.384	0.117	0.238	898	1.960
10000	1000	146.1	1.258	-21021	2.228	1.189	1.383	0.117	0.237	900	1.970
10000	1000	145.6	1.263	-21068	2.233	1.189	1.383	0.117	0.237	902	1.980
10000	1000	145.1	1.269	-21116	2.238	1.190	1.383	0.117	0.236	903	1.990
10000	1000	144.6	1.274	-21163	2.244	1.190	1.382	0.117	0.236	905	2.000
10000	1000	144.1	1.279	-21211	2.249	1.190	1.382	0.117	0.235	907	2.010
10000	1000	143.6	1.284	-21258	2.254	1.191	1.382	0.117	0.235	909	2.020
10000	1000	143.0	1.289	-21306	2.260	1.191	1.381	0.117	0.234	910	2.030
10000	1000	142.5	1.295	-21353	2.265	1.192	1.381	0.116	0.234	912	2.040
10000	1000	142.0	1.300	-21400	2.270	1.192	1.381	0.116	0.233	914	2.050
10000	1000	141.6	1.305	-21448	2.275	1.193	1.380	0.116	0.233	916	2.060
10000	1000	140.6	1.316	-21542	2.286	1.193	1.379	0.116	0.232	919	2.080
10000	1000	139.6	1.326	-21636	2.296	1.194	1.379	0.116	0.231	923	2.100
10000	1000	138.4	1.340	-21753	2.309	1.195	1.378	0.116	0.229	927	2.125
10000	1000	137.3	1.353	-21869	2.322	1.196	1.377	0.116	0.228	931	2.150
10000	1000	135.0	1.380	-22101	2.348	1.198	1.376	0.115	0.226	940	2.200
10000	1000	132.8	1.407	-22332	2.374	1.200	1.374	0.115	0.224	948	2.250

Table A.30. Gas Phase (end)

T K	P bar	ρ $kg\ m^{-3}$	H $10^1 kJ\ kg^{-1}$	G $kJ\ kg^{-1}$	S $kJ\ kg^{-1}\ K^{-1}$	C_P $kJ\ kg^{-1}\ K^{-1}$	C_P/C_V	β_T $10^{-2} bar^{-1}$	α_T $10^{-3} K^{-1}$	V_S $m\ s^{-1}$	O/U
10000	3000	630.6	0.659	-19340	1.941	1.143	1.590	0.465	0.353	736	1.750
10000	3000	612.3	0.950	-19571	1.967	1.144	1.575	0.458	0.342	749	1.800
10000	3000	595.2	1.244	-19800	1.992	1.144	1.561	0.452	0.333	762	1.850
10000	3000	587.0	1.392	-19914	2.005	1.145	1.555	0.449	0.328	768	1.875
10000	3000	579.2	1.541	-20028	2.018	1.145	1.549	0.446	0.324	774	1.900
10000	3000	573.0	1.660	-20119	2.028	1.146	1.545	0.444	0.321	779	1.920
10000	3000	567.0	1.779	-20209	2.039	1.146	1.541	0.442	0.318	784	1.940
10000	3000	564.1	1.839	-20254	2.044	1.146	1.539	0.441	0.316	786	1.950
10000	3000	561.2	1.899	-20299	2.049	1.147	1.536	0.440	0.315	789	1.960
10000	3000	558.3	1.959	-20344	2.054	1.147	1.534	0.439	0.313	791	1.970
10000	3000	555.5	2.019	-20389	2.059	1.147	1.533	0.438	0.312	793	1.980
10000	3000	552.7	2.079	-20434	2.064	1.148	1.531	0.437	0.310	796	1.990
10000	3000	549.9	2.139	-20479	2.069	1.148	1.529	0.437	0.309	798	2.000
10000	3000	547.2	2.199	-20523	2.074	1.148	1.527	0.436	0.307	800	2.010
10000	3000	544.5	2.259	-20568	2.079	1.149	1.525	0.435	0.306	803	2.020
10000	3000	541.8	2.319	-20613	2.084	1.149	1.523	0.434	0.305	805	2.030
10000	3000	539.1	2.380	-20657	2.090	1.149	1.522	0.433	0.303	807	2.040
10000	3000	536.5	2.440	-20702	2.095	1.150	1.520	0.432	0.302	809	2.050
10000	3000	533.9	2.500	-20746	2.100	1.150	1.518	0.432	0.301	812	2.060
10000	3000	528.9	2.621	-20835	2.110	1.151	1.515	0.430	0.298	816	2.080
10000	3000	523.9	2.742	-20923	2.120	1.151	1.512	0.429	0.296	821	2.100
10000	3000	517.8	2.893	-21033	2.132	1.153	1.508	0.427	0.293	826	2.125
10000	3000	511.9	3.044	-21143	2.145	1.154	1.504	0.425	0.290	832	2.150
10000	3000	500.6	3.347	-21361	2.170	1.156	1.498	0.422	0.285	842	2.200
10000	3000	489.9	3.650	-21578	2.194	1.158	1.491	0.419	0.280	853	2.250

References

1. Ackermann RJ, Rauh EG, Rand MH (1978) In:*Proc Pacific Coast Regional Meeting:Ceramic Technology.* San Diego, CA, USA
2. Ackermann RJ, Rauh EG, Rand MH (1979) In:*Proc Intern Symp on Thermodynamics of Nuclear Materials.* Julich, SM-236/17, p.11, IAEA, Vienna, Austria
3. Adamson A (1976) *Physical Chemistry of Surfaces.* Wiley Publ, New York-London
4. Adroguer B, Chatelard P, Van Dorsselaere JP *et al* (2003) Nucl Eng & Design 221:55
5. Alexander CA, Ogden JS, Cunningham GW (1967) Report BMI-1789, Battelle Memorial Institute, USA
6. Andersen HC, Chandler D, Weeks JD (1971) J Chem Phys 54:5237
7. Ashcroft NW, Lekner J (1966) Phys Rev 145:83
8. Bakhshiev HG (1972) *Spektroskopiya mezhmolekularnyh vzaimodeistviy* (The Spectroscopy of Intermolecular Interactions). Nauka Publ, Leningrad, Russia (in Russian)
9. Barin I (1993) *Thermochemical Data of Pure Substances.* Third Edition, Weinheim Publ, New York, USA
10. Barker J, Henderson D (1967) J Chem Phys 47:4714
11. Basak CB, Sengupta AK, Kamath HS (2003) J Alloys Comp 360:210
12. Baus M, Hansen JP (1980) Phys Reports 59:1
13. Baxter RJ (1970) J Chem Phys 52:4559
14. Bierrum N (1926) Kgl Dan Vidensk Mat Fys 7:1
15. Blackburn (1973) J Nucl Mater 46:244
16. Bober M, Singer J (1987) Nucl Sci & Eng 97:344
17. Booth DL (1974) *The Thermodynamic Properties of* Na, $Al_{2]}O_{3]}$, Mo, *and* UO_2 *above* 1200 K. Report TRG 1759 (R/X), UKAEA, London,UK
18. Born M, Goeppert-Mayer M (1933) *Dynamische Gittertheorie der Kristalle.* Handbuch der Physik Band 24 (part 2):623-794, Springer-Verlag, Berlin-Goettingen-Heidelberg, Germany
19. Boublik T, Nezbeda J, Hlavaty K (1980) *Statistical Thermodynamics of Simple Liquids and their Mixtures.* Academia Publ, Praha, Czechia
20. Breitung W (1975) Report KfK 2091, Karlsruhe, Germany; Bober M, Breitung W, Karow HU, Schretzmann K (1976) J Nucl Mater 60:20
21. Breitung W, Reil KO (1985) Report KfK 3939, Karlsruhe, Germany

22. Browning P, Gillan MJ, Potter PE (1977) *The Equation of State of Uranium Dioxide:A Comparison of the Corresponding States Significant Structure Theory.* Report AERE-R 8129, UKAEA, London,UK
23. Carnahan NF, Starling KE (1969) J Chem Phys 51:635
24. Chapman AT, Brynestad J, Clark GW (1980) High Temp-High Press 12:447
25. Chevalier P-Y, Fischer E, Cheynet B (2002) J Nucl Mater 303:1
26. Christensen JA (1963) J Amer Ceram Soc 40:607
27. Clausen K, Hayes W, Hutchings MT, Kjems JK, MacDonald JE, Osborn R (1985) High Temp Sci 19:189
28. Conway JB, Fincel RM Jr, Hein RA (1963) Trans Amer Nucl Soc 6:153
29. Derrington CE (1975) J Solid State Chemistry 15:171
30. Dharmadurai G (1982) J Nucl Mater 110:256
31. Drotning WD (1981) Proc 8th Symp on Thermophys Props, vol II:245, Gaithersburg, MD, USA
32. Ebeling W, Grigo M (1980) Ann Phys 37:21
33. Ebeling W, Kraeft WD, Kremp D (1976) *Theory of Bound States and Ionization Equilibrium in Plasmas and Solids.* Akademie-Verlag, Berlin, Germany
34. Ebeling W, Kraeft WD, Kremp D, Roepke G (1986) Physica A 140:160
35. Emelyanov AM, Khodeev YuS, Gorokhov LN (1970) High Temp 8:508
36. Fink JK (2000) J Nucl Mater 279:1
37. Finn PA, Sheth A, Leibowitz L (1979) J Nucl Mater 79 14; Green DW, Leibowitz L (1982) J Nucl Mater 105:184
38. Fischer EA (1979) In:*Proc Intern Symp on Thermodynamics of Nuclear Materials.* Julich, SM-236/17, p.115, IAEA, Vienna, Austria
39. Fischer EA (1987) Report KfK 4084, Karlsruhe, Germany
40. Fischer EA (1989) Nucl Sci & Eng 101:97
41. Fischer EA (1992) Report KfK 4889, Karlsruhe, Germany
42. Fischer EA, Kinsman PR, Ohse RW (1976) J Nucl Mater 59:125
43. Fisher ME, Levin Y (1993) Phys Rev Lett 71:3286
44. Fisher ME, Levin Y (1995) Physica A225:164
45. *Fizicheskie velichiny* (The Physical Constants). (1991) The Reference Book. Grigoryev IS, Meylihov EZ Eds, Energoatomizdat Publ, Moscow, Russia (in Russian)
46. Fortov VE, Dryemin AN, Leontev AA (1975) High Temp 13:1072
47. Fowler PW, Harding JH, Pyper NC (1994) J Phys B:Condens Matter 6:10593
48. Fritz IJ (1976) J Appl Phys 47:4353
49. Garbajo JJ, Yoder GL, Popov SG, Ivanov VK (2001) J Nucl Mater 299:181
50. Gillan MJ (1975) in:*Thermodynamics of Nuclear Materials.* Proceedings of the Vienna Symposium, vol 1, p.269, IAEA, Vienna, Austria
51. Gillan MJ (1983) Mol Phys 49:421
52. Grant I P, Pyper NC (1978) J Chem Soc Farad Trans II 74:1885
53. Gray CG, Gubbins KE (1984) *Theory of Molecular Fluids.* Clarendon Press, Oxford, UK
54. Green D, Leibovitz LJ (1982) J Nucl Mater 105:184
55. Grossman LN, Kaznoff AI (1968) J Amer Cer Soc 51:59
56. Gryaznov V, Iosilevski I, Yakub E, Fortov V, Hyland GJ, Ronchi C (2000) In:*Strongly Coupled Coulomb Systems.* Kalman G, Blagoev K, Rommel M Eds, p.147, Plenum Publ, New York-London, USA-UK

57. Gryaznov VK, Iosilevski IL (1976) In:*Teplofizitcheskie Svoistva Nizkotemperaturnoy Plazmy* (Thermophysical Properties of Low-Temperature Plasmas). Ievlev VM Ed, p. 25, Nauka Publ, Moscow, Russia (in Russian)
58. Gryaznov VK, Iosilevski IL, Fortov VK, *et al* (1980) *Thermophysical Properties of Substances of the Gas-Core Nuclear Engine.* Ed VM Ievlev, Atomizdat, Moscow, Russia (in Russian)
59. Gueeneau C, Baichi M, Labroche D, Chatillon C, Sundman B (2002) J Nucl Mater 304:161
60. Guggenheim A (1933) *Modern Thermodynamics by the Methods of Willard Gibbs.* Methuen and Co Publ, London, UK
61. Giussani Y, Guilliot B (1994) J Chem Phys 101:490
62. Gurvich LV, Veyts IV, Medvedev VA *et al* (1982) *Thermodynamic Properties of Individual Substances.* vol IV, Nauka Publ, Moscow, Russia (in Russian)
63. Gurvich LV, Veyts IV, Medvedev VA *et al* (1989) *Thermodynamic Properties of Individual Substances.* Hemisphere Publ, New York, USA
64. Gurvich LV, Veyts IV, Medvedev VA *et al* (1996) *Thermodynamic Properties of Individual Substances.* vol IV, Nauka Publ, Moscow, Russia (in Russian)
65. Haar L, Shenker SH (1971) J Chem Phys 35:4951
66. Halden FA, Wohlers Reinhart RH (1959) Report No SRI-6, Stanford Research Institute, CA, USA
67. Hall AR (1970) J Nucl Mater 37:314
68. Hein RA, Flagella PN (1968) Report GEMP-578 General Electric, USA
69. Hein RA, Flagella PN, Conway JB (1968) J Amer Ceram Soc 51:291
70. Hill TL (1956) *Statistical Mechanics:Principles and Applications.* Dover Publ, New York, USA
71. Hirschfelder JO, Curtiss CF, Bird RB (1954) *Molecular Theory of Gases and Liquids.* Wiley Publ, New York, USA
72. Høye JS, Lomba E, Stell G (1988) J Chem Phys 89:7462
73. Huheey JE (1983) *Inorganic Chemistry.* 3d Edition, Harper & Row Publ, New York, USA
74. Hyland GJ (1984) *Oxygen Potential Model for Stoichiometric and Non- Stocihiometric Uranium Dioxide.* Report EUR 9410 EN, European Commission, Karlsruhe, Germany
75. Ichimaru S, Yyetomi H, Tanaka S (1987) Phys Reports 149:91
76. *INTAS* Report Project 93-66 (1997) Interim Report, INTAS, Brussels
77. *INTAS* Report Project 93-66 (1999) Final Report, INTAS, Brussels
78. *INSC* Material Properties Database. http://www.insc.anl.gov/matprop
79. Iosilevski I, Chigvintsev A (1992) In:*Physics of Nonideal Plasmas.* Ebeling W, Forster A, Radtke R Eds, p.87, Teubner Verlagsgesellschaft, Leipzig-Stuttgart, Germany; J de Physique IV (2000) 10:451
80. Iosilevski I, Gryaznov V, Yakub E , Fortov VE, Hyland GJ, Ronchi C (2000) J de Physique IV 10:363
81. Iosilevski I, Hyland GJ, Ronchi C, Yakub E (1999) Trans Am Nucl Soc 81:122
82. Iosilevski I, Lappo GB, Skorohod EP *et al* (1991) In:*Fizika i Tehnika Vysokotemperaturnogo Gaza* (Physics and Technics of High-Temperature Gases). Proceedings of Moscow Aviation Institute, Moscow, Russia (in Russian)
83. Kapil SK (1976) J Nucl Mater 60:158
84. Kaplan IG (1982) *Vvedenie v teoriyu mezhmolekularnyh vzaimodeistviy* (The Introduction to the Theory of Molecular Interactions). Nauka Publ, Moscow, Russia (in Russian)

85. Kessel'man PM, Kamenezky VR, Yakub E (1976) *Svoistva perenosa real'nykh gasov* (Transport Properties of the Real Gases). Vyscha Shkola Publ, Kiev-Odessa, Ukraine (in Russian)
86. Kostrowicka Wyczalkowska A, Sengers J.V, Anisimov M.A (2004) Physica A 334:482
87. Kurosaki K, Yamada K, Uno M, Yamanaka S, Yamamoto K, Namekawa T (2001) J Nucl Mater 294:160
88. Landau D, Binder K (2000) *A Guide to Monte-Carlo Simulations in Statistical Physics.* Cambridge University Press, Cambridge, UK
89. Landau LD, Lifshitz EM (2001) *Course of Theoretical Physics, Statistical physics.* Pt.1, 3d Ed, Butterworth-Heinemann Publ, Oxford, UK
90. Larsen B (1976) J Chem Phys 65:3431
91. Lebovitz JL, Percus JK (1983) J Chem Phys 79:443
92. Lebovitz JL, Rowlinson JS (1964) J Chem Phys 41:133
93. Leibowitz L, Chasanov MG, Mishler LW (1971) J Nucl Mater 39:115
94. Leibowitz L, Mishler LW, Chasanov MG (1969) J Nucl Mater 29:356
95. Levesque D, Verlet L (1969) Phys Rev 182:307
96. Lewis WE, Singer K (1975) J Chem Soc Farady I 71:41
97. Likalter AA (1969) Sov Phys JETP 56:240
98. Magill J, Hyland GJ, Rand GJ (1988) High Temp-High Press 20:151
99. Mansoori GA, Carnahan NF, Starling KE, Leland TW (1971) J Chem Phys 54:1523
100. Marlowe MO, Kaznoff AI (1969) In:*Ceramic Nuclear Fuels.* L Owen *et al* Eds Amer Ceram Soc, pp 90-99
101. Mason EA, Spurling T H (1969) *The Virial Equation of State.* Pergamon Press, Oxford, UK
102. Menzies DC (1966) *The Equation of State of Uranium Dioxide at High Temperatures & Pressures*, Report TRG 1119 (D) UKAEA, London, UK
103. Meyer RA, Wolfe BE (1964) Trans Am Nucl Soc 7:111
104. Miller D (1965) In:*Proceedings of the Conference on Safety Fuels & Core Design in Large Fast Power Reactors.* Report USAEC ANL-7120:641, Argonne Nat Lab IL USA
105. Mistura L, Magill J, Ohse RW (1985) J Nucl Mater 135:95
106. Morita K, Fischer EA, Thurnay K (1998) Nucl Eng & Design 183:193
107. Ohse RW (1966) J Chem Phys 44:1375
108. Ohse RW, Babelot J-F, Cercignani C, Hiernaut J-P, Hoch H, Hyland GJ, Magill J (1985) J Nucl Mater 130:165
109. Padel A, de Novion Ch (1969) J Nucl Mater 33:40
110. Panagiotopoulos A Z (1992) Fluid Phase Equil 76:97
111. Peierls R (1933) Z Phys 80:763
112. Pitzer KS, Shriber DR (1987) Mol Phys 60 :1067
113. Radzig AA, Smirnov BM (1980) *Spravochnik po atomnoy i molekularnoy fizike* (The Reference Book for Atomic and Molecular Physics). Atomizdat Publ, Moscow, Russia (in Russian)
114. Rand MH, Ackermann RJ, Gronvold F, Oetting FL, Pattoret A (1978) Rev Int Haute Temp et Refract 15:355
115. Reed RC, Sherwood TK (1966) *The properties of Gases and Liquids.* McGraw-Hill Publ, New York, USA
116. Reedy GT, Chasanov MG (1972) J Nucl Mater 42:341

117. Reiss H (1956) J Chem Phys 25:400
118. Robbins EJ (1966) *Limits for the Equation of State of Uranium Dioxide.* Report TRG 1344 (R), UKAEA, London, UK
119. Ronchi C, Hiernaut J-P, Selfslag R, Hyland GJ (1993) Nucl Sci & Eng 113:1
120. Rudnyi E, Raibicheva E, Sidorov L (1992) Rapid Comm in Mass Spectr 6:356
121. Schins H, Gunnerson FS (1986) High Temp-High Press 18:249
122. Semenov AM (1968) Teplofiz Vys Temp 6:421 (in Russian)
123. Shevkunov SV, Vorontsov-Wel'aminov PN (1982) High Temp 20:1025
124. Sinanoglu O, Pitzer KS (1959) J Chem Phys 31:960
125. Sindzigre P, Gillan MJ (1988) J Phys C:Solid State 21:4017
126. Slagle OD, Nelson RP (1971) J Nucl Mater 40:349
127. *Spravochnik po fizikotehnicheskim osnovam kriogeniki* (1973) (The Reference Book on Physical and Engineering Basics of Cryogenics). Malkov MP Ed, Energiya Publ, Moscow, Russia (in Russian)
128. Stillinger FH, Lovett R (1968) J Chem Phys 48:3558
129. Stogryn DS, Hirschfelder JO (1959) J Chem Phys 31:1531
130. Tetenbaum M, Hunt PD (1970) J Nucl Mater 34:86
131. *Thermodynamic and Transport Properties of Uranium Dioxide and Related Phases* (1965) Technical Report Series No.39, p.40, IAEA, Vienna, Austria
132. Thiele E (1963) J Chem Phys 38:1959
133. Thompson SL, Lauson HS (1971) Report SC-RR-710713, Sandia Nat Lab, NM, USA
134. Tribus M (1967) *Thermostatics and Thermodynamics.* D van Nostrand Publ, New York, USA
135. Van der Waals JD (1873) *Over de continuiteit van den gas - en vloeistoftoestand.*:Thesis at the High School of Leyden, Sijthof, Leiden, (in Dutch); Engl.transl.:Van der Waals JD (1988) On the Continuity of the Gaseous and Liquid States. In: *Studies in Statistical Mechanics.* Rowlinson JS Ed vol XIV, North-Holland Publ, Amsterdam, The Netherlands
136. Verlet L, Weiss J-J, (1972) Molec Phys 24:1013
137. Voronov NM, Danilin AS, Kovalev IT (1962) In:*Proc Intern Symp on Thermodynamics of Nuclear Materials.* p.789, IAEA, Vienna, Austria
138. Vorontsov-Wel'aminov PN, Chasovskih VP (1975) High Temp 13:6
139. Wachtman JB Jr, Wheat ML, Anderson HJ, Bates JL (1965) J Nucl Mater 16:39
140. Waisman E, Lebovitz JL (1972) J Phys Chem 56:3086
141. Weinstein BK (1990) In:*Fzicheskaya Encyclopaedia* (Physical Encyclopaedia). Prohorov AM Ed, vol. 2, p.203, Moscow, Russia (in Russian)
142. Wertheim M (1963) Phys Rev Lett 8:321
143. Yakub E (1976) High Temp 14:792
144. Yakub E (1986) Zh Fiz Khim 60:2184
145. Yakub E (1992) Molec Phys 77:845
146. Yakub E (1994) Low Temp Phys 20:739
147. Yakub E (2001) J Molec Liquids 93:69
148. Yamada K, Kurosaki K, Uno M, Yamanaka S (2000) J Alloys & Compounds 307:10
149. Zel'dovich YaB (1981) Sov Phys JETP 53:1101
150. Zelener BV, Norman GE, Filinov VS (1981) *Perturbation Theory and Pseudopotential in Statistical Thermodynamics.* Nauka Publ, Moscow, Russia
151. Zernike J (1955) *Chemical Phase Theory.* Kluwer Acad Publ, Dordrecht-New York-London, USA-UK

Index

Abbreviations, 16
Adiabatic compressibility, 22, 140
Anisotropy parameter, 108, 110
Azeotropy, 62, 63, 68

Boiling curve, 8, 68, 74, 79, 80, 93, 95, 143

Charged hard sphere model, 76
Chemical potential, 14, 20, 21, 23, 32, 34–37, 56, 61–66, 71, 77, 84, 86, 88, 90–93, 97, 106
Chemically reacting fluid, 17, 18, 33, 59, 61, 64, 65, 74
Classical boiling point method, 8, 10
Clausius-Clapeyron equation, 69
Compressibility equation, 25
Compressibility factor, 25, 26, 28, 102, 104, 106, 119, 125, 126, 131
Conditions of the chemical equilibrium (CCE), 60, 62, 64, 65
Conditions of the phase equilibrium (CPE), 60–64, 71
Coulomb non-ideality, 35, 93, 94
  DHSE, 30, 44–46, 50, 53, 76, 96
  MSAE, 31, 44, 46, 53
  MSAV, 31
  parameter Γ_D, 52
Cricondenbar, 7, 70, 80
Cricondentherm, 7, 70, 80, 126
Critical point, 7, 19, 36, 37, 55, 56, 69, 70, 80, 83–86, 95–97, 102, 119, 121, 130–132

Debye-Hückel
  Debye radius, 30
  limiting law, 9, 30, 31, 45, 46, 50, 76, 94
Density, 19, 76, 80, 82, 84, 86, 88, 92, 123, 130, 131, 135
Dipole moments, 180
Dispersion coefficients, 182
Dissociation energy shift, 91

Elastic constants, 138
Electro-chemical potential, 63, 64
Electroneutrality condition, 66, 72
Enthalpy, 21–23, 123, 129, 140
EOS, 4, 6–10, 12, 13
  INTAS-99-EOS, 116, 133
  calibration, 51, 88
  caloric, 2, 8, 21, 27, 28, 66
  chemical model, 33, 34, 46, 61, 65
  empirical, 19, 29
  ideal gas, 23, 60
  Lennard-Jones fluid, 29
  local, 36, 49, 56
  model justification , 95
  near the critical point, 60, 102
  one-fluid approximation, 29
  parameter fitting, 176
  PHC, 6, 28, 84, 102
  physical model, 33
  requirements, 50
  semi-empirical, 19
  simple TPT, 102, 104, 116, 117, 119, 121–123, 130, 132
  theoretical, 19, 63

thermal, 2, 8, 10, 13, 21, 23, 25, 27, 61, 66
unified, 4, 60, 61, 95
VdW-DH-HSM, 94, 125
Excitation partition functions, 67

Forced-congruent equilibrium (FCE), 67, 69, 70, 88, 118, 119, 121, 123, 125, 126, 130–132
Forced-congruent mode (FCM), 8, 13, 68, 73, 74, 123, 143

Hard convex body model, 25, 100, 102
Hard sphere model, 24, 116
Carnahan-Starling formula, 19, 25, 102
close packing, 183
compressibility factor, 25
correlation function, 12
diameter, 24, 25, 36, 42, 44, 47, 48, 52, 76, 77, 80, 81, 85, 93, 94, 102–104, 107, 109–111, 114, 116, 183–186
free energy, 26
mixture, 45, 75, 96
Percus-Yevick approximation, 31
Heat capacity, 3, 8, 21, 22, 27, 28, 49, 51, 54, 115, 116, 120, 122, 123, 138, 140, 141
Heat of vaporisation, 52, 119, 124–126, 131, 132
Helmholtz free energy, 65, 100, 101

Isothermal compressibility, 14, 22, 49, 50, 121

Law of rectilinear diameters, 4, 83, 119
Lennard-Jones fluid, 9, 130
Lennard-Jones potential, 27, 100, 103, 104, 107, 114, 183
Liquid Metal Fast Breeder Reactor, 1

Maxwell construction, 61, 68
MIX Models
MIX-I, 37, 81, 89
MIX-II, 38, 67, 95

Neutral models
NM-0, 34
NM-I, 35
NM-II, 36
Restricted neutral model (RNM), 35
RPNM, 35
Non-congruency, 4, 75, 88, 94

Oxygen potential, 5, 9, 34–37, 47, 56, 62, 77, 81, 85, 86, 88, 90–95, 115

Packing fraction, 52, 104
Partition function cut-off, 67
Physical cluster theory, 33
Polarisability, 179
Principle of corresponding states, 4, 19
Pure ionic models
compatibility conditions, 51
IM-I, 43, 55
IM-II, 37, 46–48, 56
IM-III, 50, 51, 54–56

Retrograde condensation, 8

Saturation curve, 7, 68, 80, 143
Significant structure theory, 4
Single substance approximation (SSA), 67, 75, 95, 127
Slater radius, 178, 184
Soft sphere model, 26, 100
Speed of sound, 15, 22, 42, 138
Stillinger-Reiss theorem, 7, 9, 96, 99
Stockmayer potential, 107
Strongly coupled plasma, 96

Thermal expansion, 12, 18, 22, 49–51, 122, 135, 137
Thermodynamic perturbation theory (TPT), 9, 12, 28, 99–102, 104, 105, 111
Andersen, Chandler and Weeks, 27, 99, 103, 104
Barker and Henderson, 26, 99, 102, 103, 109
Total vapour pressure, 142

Van der Waals
attraction parameter, 9, 27, 31, 37, 49, 51, 52, 56, 103, 105
critical point, 250
EOS, 27, 60, 75, 76, 99, 102, 117, 176
heat capacity, 9
repulsive parameter, 10, 102
Van der Waals
critical point, 61

Printing: Mercedes-Druck, Berlin
Binding: Stein + Lehmann, Berlin